AF720674

**Carbohydrate Chemistry in the Total Synthesis of Naturally Occurring Glycosides**

# Carbohydrate Chemistry in the Total Synthesis of Naturally Occurring Glycosides

*Biao Yu and Xiaoyu Yang*

WILEY-VCH

**Authors**

***Prof. Biao Yu***
State Key Laboratory of Bioorganic and Natural Products Chemistry
Shanghai Institute of Organic Chemistry
Chinese Academy of Sciences
Shanghai, 200032
China

***Dr. Xiaoyu Yang***
School of Physical Science and Technology
ShanghaiTech University
China

**Cover:** © aquapix/Shutterstock

All books published by **WILEY-VCH** are carefully produced. Nevertheless, authors, editors, and publisher do not warrant the information contained in these books, including this book, to be free of errors. Readers are advised to keep in mind that statements, data, illustrations, procedural details or other items may inadvertently be inaccurate.

**Library of Congress Card No.:** applied for

**British Library Cataloguing-in-Publication Data**
A catalogue record for this book is available from the British Library.

**Bibliographic information published by the Deutsche Nationalbibliothek**
The Deutsche Nationalbibliothek lists this publication in the Deutsche Nationalbibliografie; detailed bibliographic data are available on the Internet at <http://dnb.d-nb.de>.

© 2024 WILEY-VCH GmbH, Boschstraße 12, 69469 Weinheim, Germany

All rights reserved (including those of translation into other languages). No part of this book may be reproduced in any form – by photoprinting, microfilm, or any other means – nor transmitted or translated into a machine language without written permission from the publishers. Registered names, trademarks, etc. used in this book, even when not specifically marked as such, are not to be considered unprotected by law.

**Print ISBN:** 978-3-527-34519-9
**ePDF ISBN:** 978-3-527-81792-4
**ePub ISBN:** 978-3-527-81790-0
**oBook ISBN:** 978-3-527-81789-4

**Typesetting:** Straive, Chennai, India
**Printing and Binding:** CPI Group (UK) Ltd, Croydon, CR0 4YY

C9783527345199_140324

# Contents

# Preface

Naturally occurring glycosides are prevalent in plants, microorganisms, and low animals, where they primarily function as signaling and defense agents. These compounds, spanning various categories of natural products such as polyketides, steroids, triterpenes, flavonoids, alkaloids, peptides, and lipids, are produced by stepwise glycosylation under the catalysis of diverse glycosyltransferases. Boasting a wide spectrum of pharmacological activities, including antitumor, anti-infective, and immunomodulatory effects, some glycosides, especially antibiotics, nucleosides, and cardiac glycosides, have found extensive use as therapeutic agents, while many others persist in folkloric usage. The saccharide residues in these compounds play an essential role in shaping their pharmacophore and significantly influence their pharmacokinetic and pharmacodynamic properties.

Given the microheterogeneity and the limited availability of naturally occurring glycosides, chemical synthesis emerges as a practical approach to the availability of these compounds, especially when conducting structure–activity relationship studies for pharmaceutical development. Consequently, the synthesis of natural glycosides has garnered significant research interest. Although it falls into the discipline of total synthesis of natural products, the total synthesis of naturally occurring glycosides encompasses an additional content that involves the installation of the saccharide residues with various glycosylation tactics and methods and protecting group manipulations. While certain elegant total synthesis of naturally occurring glycosides are included in books of natural product synthesis, to the best of our knowledge, no monograph is available that comprehensively discusses the total synthesis of naturally occurring glycosides, especially focusing on the carbohydrate chemistry involved in the total syntheses.

For nearly three decades, we (the Yu group) have dedicated our efforts to the total synthesis of naturally occurring glycosides and successfully accomplished the synthesis of a large variety of natural glycosides, such as landomycin A, periploside A, and luzonicosides A. Throughout the course of our research, it becomes evident that a comprehensive book focusing on the carbohydrate chemistry involved in the total synthesis should be of big value to those interested in this topic and, more generally, in the synthetic carbohydrate chemistry. When approached by Dr. Lifen Yang to explore the idea of documenting those research in a book, the notion immediately resonated with us. Having gone through difficulties in compiling a book, including those unexpectedly presented by the COVID-19 pandemic, it ultimately took us six years to make a conclusion. Within its pages, we have detailed the comprehensive evolution of carbohydrate chemistry for the total synthesis of naturally occurring glycosides, categorizing the development across ten different types of aglycones to which saccharide residues are assembled. Our aim for this book is to provide readers with a comprehensive and clear understanding of the progress in the total synthesis of naturally occurring glycosides, from simple to complex molecules and from intricate strategies to highly efficient and economical methodologies. We anticipate that this book will serve as a valuable handbook for new researchers entering the field of carbohydrate synthesis, where they can find widely utilized assembly tactics, glycosylation methods, and information on temporary and permanent protecting groups essential for the synthesis of specific types of glycosides.

We express our sincere gratitude to the Wiley group for affording us this invaluable opportunity to author this book and for their enduring patience and unwavering support. Specifically, we extend our deep appreciation to Dr. Lifen Yang for her steadfast management of this project as well as to Pinky Sathishkumar, Katherine Wong, Katrina Maceda, and Shwathi Srinivasan, each of whom served as managing editor during different periods. We are also grateful to Ashok Ravi for his meticulous proofreading of the manuscript. Lastly, we wish to convey our heartfelt thanks to our coworkers and our families for their exceptional patience and unwavering support, which has been indispensable in making the completion of this book possible.

December 2023

*Dr. Biao Yu*
*Dr. Xiaoyu Yang*
*Shanghai*

# List of Abbreviations

| | |
|---|---|
| Ac | acetyl |
| AIBN | azodiisobutyronitrile |
| All | allyl |
| Alloc | allyloxycarbonyl |
| AzMB | *o*-azidomethylbenzoyl |
| Bn | benzyl |
| Boc | *t*-butoxycarbonyl |
| Bom | benzyloxymethyl |
| b.r.s.m. | based on recovered starting material |
| BSTFA | bis(trimethylsilyl)trifluoroacetamide |
| Bz | benzoyl |
| CAN | cerium(IV) ammonium nitrate |
| Cbz | carbobenzyloxy |
| ClAc | chloroacetyl |
| Cp | cyclopentadienyl |
| CSA | camphorsulfonic acid |
| DABCO | 1,4-diazabicyclo[2.2.2]octane |
| DBAD | dibenzyl azodicarboxylate |
| dba | dibenzylideneacetone |
| DBU | 1,8-diazabicyclo[5,4,0]undec-7-ene |
| DCC | dicyclohexylcarbodiimide |
| DCM | dichloromethane |
| DDQ | 1,2-dichloro-4,5-dicyanobenzoquinone |
| DEAD | diethyl azodicarboxylate |
| DEIPS | diethylisopropylsilyl |
| DHQD | dihydroquinidine |
| DIAD | diisopropyl azodicarboxylate |
| DIBAL-H | diisobutylaluminium hydride |
| DIPEA | ethyldiisopropylamine |
| DMA | *N*,*N*′-dimethyl-1,2-ethanediamine |
| DMAP | 4-*N*,*N*-dimethylaminopyridine |
| DMDO | dimethyldioxirane |
| DMF | *N*,*N*-dimethylformamide |

| | |
|---|---|
| DMP | Dess–Martin periodinane |
| DMS | dimethyl sulfide |
| DMSO | dimethylsulfoxide |
| DNP-Br | 2,4-di-nitrophenylbromide |
| dr | diastereomeric ratio |
| DTBMP | 2,6-di-*tert*-butyl-4-methylpyridine |
| DTT | DL-1,4-dithiothreito |
| EDCI | 1-ethyl-3-(3-dimethylaminopropyl) carbodiimide |
| ee | enantiomeric excess |
| FDPP | perfluorophenyl diphenylphosphinate |
| Fmoc | fluorenylmethyloxycarbonyl |
| HFIP | hexafluoroisopropanol |
| HMPA | hexamethylphosphoramide |
| LDA | lithium diisopropylamide |
| Lev | levulinyl |
| HATU | 2-(1*H*-7-azabenzotriazole-1-yl)-1,1,3,3-tetramethyluronium hexafluorophosphate |
| HMDS | bis(trimethylsilyl)amide |
| HOAt | 1-hydroxy-7-azabenzotriazole |
| LiDBB | lithium 4,4′-ditert-butyl-1,1′-biphenyl |
| MBz | 4-methoxybenzoyl |
| mCPBA | *m*-chloroperoxybenzoic acid |
| MMTr | 4-methoxtrityl |
| MOM | methoxymethyl |
| Ms | methanesulfonyl |
| MS | molecular sieve |
| Nap | 2-naphthylmethy |
| NBS | *N*-bromosuccinimide |
| NCS | *N*-chlorosuccinimide |
| NIS | *N*-iodosuccinimide |
| NMO | *N*-methylmorpholine-*N*-oxide |
| OTf | trifluoromethanesulfonate |
| PCC | pyridinium chlorochromate |
| PDC | pyridinium dichromate |
| Phth | phthaloyl |
| Piv | pivaloyl |
| PMB | para-methoxybenzyl |
| PMP | para-methoxyphenyl |
| pNs | para-nitrobenzenesulfonyl |
| PPTS | pyridinium *p*-toluenesulfonate |
| Pyr | pyridine |
| RCM | ring-closing metathesis |
| rt | room temperature |
| S-Phos | 2-dicyclohexylphosphino-2′,6′-dimethoxybiphenyl |
| TASF | tetra-*n*-butylammonium fluoride-sulfur tetrafluoride |

| | |
|---|---|
| TBAB | tetrabutylammonium bromide |
| TBAI | tetrabutylammonium iodide |
| TBAF | tetra-*n*-butylammonium fluoride |
| TBDPS | *tert*-butyldiphenylsilyl |
| TBHP | *tert*-butyl hydroperoxide |
| TBS | *t*-butyldimethylsilyl |
| TDS | hexyldimethylsilyl |
| TEA | triethylamine |
| TES | triethylsilyl |
| Teoc | 2-(trimethylsilyl)ethoxycarbonyl |
| TFA | trifluoroacetic acid |
| THF | tetrahydrofuran |
| TIPS | triisopropylsilyl |
| TMS | trimethylsilyl group |
| TMU | tetramethylurea |
| Tol | toluene |
| Ts | toluenesulfonyl |
| TTBP | 2,4,6-tri-*tert*-butylpyrimidine |

# 1

# Introduction

Glycosides occur ubiquitously in nature, being especially abundant in plants, microorganisms, and low animals, which are evolved as secondary metabolites to function mainly as signal and defense chemicals [1–3]. A recent survey concluded that about 16.2% of the reported natural products are glycosides [4]. The aglycones comprise all types of natural products, such as polyketides, steroids, triterpenes, flavonoids, nucleobases, peptides, and lipids (Figure 1.1). The sugar moieties are usually added onto the aglycones and elongated subsequently via stepwise glycosylation under the action of various glycosyltransferases [5, 6]. Depending on the type of aglycones, the saccharide parts are highly characteristic and conservative in monosaccharide composition and glycosidic linkages. Enormous microheterogeneity occurs due to the tolerance of the glycosyltransferases for variation of the monosaccharide units and the aglycones, the incompleteness of the enzymatic reactions, as well as the subsequent modifications, such as acylation, oxidation, and degradation.

These naturally occurring glycosides have shown various pharmacological activities, especially antitumor, anti-infective, and immunomodulatory effects [7]. Some have been long and widely used as therapeutic agents, including, most importantly, antibiotics, nucleosides, and cardiac glycosides [8]. Many others still remained in folkloric usage, such as the saponin extracts from ginseng, licorice, ivy leaves, primula roots, and senega roots [9]. The saccharide residues can be an indispensable part of the pharmacophore or contribute critically to the pharmacokinetic and pharmacodynamic properties of the glycosides.

Chemical synthesis of a glycoside demands integration of the synthetic chemistry of the particular aglycone and the saccharide, involving especially a condensation of the two distinct parts and an overall protecting-group arrangement. Based on the stage at which the glycosidic bond between the saccharide and the aglycone is constructed, five tactics can be applied to the synthesis (Figure 1.2) [10]. The most straightforward and convergent tactic is a direct late-stage glycosylation of the aglycone with a prefabricated saccharide donor, followed by global deprotection (Tactic I). Glycosylation of the aglycone with a fully developed oligosaccharide donor might be problematic; then, the sugar units can be assembled in a linear manner (Tactic II). This tactic could warrant a stereospecific and high-yielding formation of the glycosidic bond to the aglycone but demands manipulation of temporary

*Carbohydrate Chemistry in the Total Synthesis of Naturally Occurring Glycosides*,
First Edition. Biao Yu and Xiaoyu Yang.
© 2024 WILEY-VCH GmbH. Published 2024 by WILEY-VCH GmbH.

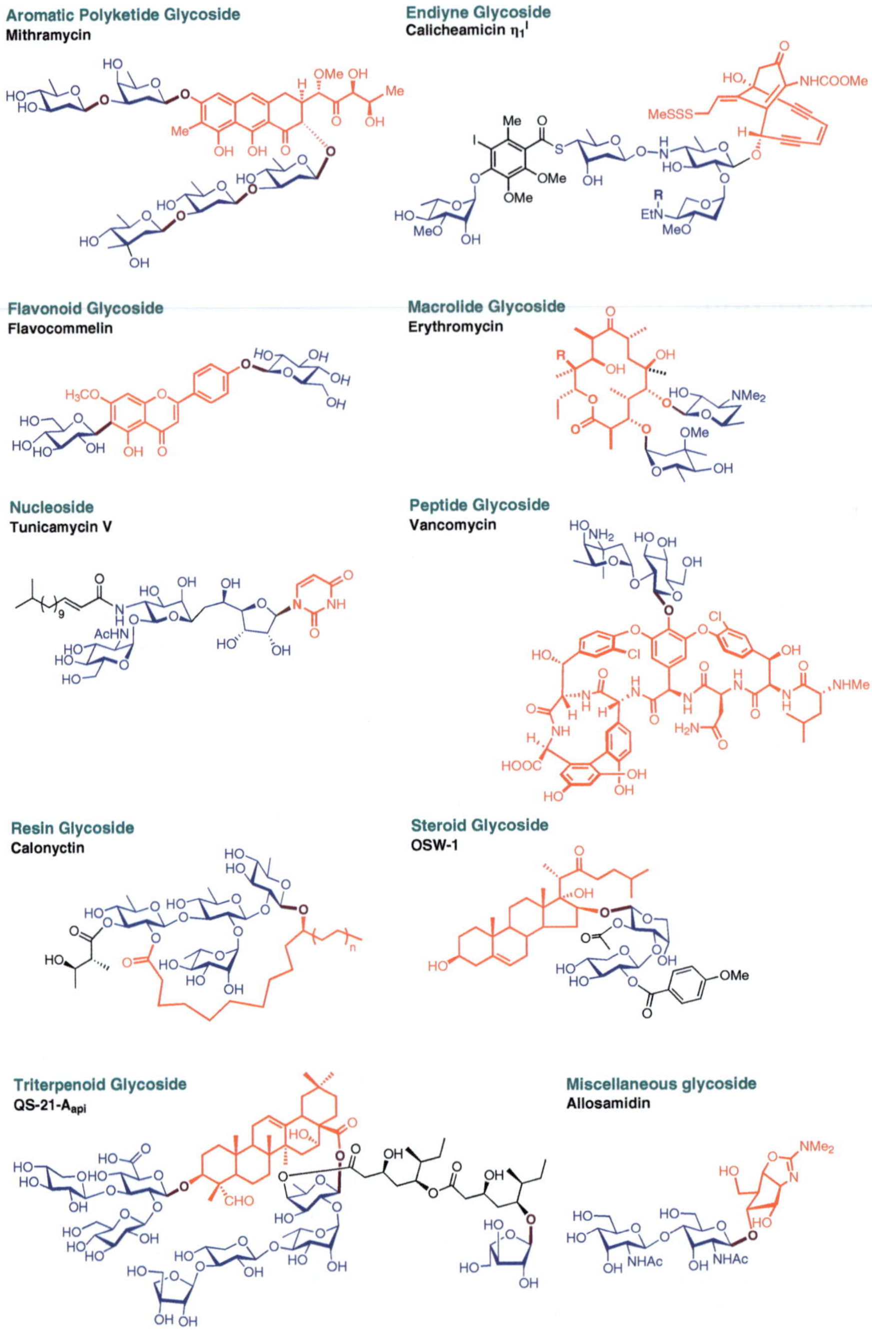


**Figure 1.1** Representative molecules in the major categories of the naturally occurring glycosides, which have been synthesized. In red are the algycones, and in blue are the sugar residues.

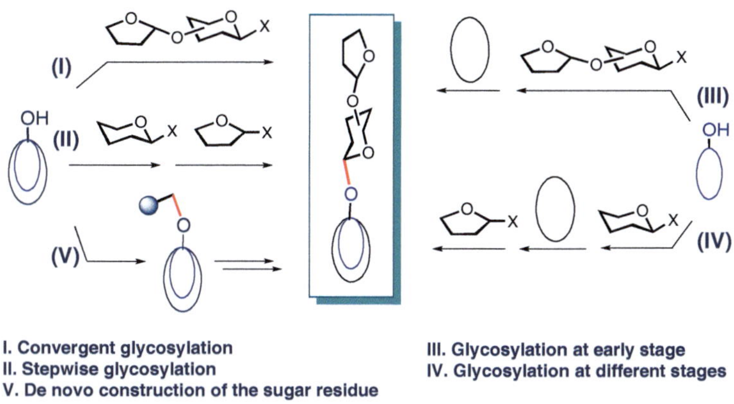


**Figure 1.2** Five general tactics for the synthesis of complex glycosides based on the stage of incorporating the saccharide onto the aglycone.

protecting groups in between each glycosylation step. When glycosylation of the aglycone, even with a compromised monosaccharide donor, is unsuccessful, then this glycosidic linkage should be built before elaboration of the full aglycone (Tactic III). The fourth alternative for the assembly of a complex glycoside involves the elaboration of both the aglycone and the glycan after the construction of the glycosidic linkage (Tactic IV). Tactic V involves de novo synthesis of the sugar residue on the aglycone so as to bypass the glycosylation reaction [11, 12].

Installation of a saccharide onto an aglycone usually demands judicious choice of a glycosylation reaction. Numerous glycosylation protocols employing a wide variety of glycosyl donors have been developed (Figure 1.3). The major types that have been applied in the synthesis of natural glycosides include glycosyl bromides [13], fluorides [14], iodides [15], trichloroacetimidates [16, 17], *N*-phenyl trifluoroacetimidates [18, 19], thioglycosides [20–22], sulfoxides [23], heteroaryl thioglycosides [24, 25], 1-hydroxyl sugars [26, 27], 1-*O*-acetates [28], and *ortho*-alkynylbenzoates [29, 30]. Listed are also the commonly used promoters for each type of donor, which are determined by the nature of the leaving groups. It should be noted that the reactivity of a donor is also dependent on the sugar type and protecting group pattern.

Besides the coupling yield, another critical issue for a glycosylation reaction is stereoselectivity. In general, activation of a glycosyl donor by a promoter yields a continuum of species relevant to the sugar oxocarbenium intermediate (Figure 1.4) [31–35]. Each of these interconvertible species can react with an acceptor (an *O*-, *N*-, *S*-, or *C*-nucleophile) to give the glycoside, but with a different stereo-preference. Thus, the relative abundance of these transient species and their kinetic preference for glycosylation determine the overall outcome of the stereoselectivity. Usually, 1,2-*trans*-glycosides can be confidently synthesized with donors equipped with a neighboring (or remote) participating group (path a). This also constitutes a reliable approach to the synthesis of 2-deoxy-glycosides, in that the neighboring participating group needs to be removed afterwards [36–38]. Glycosylation through path b (via the oxocarbenium species) erodes the stereoselectivity. Direct stereoselective synthesis of the 1,2-*cis*-glycosides and the 2-deoxy-glycosides must resort to

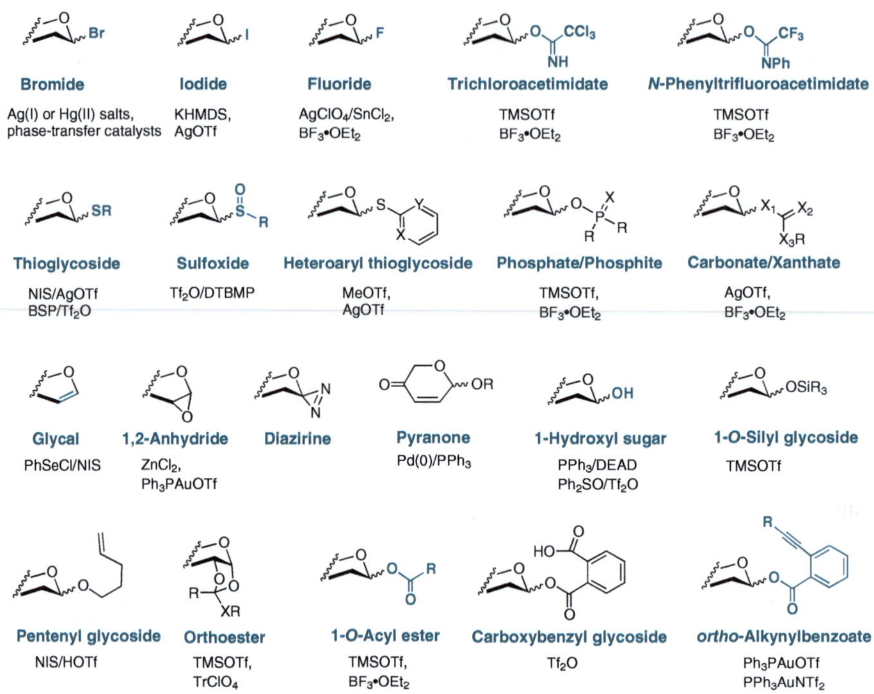


**Figure 1.3** The major types of glycosyl donors and their promoters.

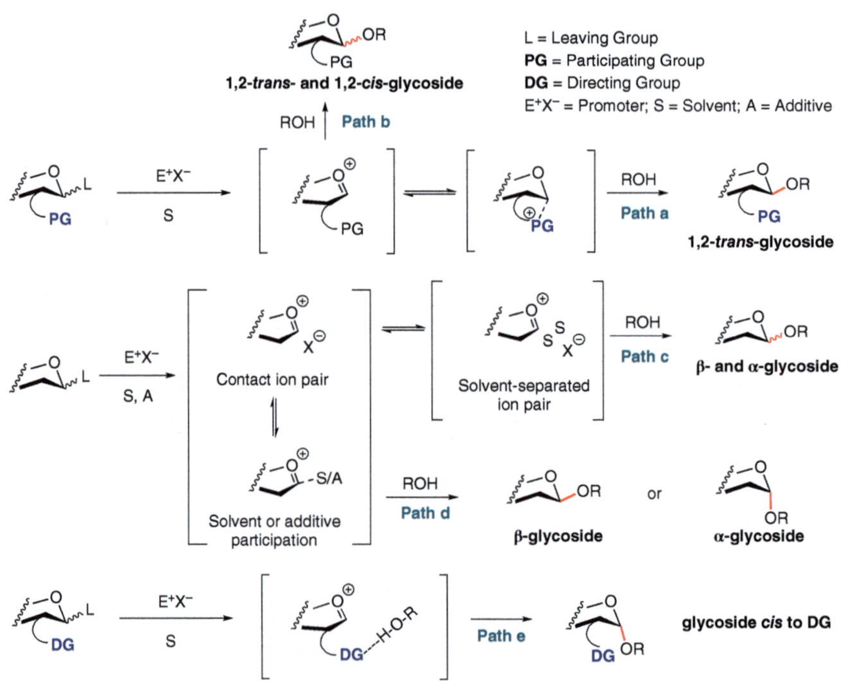


**Figure 1.4** A general mechanistic scheme for the stereochemical outcomes in the glycosylation reactions.

fine-tuning of the reaction parameters, to force the glycosylation to proceed via a contact ion pair or a solvent-participating intermediate (path d) [39, 40]. Glycosylation through path c (via the solvent-separated ion pair) usually leads to a mixture of α/β glycosides; however, controlling the conformation of the sugar oxocarbenium intermediate (mainly by the protecting groups) could also lead to stereoselective glycosylation [41–44]. Recently, some directing groups have been developed, which could deliver stereoselective glycosylation via a H-bonding intermediate (path e) [45–47]. It is noteworthy that the inherent stereochemistry of the aglycone would influence strongly the stereoselectivity of the glycosylation reaction. In addition, the glycosidic linkages, especially the abundantly occurring deoxy-glycosides, might undergo anomerization or cleavage under acidic conditions.

In this book, we compile the successful synthesis of the representative and complex natural glycosides. These syntheses are presented in 10 chapters based on the types of target glycosides, namely, aromatic polyketide glycosides, enediyne glycosides, flavonoid glycosides, macrolide glycosides, nucleosides, peptide glycosides, resin glycosides, steroid glycosides, triterpenoid glycosides, and miscellaneous glycosides (Figure 1.1). For each glycoside, a brief introduction has been provided about its origin, structural features, and biological activities. In each synthesis, we focus on the glycosylation steps, especially the step for the construction of the glycosidic bond connecting to the aglycone. The glycosylation yields and stereoselectivity are given and highlighted. The subsequent transformations toward the final targets are also depicted; those include elongation of the glycans, elaboration of the aglycone, and/or manipulation of the protecting groups. Especially, the steps and yields for the final cleavage of the protecting groups are highlighted. These late-stage synthetic steps demonstrate the compatibility of the chemical transformations demanded in the presence of the complex aglycone and the saccharide residues.

## References


1 Ikan, R. (1999). *Naturally Occurring Glycosides*. Chichester, England: Wiley.
2 Dembitsky, V.M. (2004). *Chem. Biodivers.* 1: 673–781.
3 Elshahawi, S.I., Shaaban, K.A., Kharel, M.K., and Thorson, J.S. (2015). *Chem. Soc. Rev.* 44: 7591–7697.
4 Chen, Y., Liu, Y., Chen, N. et al. (2023). *Nat. Prod. Rep.* 40: 1464–1478.
5 Thibodeaux, C.J., Melançon, C.E. III, and Liu, H.W. (2008). *Angew. Chem. Int. Ed.* 47: 9814–9859.
6 Gantt, R.W., Peltier-Pain, P., and Thorson, J.S. (2011). *Nat. Prod. Rep.* 28: 1811–1853.
7 Wong, C.-H. (2003). *Carbohydrate-Based Drug Discovery*, vol. 1–2. Weinheim: Wiley-VCH.
8 Cao, X., Du, X., Jiao, H. et al. (2022). *Acta Pharm. Sin. B* 12: 3783–3821.
9 Hostettmann, K. and Marston, A. (ed.) (1995). *Saponins*. Cambridge, UK: Cambridge University Press.
10 Yu, B., Sun, J., and Yang, X. (2012). *Acc. Chem. Res.* 45: 1227–1236.
11 Babu, R.S., Zhou, M., and O'Doherty, G.A. (2004). *J. Am. Chem. Soc.* 126: 3428–2439.

**12** Zhu, D., Geng, M., and Yu, B. (2022). *Angew. Chem. Int. Ed.* 61: e202203239.
**13** Koenigs, W. and Knorr, E. (1901). *Ber. Dtsch. Chem. Ges.* 34: 957–981.
**14** Mukaiyama, T., Murai, Y., and Shoda, S. (1981). *Chem. Lett.* 10: 431–432.
**15** Helferich, B. and Gootz, R. (1929). *Ber. Dtsch. Chem. Ges.* 62: 2788–2792.
**16** Schmidt, R.R. and Michel, J. (1980). *Angew. Chem. Int. Ed. Engl.* 19: 731–732.
**17** Schmidt, R.R. (1986). *Angew. Chem. Int. Ed. Engl.* 25: 212–235.
**18** Yu, B. and Tao, H. (2001). *Tetrahedron Lett.* 42: 2405–2407.
**19** Yu, B. and Sun, J. (2010). *Chem. Commun.* 46: 4668–4679.
**20** Ferrier, R.J., Hay, R.W., and Vethaviyasar, N. (1973). *Carbohydr. Res.* 27: 55–61.
**21** Codee, J.D.C., Litjens, R.E.J.N., van den Bos, L.J. et al. (2005). *Chem. Soc. Rev.* 34: 769–782.
**22** Lian, G., Zhang, X., and Yu, B. (2015). *Carbohydr. Res.* 403: 13–22.
**23** Kahne, D., Walker, S., Cheng, Y., and van Engen, D. (1989). *J. Am. Chem. Soc.* 111: 6881–6882.
**24** Haneasian, S., Bacquet, C., and Lehong, N. (1980). *Carbohydr. Res.* 80: C17–C22.
**25** El Ashry, E.S.H., Awad, L.F., and Atta, A.I. (2006). *Tetrahedron* 62: 2943–2998.
**26** Fischer, E. (1893). *Ber. Dtsch. Chem. Ges.* 26: 2400–2412.
**27** Garcia, B.A., Poole, J.L., and Gin, D.Y. (1997). *J. Am. Chem. Soc.* 119: 7597–7598.
**28** Helferich, B. and Schmitz-Hillebrecht, E. (1933). *Chem. Ber.* 66: 378–383.
**29** Li, Y., Yang, Y., and Yu, B. (2008). *Tetrahedron Lett.* 49: 3604–3608.
**30** Tang, Y., Li, J., Zhu, Y. et al. (2013). *J. Am. Chem. Soc.* 135: 18396–18405.
**31** Rhind-Tutt, A.J. and Vernon, C.A. (1960). *J. Chem. Soc.* 4637–4644.
**32** Lemieux, R.U., Hendrika, K.B., Stick, R.V., and James, K. (1975). *J. Am. Chem. Soc.* 97: 4056–4062.
**33** Crich, D. (2010). *Acc. Chem. Res.* 43: 1144–1153.
**34** Mydock, L.K. and Demchenko, A. (2010). *Org. Biomol. Chem.* 8: 497–510.
**35** Bohe, L. and Crich, D. (2015). *Carbohydr. Res.* 403: 48–59.
**36** De Lederkremer, R.M. and Marino, C. (2007). *Adv. Carbohydr. Chem. Biochem.* 61: 143–216.
**37** Hou, D. and Lowary, T.L. (2009). *Carbohydr. Res.* 344: 1911–1940.
**38** Borovika, A. and Nagorny, P. (2012). *J. Carbohydr. Chem.* 31: 255–283.
**39** Nigudkar, S.S. and Demchenko, A.V. (2015). *Chem. Sci.* 6: 2687–2704.
**40** Satoh, H., Hansen, H.S., Manabe, S. et al. (2010). *J. Chem. Theory Comput.* 6: 1783–1797.
**41** Nukada, T., Bérces, A., and Whitfield, D.M. (2002). *Carbohydr. Res.* 337: 765–774.
**42** Yang, M.T. and Woerpel, K.A. (2009). *J. Org. Chem.* 74: 545–553.
**43** Jensen, H.H., Nordstrøm, L.U., and Bols, M. (2004). *J. Am. Chem. Soc.* 126: 9205–9213.
**44** Moumé-Pymbock, M., Furukawa, T., Mondal, S., and Crich, D. (2013). *J. Am. Chem. Soc.* 135: 14249–14255.
**45** Yasomanee, J.P. and Demchenko, A.V. (2014). *Angew. Chem. Int. Ed.* 53: 10453–10456.
**46** Yasomanee, J.P. and Demchenko, A.V. (2012). *J. Am. Chem. Soc.* 134: 20097–20102.
**47** Liu, X., Lin, Y., Liu, A. et al. (2022). *Chin. J. Chem.* 40: 443–452.

# 2

# Aromatic Polyketide Glycosides

Aromatic polyketides are a diverse family of natural products with considerable structural variation [1, 2]. The majority of these compounds contain highly substituted, fused-ring polyphenols that are derived from poly-β-ketone precursors and decorated with deoxysaccharide residues via *C*- and *O*-linkages (Figure 2.1). The aglycones exhibit three key structural features, including the number of rings, the cyclization pattern dictated by the regioselective cyclization of the linear polyketide precursors, and the ring topology resulting from oxidative rearrangement reactions. Consequently, aromatic polyketides fall into different categories, such as the anthracycline family (e.g. doxorubicin **1**) [3], angucycline family (e.g. landomycin A **2**) [4–6], tetracycline family (e.g. SF2575 **3**), and aureolic acids family (e.g. mithramycin **4**), among others. Many members of the aromatic polyketide family have potent activities, such as antibiotic, immunosuppressive, antiparasitic, cholesterol-lowering, and antitumor activities. Clinical use is seen for some of these compounds, such as doxorubicin (adriamycin, **1**) and mithramycin (**4**), as anticancer drugs.

## Landomycin A

The landomycins represent the largest subfamily of angucycline antibiotics, which were initially isolated from *Streptomyces cyanogenus* by Rohr and coworkers in 1990 [7]. Landomycin A (**2**) is the most significant member of this group (Figure 2.2), which features an angular tetracyclic core containing a dearomatized B-ring and a hexasaccharide composed of two repeat trisaccharide, i.e. (α-L-rhodinose-(1 → 3)-β-D-olivose-(1 → 4)-β-D-olivose). It exhibits potent antitumor activities by inhibiting DNA synthesis and $G_1$/S cell cycle progression, although the precise mechanism of action remains unclear. Structure–activity relationship (SAR) studies indicate that the cytotoxic activity of landomycins depends on the length of the glycan, and landomycin A with the longest sugar chain is the most active congener. Due to its unique structure and potent antitumor activities, there has been significant interest in the total synthesis of landomycin A among organic chemists. Indeed, several research groups have achieved the synthesis of the hexasaccharide fragment of landomycin A using various glycosylation

*Carbohydrate Chemistry in the Total Synthesis of Naturally Occurring Glycosides*,
First Edition. Biao Yu and Xiaoyu Yang.
© 2024 WILEY-VCH GmbH. Published 2024 by WILEY-VCH GmbH.

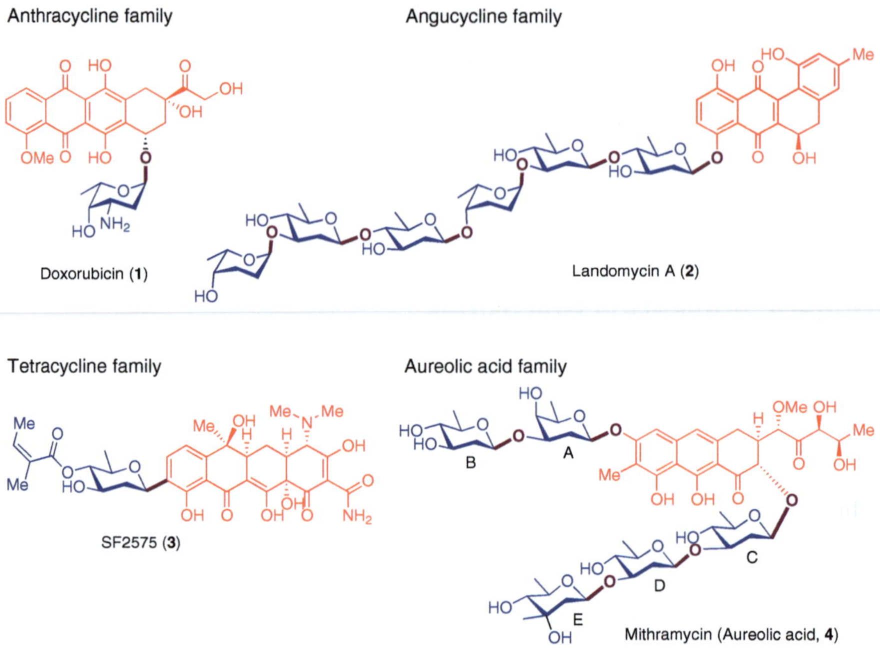


**Figure 2.1** Representative aromatic polyketide glycosides.

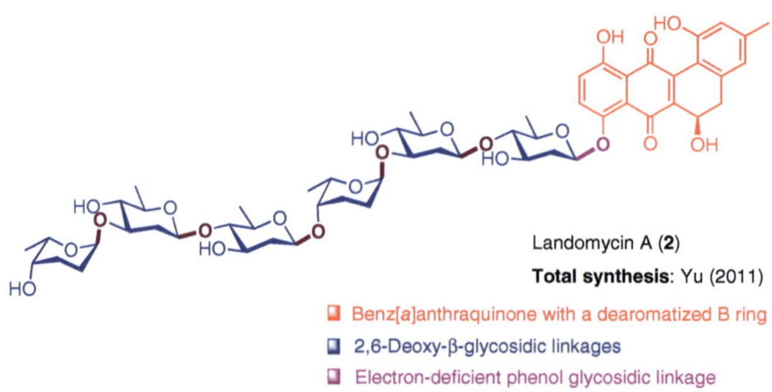


**Figure 2.2** Landomycin A (2) and its structural features.

methods. Nevertheless, only Yu and coworkers accomplished the total synthesis of landomycin A in 2011 [8].

In 1997, Sulikowski and coworkers reported the first approach to the synthesis of the hexasaccharide fragment of landomycin A (Scheme 2.1) [9]. The desired hexasaccharide **7** was obtained in a moderate yield of 42% and a β/α stereoselectivity of ~1 : 1 through a late-stage coupling of trisaccharide acceptor **6** with 2-deoxy-trisaccharide phosphite **5** under the catalysis of TMSOTf.

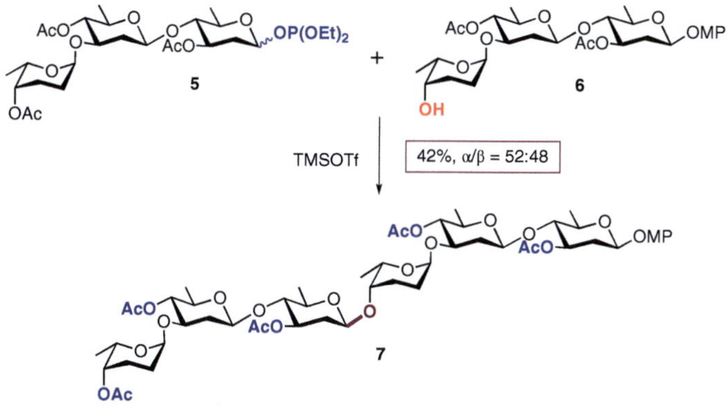


**Scheme 2.1** The key glycosylation step employed in the synthesis of landomycin A hexasaccharide by the Sulikowski group.

In 2000, Roush et al. reported a stereoselective synthesis of the hexasaccharide of landomycin A (Scheme 2.2) [10]. The construction of the 2-deoxy-β-glycosidic linkages exhibited excellent stereoselectivity (>95 : 5) using 2-iodo-glycosyl donors, in which the 2-iodo group participated in the glycosylation. Glycosylation of trisaccharide acceptor **9** with trichloroacetimidate donor **8** in the presence of TBSOTf at −78 °C resulted in the desired β-linked hexasaccharide **10** with high selectivity and in yields ranging from 73% to 80%.

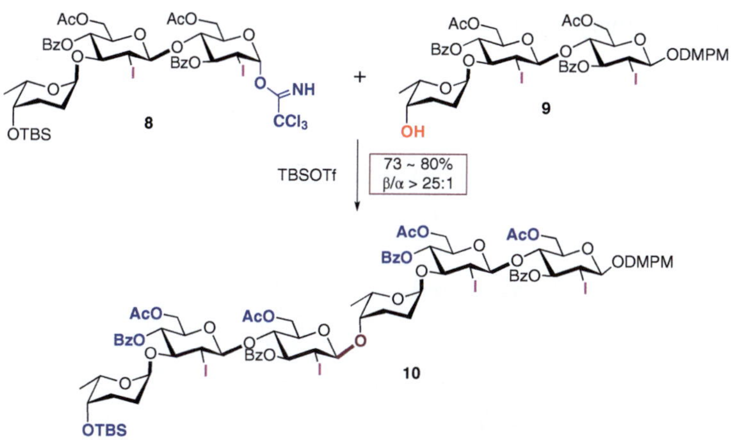


**Scheme 2.2** The key glycosylation step employed in the synthesis of landomycin A hexasaccharide by the Roush group.

Using phenyl 2,3-*O*-thionocarbonyl-1-thioglycosides as 2-deoxy-β-glycoside precursors, which can undergo a 1→2 sulfide migration and concurrent glycosidation, Yu et al. achieved the synthesis of the hexadeoxysaccharide fragment of landomycin A in 2002 (Scheme 2.3) [11]. A late-stage glycosylation of trisaccharide acceptor **12** with disaccharide sulfide donor **11** in the presence of MeOTf and 2,6-di-*tert*-butyl-4-methylpyridine (DTBMP) yielded pentasaccharide **13** with

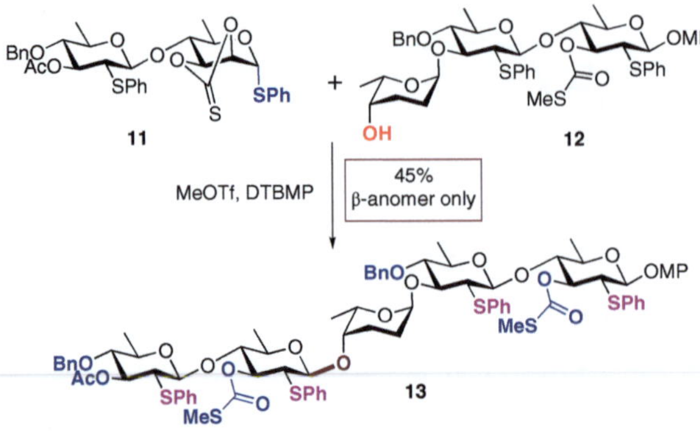

Scheme 2.3 The key glycosylation step employed in the synthesis of landomycin A hexasaccharide by the Yu group.

complete stereocontrol and a yield of 45%. The coupling partners, **11** and **12**, were prepared using similar methodologies.

In 2010, Takahashi and coworkers reported a stereoselective synthesis of a panel of deoxyhexasaccharides relevant to the landomycin A glycan, in which the β-selective glycosylation was realized using 2-deoxyglycosyl imidate donors featuring an electron withdrawing sulfonate at the *O*-4 position and being activated with $I_2$ and $Et_3SiH$ at an extremely low temperature (Scheme 2.4) [12]. Glycosylation of trisaccharide acceptor **15** with trisaccharide imidate donor **14** was performed at −95 °C in the presence of $I_2$ and $Et_3SiH$, yielding hexasaccharide **16** in 93% yield and with good β/α selectivity (82 : 18).

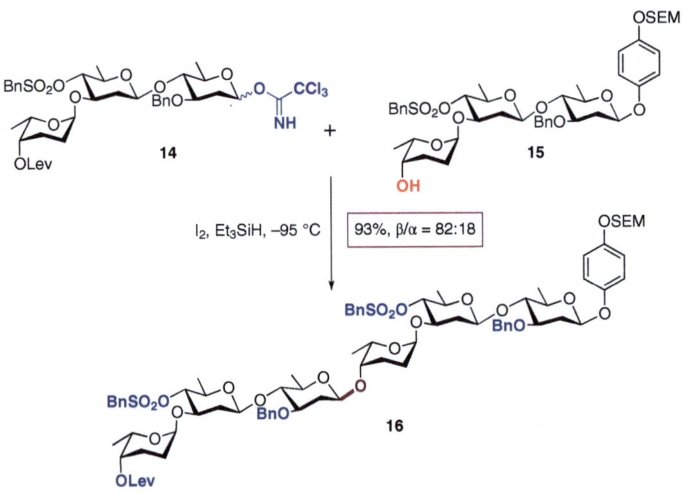

Scheme 2.4 The key glycosylation step employed in the synthesis of landomycin A hexasaccharide by the Takahashi group.

**Total synthesis of landomycin A (2) by the Yu group.** Although various approaches for synthesizing the hexasaccharide and shorter deoxysaccharides [13–15] of landomycins have been developed, the synthesis of the aglycone was only achieved by Roush et al. in 2004 [16], and the total synthesis of landomycin A was not achieved until 2011 [8, 17, 18]. Due to the challenging linkage between the landomycin aglycone and the saccharide chain, the glycosylation of aglycone was carried out with monosaccharide donors. The remaining pentasaccharide building block was synthesized as shown in Scheme 2.5. Glycosylation of acceptor **17** with L-rhodinosyl acetate **18** under the catalysis of TBSOTf at −78 °C afforded the thermodynamically favored α-disaccharide **19** in 94% yield. Selective removal of the C4′-*O*-TBS group in **19** provided disaccharide acceptor **20**. Coupling of acceptor **21** with 2,6-dideoxy-2-iodo-glucopyranosyl trifluoroacetimidate **22** under the action of TBSOTf at −78 °C afforded β-disaccharide **23** in excellent yield (91%), along with the corresponding α-disaccharide in ~5% yield. After removal of the C3-*O*′-TBS group, glycosylation with L-rhodinosyl acetate **24** enabled by TBSOTf at −50 °C gave trisaccharide **25** in a high yield of 89%. Trisaccharide **25** was converted to trifluoroacetimidate **26** in two steps, which was then coupled with disaccharide acceptor **20** under the catalysis of TBSOTf at −78 °C, yielding the desired β-pentasaccharide **27** in a satisfactory yield of 88%. Pentasaccharide acetate **27** was then transformed to trifluoroacetimidate **28** in a similar manner to the **25** → **26** transformation.

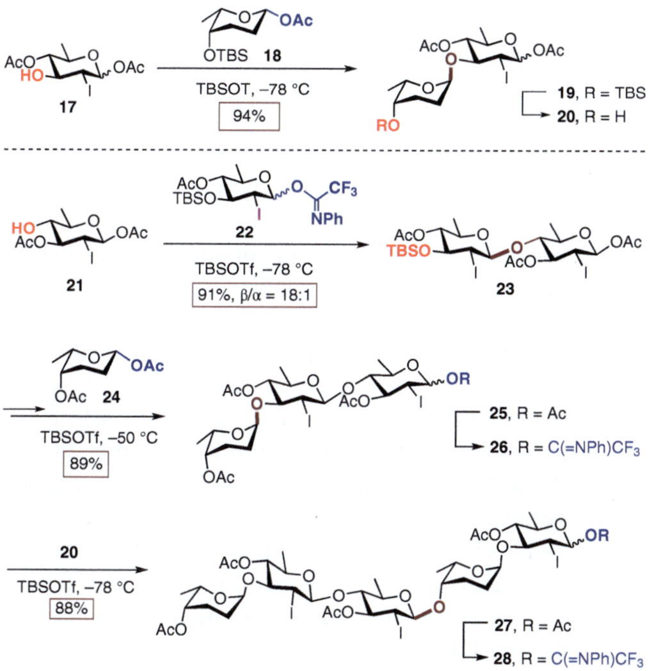


**Scheme 2.5** Synthesis of pentasaccharide donor **28** by Yu and coworkers.

The synthesis of landomycin aglycone started with the preparation of naphthalene **31** via Hauser annulation between cyanophthalide **29** and aryl enone **30** under basic conditions (Scheme 2.6). Compound **31** was then converted to naphthoquinone **32** by stereoselective reduction of the C-6 carbonyl group,

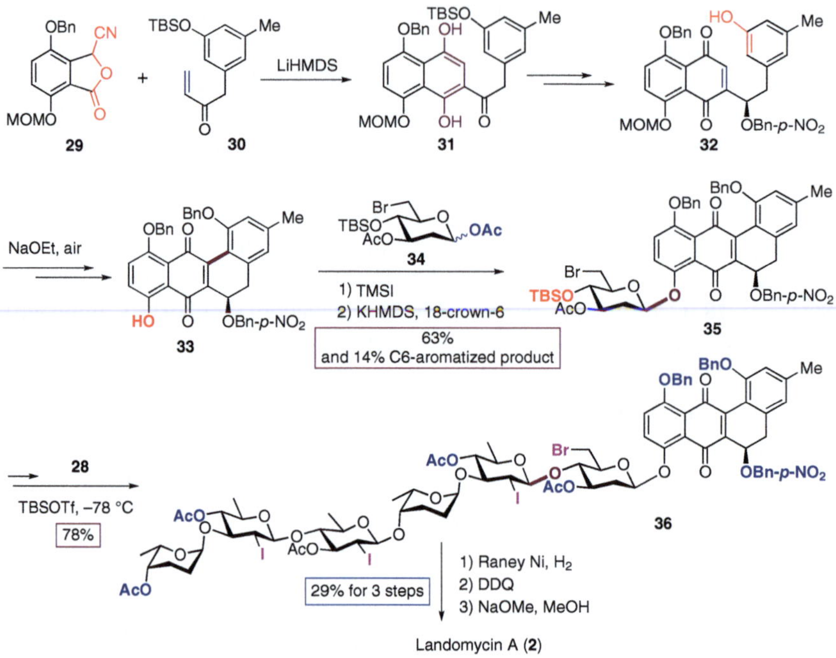

**Scheme 2.6** Total synthesis of landomycin A (**2**) by Yu and coworkers.

protection of the resultant hydroxyl with *p*-nitrobenzyl group, removal of the *O*-TBS group, and oxidation of the hydroquinone. The C-ring was formed via intramolecular Michael addition, followed by oxidation of **32** through treatment with NaOEt and air [16]; subsequent protecting group manipulation provided C8-OH derivative **33**. After extensive studies, glycosylation of aglycone **33** was achieved by using a $S_N2$-type reaction with the *in situ* generated α-glycosyl iodide from acetate **34** under basic conditions, furnishing β-glycoside **35** in 63% yield, with the corresponding C6-aromatizaed product being isolated in 14% yield. Subsequently, the C4′-*O*-TBS group in landomycinone glycoside **35** was selectively cleaved, and coupling with pentasaccharide trifluoroacetimidate donor **28** was achieved under the catalysis of TBSOTf (3 mol%) at −78 °C, yielding hexasaccharide **36** in a good yield of 78%. Stronger reaction conditions such as greater equivalents of TBSOTf or higher temperature led to cleavage of the 2,3,6-trideoxy-α-glycosidic and the C8-*O*-2′-deoxy-β-glycosidic linkages. Finally, removal of the bromide, iodide, and benzylic protecting groups (with $H_2$, Raney-Ni), regeneration of the quinone motif (with 2,3-dichloro-5,6-dicyano-1,4-benzoquinone [DDQ]), and removal of the five acetyl groups (with NaOMe/MeOH) accomplished the total synthesis of landomycin A (**2**) in 29% yield for the last three steps.

In 2021, Mong and coworkers reported the total synthesis of landomycin Q and landomycin R, which are the trisaccharide and disaccharide derivatives of the B-ring unsaturated anhydrolandomycinone congeners [19]. The construction of the key linkage between aglycone and the saccharide chain was also achieved using an $S_N2$-type glycosylation with glycosyl iodide under basic conditions.

## Olivomycin A

Olivomycins, member of the aureolic acid family, were first isolated from *Streptomyces olivoreticuli* by Brazhnikova et al. in 1962 [20]. Olivomycin A (**37**), chromomycin A3 (**38**), and mithramycin (**4**) are the most well-known antibiotics of this family (Figure 2.3) [21]. Structurally, these compounds contain a tricyclic ring system and a unique dihydroxy-methoxy-oxo-pentyl aliphatic side chain at C-3 position. Olivomycin A (**37**) and chromomycin A3 (**38**) each contain two oligosaccharide chains attached to the aromatic and aliphatic rings, respectively. These chains are composed of 2,6-dideoxysugars, such as D-olivose, D-oliose, D-mycarose, L-chromose B, and their *O*-methyl or *O*-acetyl derivatives, linked via β-(1 → 3)-glycosidic linkages. These aureolic acids can inhibit DNA-dependent RNA polymerase, and selectively form 2 : 1 antibiotic/$Mg^{2+}$ complexes in the DNA minor groove with a preference for GC-rich sequences. SAR studies have demonstrated that the two oligosaccharide chains are critical for the DNA binding and biological activities.

**Synthesis of the oligosaccharide fragments of mithramycin by the Thiem group.** In 1987, Thiem et al. reported the synthesis of the A–B disaccharide and C–D–E trisaccharide fragments of mithramycin (Scheme 2.7) [22, 23]. Glycosylation of sugar acceptor **39** with 2-bromo-2-deoxyglycosyl bromide **40** in the presence of AgOTf afforded disaccharide **41** in 56% yield with moderate β-selectivity (β/α = 3 : 1). Reductive dehalogenation and cleavage of the acyl protecting groups yielded the desired A–B disaccharide **42**. Coupling between sugar acceptor **43** and glycosyl bromide **40** enabled by AgOTf provided disaccharide **44** in excellent yield (92%). After selective removal of the formyl group, iterative glycosylation with **40** in the presence of AgOTf furnished trisaccharide **45** in 69% yield with high β-selectivity (β/α = 10 : 1). Finally, the C3″-tertiary alcohol motif was constructed through several steps of transformations, along with the reductive dehalogenation and cleavage of the acyl protecting groups, resulting in the formation of C–D–E trisaccharide **46**.

**Synthesis of the oligosaccharide fragments of olivomycin A by the Toshima group.** In 1995, Toshima et al. reported a highly stereocontrolled synthesis of the C–D–E trisaccharide fragment of olivomycin A using 2,6-anhydro-2-thio

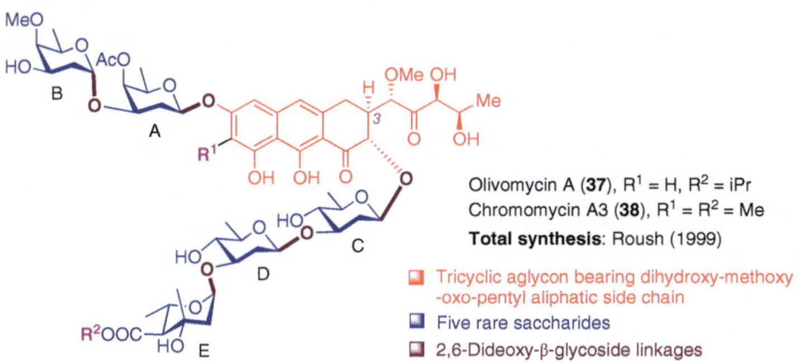


**Figure 2.3** Olivomycin A (**37**) and chromomycin A3 (**38**) and their structural features.

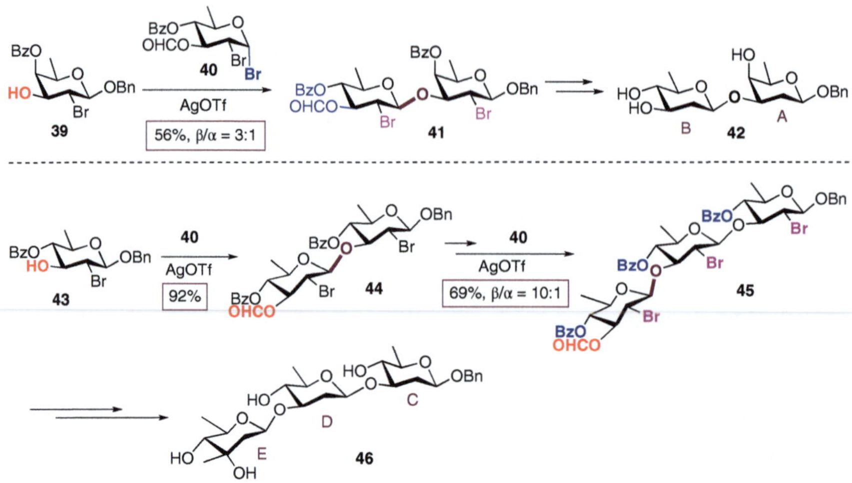


**Scheme 2.7** Synthesis of the di- and trisaccharides (**42** and **46**) of mithramycin by Thiem and coworkers.

sugars as donors (Scheme 2.8) [24]. Glycosylation of cyclohexanol with 2,6-anhydro-2-thioglycosyl acetate **47** in the presence of TMSOTf provided glycoside **48** in 90% yield with exclusive β-selectivity. After C3 hydroxy group inversion and oxidation of the thioether, glycosylation of the resultant **49** with **47** led to β-disaccharide **50** in 89% yield without detection of the α-anomer. Disaccharide **50** was subjected to similar transformations as **48**→**49** to give disaccharide acceptor **51**, which was then coupled with 2,6-anhydro-2-thio-glycoside **52** in the presence of N-bromosuccinimide (NBS) and furnished the desired trisaccharide **53** in 89% yield with exclusive α-selectivity. Finally, C–D–E trisaccharide **54** was obtained through hydrogenation of **53** using Raney-Ni as the catalyst, followed by desilylation.

**Total synthesis of olivomycin A by the Roush group.** In 1999, Roush and coworkers reported the total synthesis olivomycin A (**37**) [25–29]. Previous studies showed poor efficacy in the glycosylation of a protected aureolic acid aglycone with fully elaborated C–D–E trisaccharides donors, resulting in <15% yield of the desired β-glycoside. Therefore, an alternative approach was used in which the C saccharide was first coupled to the aglycone, followed by sequential installation of the A–B disaccharide **58** and D–E disaccharide **62**. The A–B disaccharide was prepared by coupling of 2-phenylseleno-galactosyl acetate **55** with glycal **56** under the promotion of $I(coll)_2ClO_4$ in dichloromethane (DCM), yielding α-disaccharide **57** in 63% yield along with 6% of the isomer with an equatorial iodide in the B residue (Scheme 2.9). Two steps were then performed to convert **57** into reducing disaccharide **58**. Meanwhile, coupling of glycal derivative **59** with olivomycosyl acetate **60** in the presence of TMSOTf in DCM at −78 °C led to the desired α-D–E disaccharide glycal **61** in 74% yield, which was then converted in two steps into the corresponding 2-SPh-disaccharide trichloroacetimidate donor **62**.

The synthesis of the protected aglycone began with the Horner–Wadsworth–Emmons (HWE) reaction of phosphate **63** with aldehyde **64**, resulting in the coupled

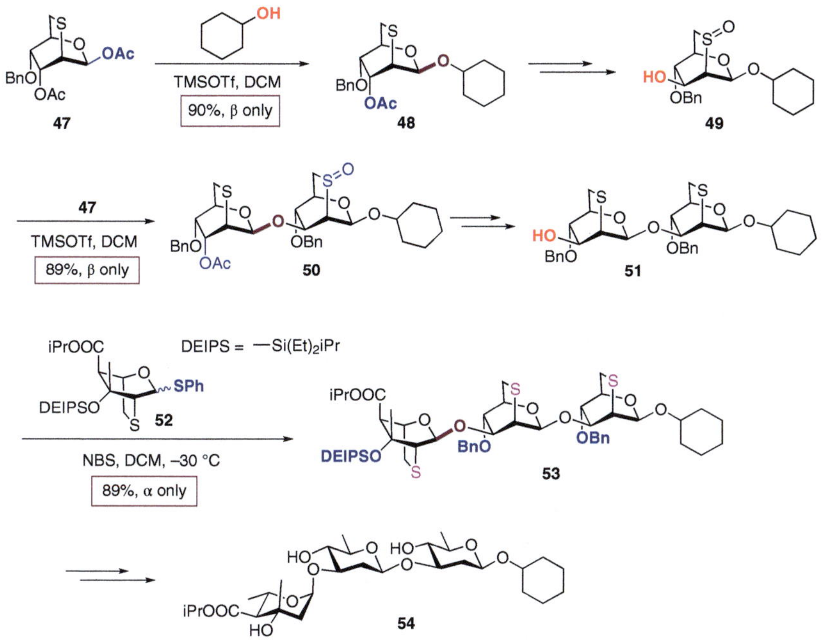


**Scheme 2.8** Synthesis of the C–D–E trisaccharide fragment **54** of olivomycin by Toshima and coworkers.

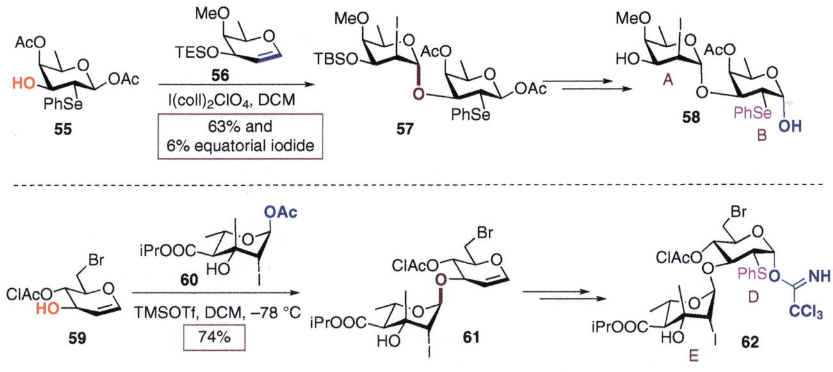


**Scheme 2.9** Synthesis of the A–B and D–E disaccharide fragments of olivomycin A.

enone **65** in 80% yield (Scheme 2.10). After steps of functional group transformations, addition of lithiated methyl acetate to compound **66** furnished naphthene derivative **67**, which was further converted to diester **68**. Dieckman condensation of **68** under basic conditions afforded the tricyclic aglycone derivative **69**. Subsequently, glycosylation of **69** with C2-SPh-glycosyl imidate **70** afforded glycoside **71** in 58% yield with good β-selectivity (β/α = 8 : 1). After protecting group manipulation, the coupling of the monosaccharide acceptor with C2-SPh substituted D–E disaccharide imidate donor **62** in the presence of TBSOTf (0.3 equiv) afforded

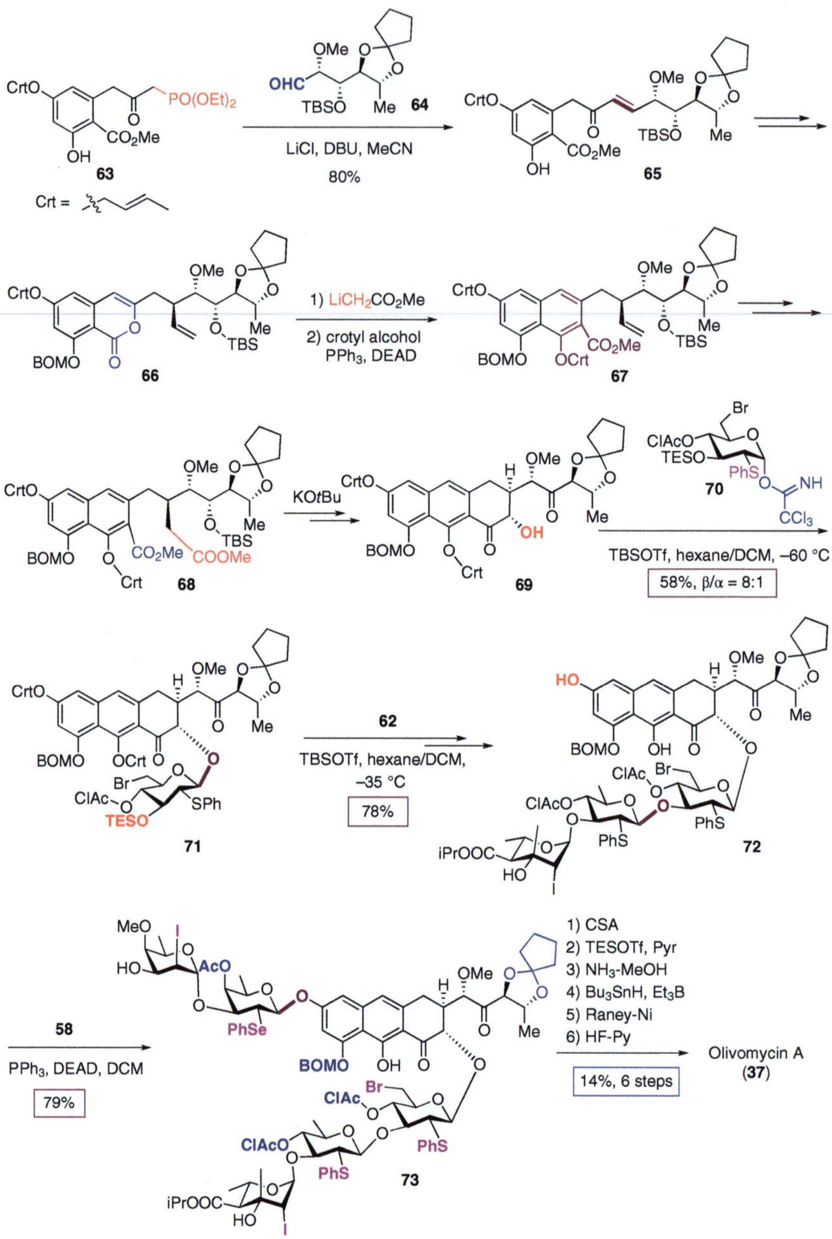


**Scheme 2.10** Total synthesis of olivomycin A (**37**) by Roush and coworkers.

trisaccharide **72** in 78% yield. Following this, glycosylation of the phenol group was carried out with the reducing A–B disaccharide **58** under Mitsunobu conditions ($PPh_3$, DEAD) to provide glycoside **73** in 79% yield. The final functional group manipulations began with acid-catalyzed cleavage of the cyclopentylidene ketal and were followed by per-triethylsilylation to give intermediates with improved solubility. Subsequently, the two chloroacetate groups were removed, followed by radical reduction of the iodo-, bromo-, and selenophenyl substituents, further cleavage of the thiophenyl and benzyloxymethyl (BOM) groups with Raney-Ni, and final removal of the three triethylsilyl (TES) ethers with HF-pyridine furnished olivomycin A (**37**) in 14% yield (for the last six steps).

## Ciclamycin

Ciclamycin 0 (**74**, Figure 2.4), an anthracycline trisaccharide originally isolated from *Streptomyces capoamus* in 1964, shows potent activity against tumors both *in vitro* and *in vivo* [30]. While the majority of ciclamycin compounds share the same aglycone, variance in their toxicity levels can be attributed to dissimilarities in their oligosaccharide fragments. Consequently, development of a synthetic route to this type of anthracyclines with varied oligosaccharides is necessary to elucidate the role of the glycan in the antitumor activities.

**Total synthesis of ciclamycin 0 by the Danishefsky group.** In 1990, Danishefsky and coworkers reported the synthesis and structure corroboration of ciclamycin 0 (**74**) by utilizing substituent-controlled oxidative coupling of glycals to assemble the glycan (Scheme 2.11) [31]. The coupling of L-fucal derivative **75** with 3,6-dideoxy-glycal **76** under the action of iodinium salt ($I^+$(sym-collidine)$_2ClO_4^-$) afforded the corresponding iododisaccharide, which was then reduced with $Ph_3SnH$ and AIBN to yield the desired disaccharide **77** in 57% yield. It is worth noting that the benzoyl group at C3 of glycal **75** has apparently deactivated this glycal as compared to **76** in the presence of $I^+$. After two steps of protecting group manipulations, iterative glycosylation of **78** with glycal **75**, followed by iodide removal with $Ph_3SnH$, provided trisaccharide **79** in 54% yield. Functional group adjustments were then performed to furnish keto-group bearing trisaccharide glycal **80**. The reaction of glycal **80** and ε-pyrromycinone (**81**), which was obtained from the hydrolysis of the natural bohemic acid complex, was carried out in the presence of $I^+$(sym-collidine)$_2ClO_4^-$ to produce a mixture of three coupled glycosides. Two major products (35% for **82a** and 25% for **82b**) possessed the required axial α-glycosidic linkage to the C7 oxygen but differed in the stereochemistry of the

**Figure 2.4** Ciclamycin 0 (**74**) and its structural features.

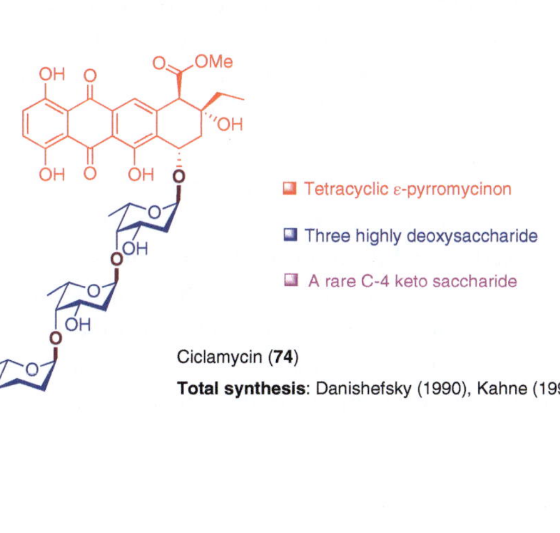


Ciclamycin (**74**)

**Total synthesis**: Danishefsky (1990), Kahne (1999)

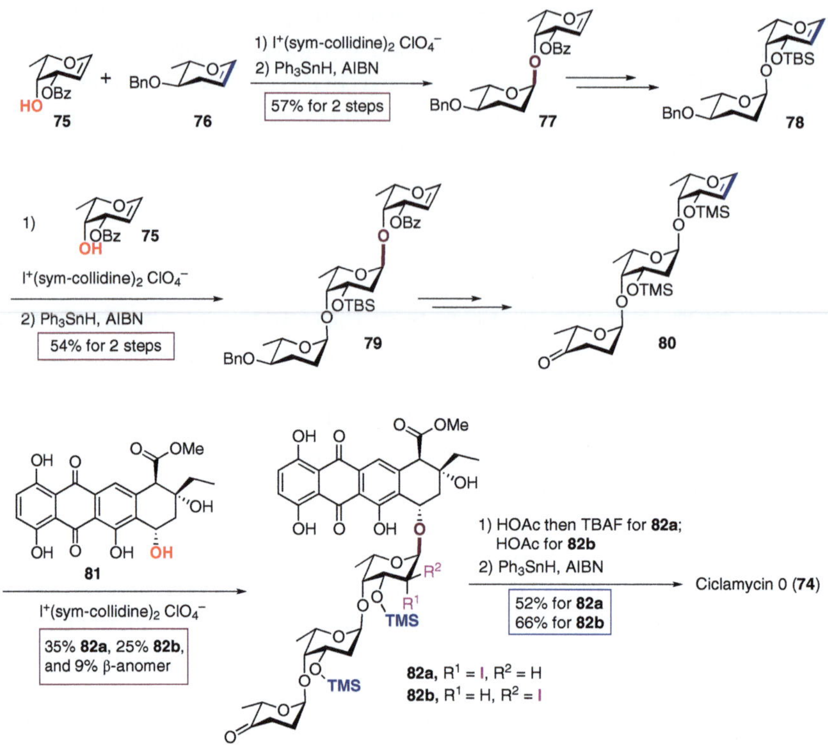

**Scheme 2.11** Total synthesis of ciclamycin 0 (**74**) by Danishefsky and coworkers.

iodo substituent, while the minor product was a β-anomer (9%). Finally, removal of the TMS groups followed by deiodonation with $Ph_3SnH$/AIBN accomplished the total synthesis of ciclamycon 0 (**74**) in 52% (from **82a**) and 66% (from **82b**) yield, respectively.

**Total synthesis of ciclamycin 0 by the Kahne group.** In 1993, Kahne and coworkers reported a stereoselective one-pot synthesis of ciclamycin 0 trisaccharide **88** using glycosyl sulfoxide donors (Scheme 2.12) [32]. During the preliminary studies, the authors observed two key findings: (i) for perbenzyl glucosyl sulfoxides, the *p*-methoxyphenyl sulfoxide could be selectively activated in the presence of the unsubstituted phenyl sulfoxide by employing 0.5 equiv of the activating agent; (ii) silyl ethers reacted at a much slower rate than alcohols under the catalysis of TfOH. Based on these findings, the assembly of trisaccharides **87** was realized through one-pot reaction of monosaccharide sulfoxides **83**, **84**, and sulfide **85** in the presence of HOTf (5 mol%). The desired trisaccharide **87** was isolated in 25% yield, while disaccharide **86** in 14%, indicating that the glycosylation process occurred in a sequential manner, with *p*-methoxyphenyl sulfoxide **84** being activated more quickly than phenyl sulfoxide **83** and alcohol **85** reacted faster than silyl ether **84**. Oxidation of the trisaccharide sulfide with *m*CPBA led to the corresponding glycosyl sulfoxide, which was then coupled with aglycone **81** in the presence of $Tf_2O$ to provide the desired glycoside **88**, albeit in a low 16% yield. Furthermore,

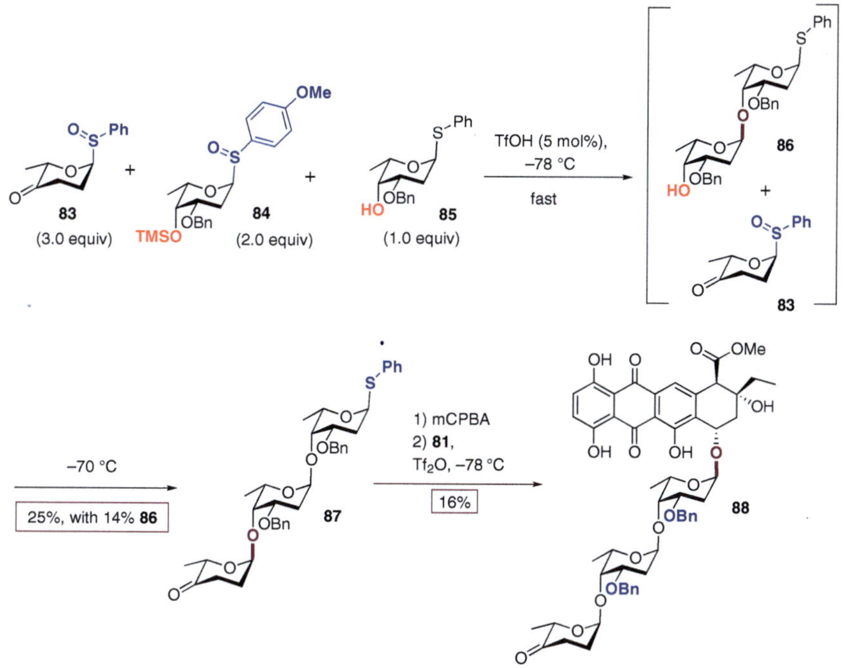

**Scheme 2.12** One-pot synthesis of ciclamycin 0 trisaccharide **87** by Kahne and coworkers.

deprotection of the benzyl group under hydrogenation conditions was unsuccessful due to the simultaneous cleavage of the benzylic linkage of the aglycone to the glycan.

Kahne and coworkers concluded the total synthesis of ciclamycin 0 (**74**) in 1999 (Scheme 2.13) [33]. Based on the previous studies, they rationalized that glycosylation with glycosyl sulfoxide under the action of $Tf_2O$ could generate phenylsulfenyl triflate which could also activate anomeric sulfides. To avoid this undesired activation, two alternatives were proposed: (i) the addition of an alkene to trap the phenylsulfenyl triflate specie, with 4-allyl-1,2-dimethoxybenzene providing an optimal one; (ii) the use of hindered sulfides such as 2,6-dichlorophenyl sulfide that were activated sluggishly by phenylsulfenyl triflate. Therefore, glycosylation of thioglycoside **89** with glycosyl sulfoxide **90** under inverse addition conditions ($Tf_2O$, DTBMP, −78 °C) in the presence of 4-allyl-1,2-dimethoxybenzene led to disaccharide **91** in 82% yield, with no detection of trisaccharide. After deprotection of the acetyl group, the resulting disaccharide acceptor was coupled with sulfoxide **92** under similar conditions ($Tf_2O$, DTBMP, −78 °C) to provide trisaccharide **93** in 68% yield with good α-selectivity (α/β = 5 : 1). Treatment of sulfide **93** with DMDO (dimethyldioxirane) led to the corresponding sulfoxide, which was subjected to the glycosylation with aglycone **81** and the coupled glycoside **94** was obtained in 68% yield under similar conditions as for the previous glycosylations. Finally, the removal of the two PMB groups with DDQ completed the total synthesis of ciclamycin 0 (**74**) in 60% yield.

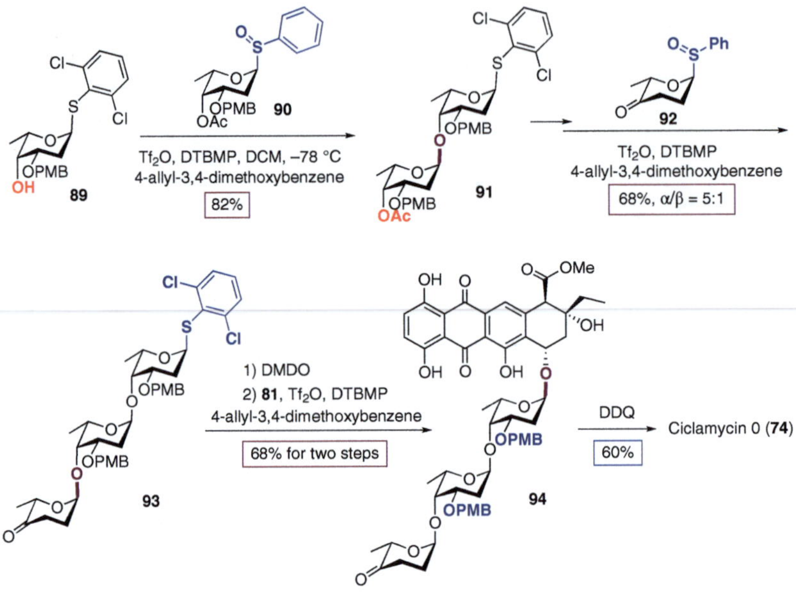


**Scheme 2.13** Total synthesis of ciclamycin 0 (**74**) by Kahne and coworkers.

## Vineomycin B2

Vineomycin B2 (**96**, Figure 2.5), an anthracycline antibiotic active against gram-positive bacteria and sarcoma-180 solid tumors, was isolated by Ōmora et al. from the culture broth of *Streptomyces matensis* subsp. *vineus* in 1977 [34]. It consists of a *C*-linked trisaccharide at the aromatic ring and an *O*-linked disaccharide at the β-oxo-*tert*-hydroxyl group of the side chain. It is noteworthy that the trisaccharide contains the disaccharide, namely α-L-acurosyl-α-L-rhodinose, as the terminal residue, and this trisaccharide also occurs in angucycline natural product PI-080 [35]. When subjected to acid-catalyzed methanolysis, vineomycin B2 (**96**) undergoes cleavage of the *O*-glycosidic linkages to give vineomycinone B2 methyl

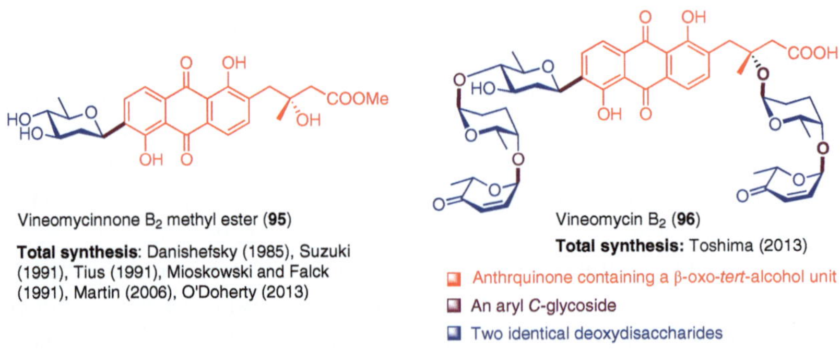


**Figure 2.5** Vineomycinone B2 methyl ester (**95**) and vineomycin B2 (**96**) and the structural features.

ester (**95**), which can be classified as the *C*-glycoside (β-D-olivoside) of the methyl ester of fridamycin E.

**Total synthesis of vineomycinone B2 methyl ester (95).** In 1985, Danishefsky and coworkers reported the first total synthesis of vineomycinone B2 methyl ester (**95**, Scheme 2.14) [36]. The key *C*-glycoside was constructed via a hetero-Diels-Alder reaction, whereby the addition of aldehyde **98** and diene **97** under the catalysis of $Eu(fod)_3$ gave **99** (95%). Compound **99** was subjected to hydroboration ($BH_3$·DMS) and subsequent oxidation, providing 2,6-dideoxyglycoside **100**. Lastly, the target molecule (**95**) was obtained via removal of the methyl and silyl ethers, followed by addition of (carbomethoxy)methylmagnesium to elaborate the side chain, albeit as a racemic and diastereomeric mixture.

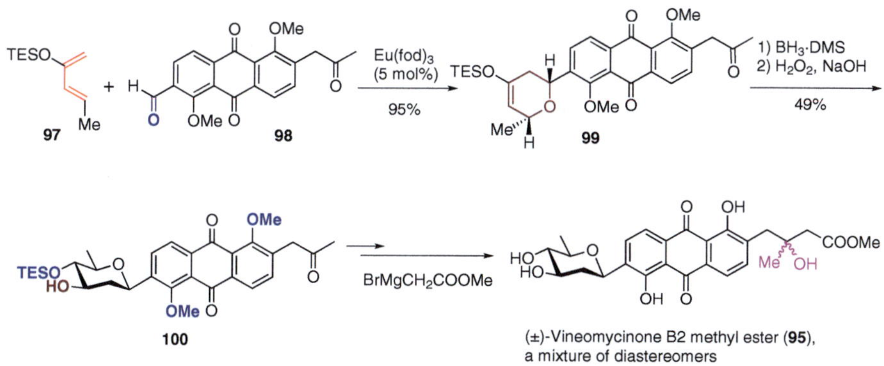


**Scheme 2.14** Synthesis of vineomycinone B2 methyl ester (**95**) by Danishefsky and coworkers.

In 1991, Mioskowski, Falck, and coworkers reported the total synthesis of vineomycinone B2 methyl ester (**95**) through a double Bradsher cyclization (Scheme 2.15) [37]. After the activation of benzoisoquinoline **101** with 2,4-di-nitrophenylbromide (DNP-Br), a Bradsher cyclization with glycosyl vinyl ether **102** led to the desired β-*C*-glycoside **103** in 47% yield. Dakin oxidation followed by singlet oxygen addition furnished anthraquinone **104**. Finally, oxidative construction of the methyl ester motif and cleavage of the BMP (4-PhPh$CH_2$–) protecting groups by catalytic hydrogenation accomplished the total synthesis of vineomycinone B2 methyl ester (**95**) in 52% yield.

In the same year, Tius and coworkers reported their results on the total synthesis of vineomycinone B2 methyl ester (**95**, Scheme 2.16) [38]. The *C*-glycosidic bond was constructed by Pd-catalyzed coupling of anthracene iodide **105** with glycalzinc **106**, yielding **107** in 78% yield. Stereoselective reduction of the glycal motif provided the desired β-*C*-glycoside. The introduction of a tributylstannane residue led to **108**, which was coupled with bromide **109** under Pd-catalysis to provide **110** in 50% yield. Stereoselective introduction of the methyl group and oxidative generation of the anthraquinone moiety led to **111**, which was treated with HCl in MeOH to cleave the MOM (methoxymethyl) ether, silyl ethers, and acetal group, and to form the methyl ester in a one-step process (76%), thus completing the total synthesis.

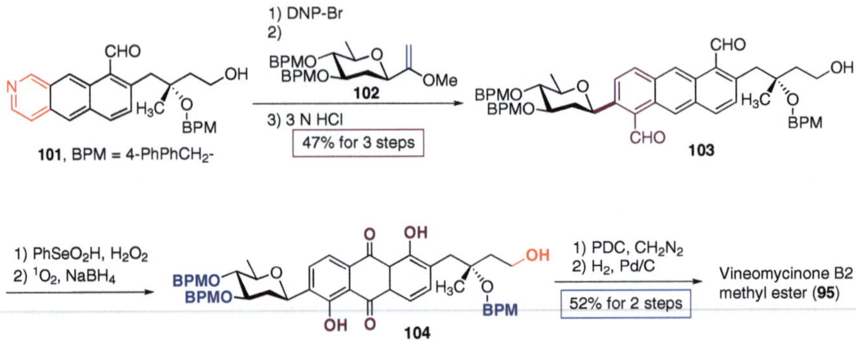


**Scheme 2.15** Total synthesis of vineomycinone B2 methyl ester (**95**) by Mioskowski, Falck and coworkers.

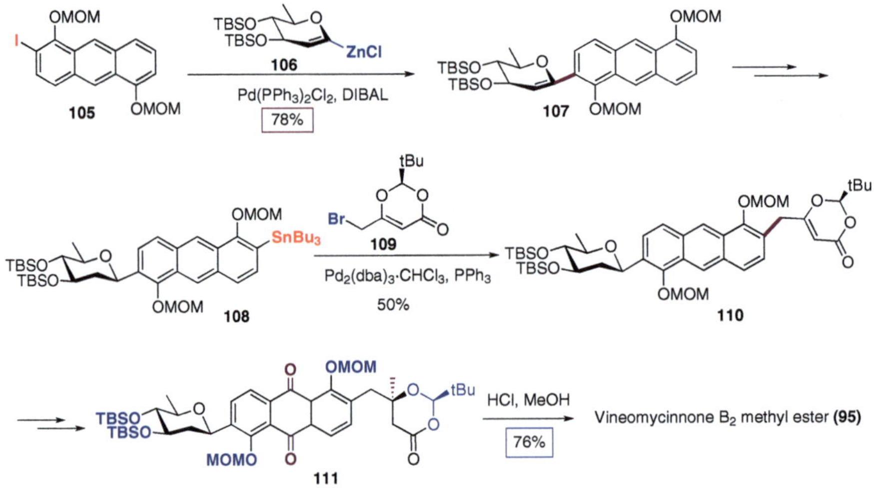


**Scheme 2.16** Total synthesis of vineomycinone B2 methyl ester (**95**) by Tius and coworkers.

In 1991, Suzuki and coworkers reported their total synthesis of vineomycinone B2 methyl ester (**95**, Scheme 2.17) [39]. The glycosylation of anthrol **112** with glycosyl fluoride **113** in the presence of $Cp_2HfCl_2$-$AgClO_4$ provided the desired β-*C*-glycoside **114** in 86% yield with complete β-selectivity. This reaction was proposed to undergo through an $O \rightarrow C$-glycoside rearrangement. After manipulating the protecting group and introducing the trimethylstannane residue, compound **115** was lithiated and then added onto aldehyde **116** to provide **117**. Finally, selective deoxygenation, oxidative cleavage of the double bond, and removal of all the methyl and silyl protecting groups accomplished the total synthesis.

In 2013, O'Doherty and coworkers accomplished the total synthesis of vineomycinone B2 methyl ester (**95**) by employing Suzuki's *C*-glycosylation method, in which a D-olivose derivative was prepared from commercially available acyl furan through a Pd-catalyzed asymmetric synthesis route [40].

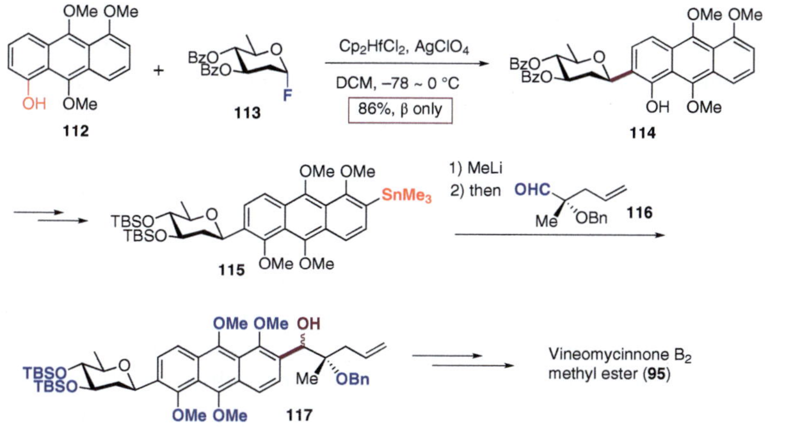


**Scheme 2.17** Total synthesis of vineomycinone B2 methyl ester (**95**) by Suzuki and coworkers.

In 2006, Martin and coworkers reported a novel and highly convergent total synthesis of vineomycinone B2 methyl ester (**95**, Scheme 2.18) [41]. Their synthesis began with the addition of lithiated furan **119** onto lactone **118**, which was followed by reduction with $NaCNBH_3$ to give β-glycosylfuran **120** in 80% yield. Furan **120** was converted into silicon-substituted alcohol **121**, which was subjected to Mitsunobu reaction with phenol **122** to provide the Diels-Alder precursor **123**. Next,

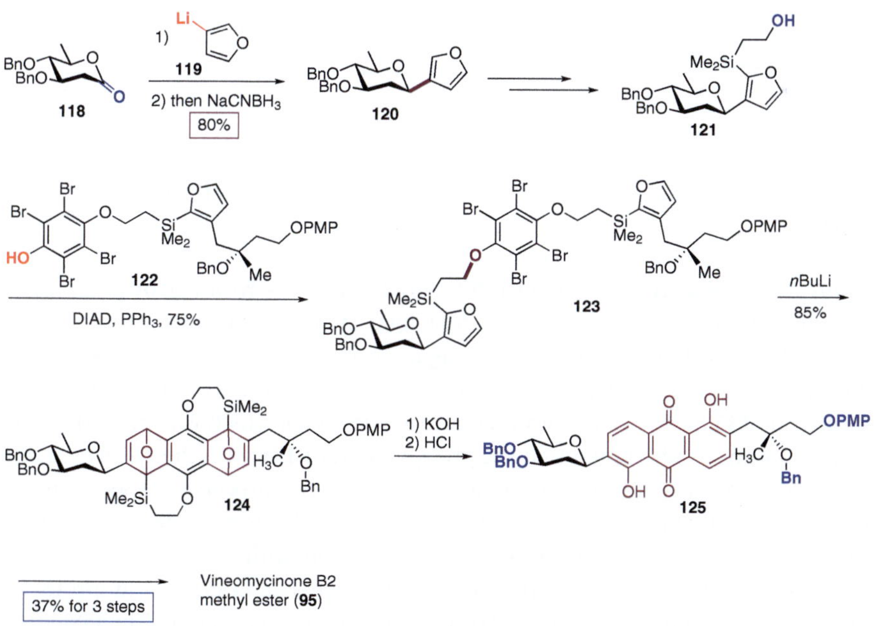


**Scheme 2.18** Total synthesis of vineomycinone B2 methyl ester (**95**) by Martin and coworkers.

the pivotal domino benzyne-furan cycloaddition was enabled by *n*BuLi, affording the biscycloaddition product **124** in 85% yield, which was then transformed into anthraquinone derivative **125** via treatment with KOH and subsequently HCl/EtOH. Finally, sequential removal of the *para*-methoxyphenyl (PMP) group, oxidation of the alcohol to carboxylic acid, removal of the Bn groups, and formation of the methyl ester accomplished the total synthesis of vineomycinone B2 methyl ester (**95**) in 37% yield.

**Synthesis of vineomycinone B2 trisaccharide by the O'Doherty group.** In 2008, O'Doherty and coworkers reported a de novo synthesis of a trisaccharide derivative (**130**) of vineomycinone B2 (Scheme 2.19) [42]. The coupling of olivoside **126** with α-L-pyranone **127** under Pd-catalyzed conditions afforded disaccharide **128** with complete stereocontrol at the anomeric center. Subsequent functional group transformations led to diol **129**, which was subjected to a regioselective glycosylation with α-L-pyranone **127** under the previous Pd-catalyzed conditions to provide trisaccharide **130** in 35% yield, along with 61% of the recovered starting material (89% brsm). The low conversion was found necessary to achieve the regioselectivity with the reaction being performed at a dilute concentration.

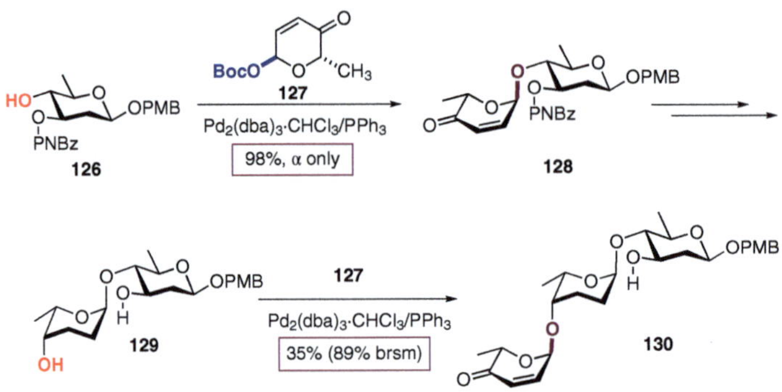


**Scheme 2.19** Synthesis of vineomycinone B2 trisaccharide **130** by O'Doherty and coworkers.

**Total synthesis of vineomycin B2 (96) by the Toshima group.** In 2013, Toshima and coworkers reported the first total synthesis of vineomycin B2 (**96**, Scheme 2.20) [43]. Glycosylation of sugar acceptor **131** with 2,3-unsaturated glycosyl acetate **132**, under the action of TBSOTf at −98 °C, gave the desired disaccharide **133** in 74% yield with exclusive α-selectivity, which was converted into **134** in two steps. It was noted that the glycosylation of **131** with the corresponding 2,3-unsaturated-4-keto glycosyl acetate as donor did not proceed at −98 °C, while the reaction at a higher temperature, such as at −78 °C, led to considerable amounts of byproducts. The elaboration of the aglycone began with a stereo- and regioselective *C*-glycosylation of 5,9,10-trimethoxyanthracen-1-ol (**135**) with unprotected D-olivose (**136**) in the presence of TMSOTf, providing *C*-glycoside **137** in 45% yield. Steps of transformation were employed to convert **137** into aryl stannane derivative **138**, which was then coupled with enantio-enriched aldehyde **139** in the presence

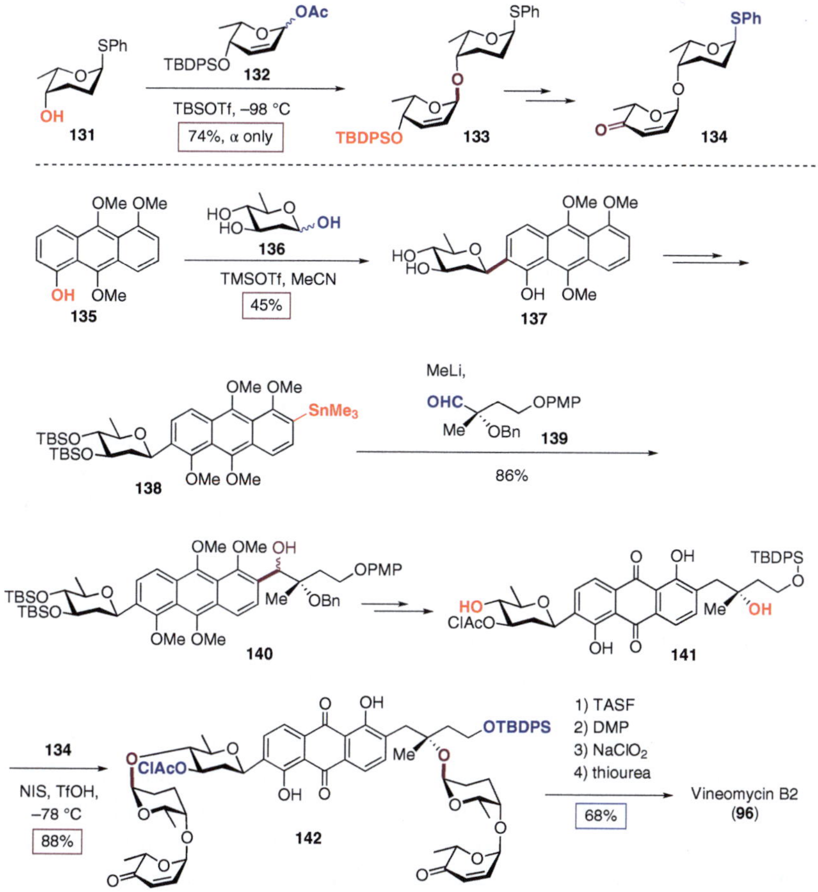

Scheme 2.20 Total synthesis of vineomycin B2 (**96**) by Toshima and coworkers.

of MeLi to give **140** in 86% yield. Removal of the benzylic alcohol and adjustment of protecting groups afforded diol **141**, which was coupled with thiodisaccharide **134** in the presence of NIS/TfOH to give the desired pentasaccharide **142** in 88% yield. Notably, it was found that a high concentration of substrates (25 mM for **141**) and an excess amount of disaccharide donor **134** (4 equiv) were critical to ensure the high yield of this glycosylation reaction. Finally, deprotection of the TBDPS group with TASF, followed by oxidation of the resulting primary alcohol to carboxylic acid, and cleavage of the ClAc group with thiourea furnished vineomycin B2 (**96**) in 68% yield (four steps).

## Trioxacarcin

The trioxacarcin family natural products were first isolated from *Streptomyces bottropensis* DO-4 by Tomita et al. in 1981 [44], with trioxacarcin A (**143**, Figure 2.6)

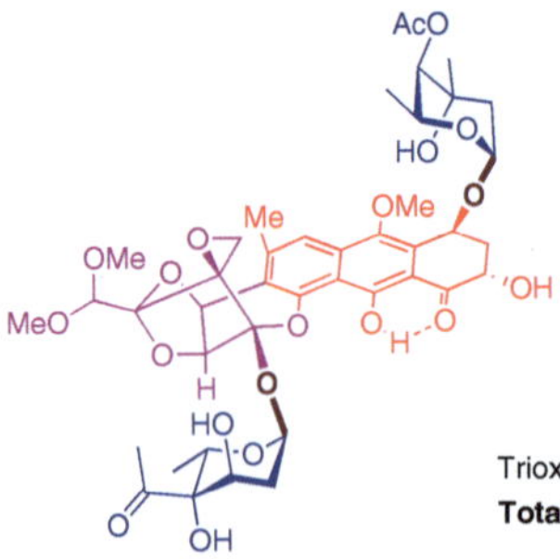


**Figure 2.6** Trioxacarcin A (**143**) and its structural features.

being the most toxic member, exhibiting subnanomolar 70% growth inhibition ($GI_{70}$) values against various human cancer cell lines. Structurally, trioxacarcin A (**143**) consists of a rigid and highly oxygenated polycyclic skeleton bearing a fused spiro epoxide function, five ketal/hemiketal residues (three of them within a span of four contiguous carbon atoms), and two unusual glycosidic residues at hindered hydroxyl groups. The remarkable cytotoxicity of trioxacarcin A could be attributed to its ability to efficiently and irreversibly alkylate the N7 of G residues in duplex DNA via opening of the spiro epoxide residue.

**Total synthesis of trioxacarcin A (143) by the Myers group.** In 2013, Myers and coworkers reported the first total synthesis of trioxacarcin A (**143**) using a convergent approach with five modular components (Scheme 2.21) [45, 46]. The synthesis of the aglycone commenced with a Kraus-Sugimoto annulation between cyanophthalide **144** and enone **145** in the presence of *t*BuOLi, providing anthrone methyl ether **146** in 70% yield, after methylation with dimethylsulfate. Compound **146** was converted into aldehyde **147**, which was subjected to condensation with epoxy diazo diketone **148** under the catalysis of rhodium(II) acetate. The reaction, leading to a mixture of diastereomeric cycloadducts, went through a 1,3-dipolar cycloaddition between an *in situ* generated five-membered cyclic carbonyl ylide and the aromatic aldehyde with high endo-selectivity. Cleavage of the cyclic di-tertbutylsiloxane protecting group facilitated the separation of the resulting stereoisomers, giving the desired polycyclic product **149** in 26% yield (two steps). Next, glycosylation of **149** with trioxacarcinose B acetate **150** in the presence of $TMSNTf_2$ provided the desired glycoside **151** in 78% yield, with complete α-selectivity. It was noted that attempts to glycosylate **149** with other trioxacarcinose B donors, such as fluoride and sulfide donors, were unsuccessful under various conditions. Removal of the PMB ether, followed by glycosylation with 1-phenylthiotrioxacarcin A **152** in the presence of $AgPF_6$ and DTBMP led to α,α-bisglycoside **153** in 74% yield. This glycosylation sequence was found necessary because the preinstalled trioxacarcinose A residue could not survive under the glycosylation conditions for installation of the trioxacarcinose B residue. Finally, selective removal of the acetate on the trioxacarcinose B residue (with $K_2CO_3$ in MeOH) followed by cleavage of the silyl ether (with $3HF{\cdot}Et_3N$) accomplished the synthesis of trioxacarcin A (**143**) in 55% yield over two steps.

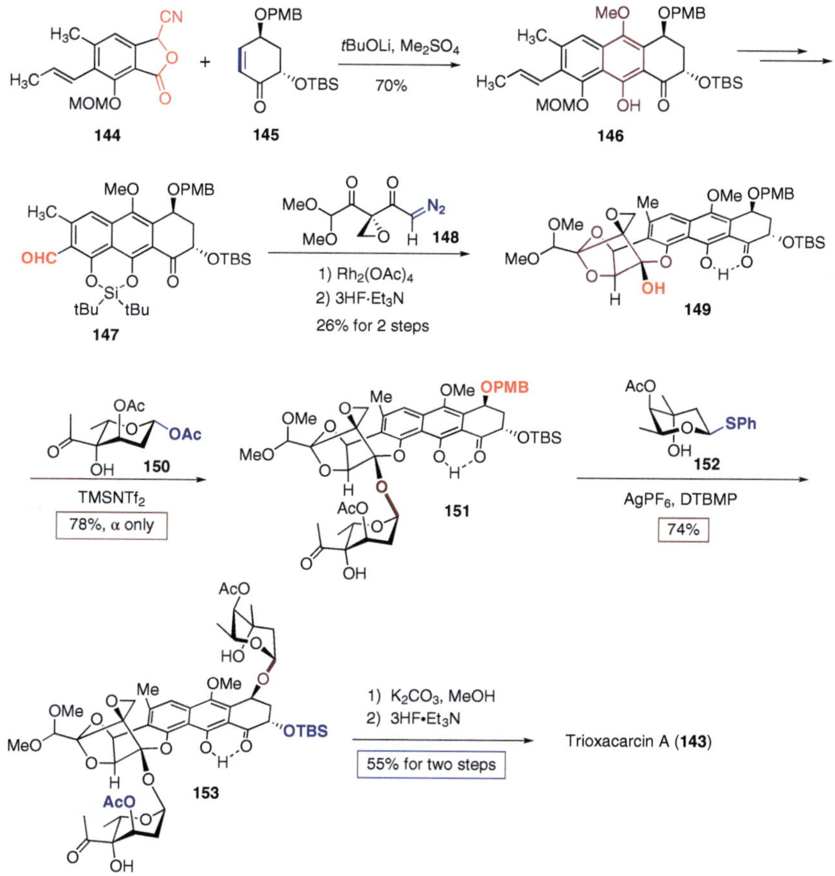


**Scheme 2.21** Total synthesis of trioxacarcin A (**143**) by Myers and coworkers.

**Total synthesis of trioxacarcin A (143) by the Nicolaou group.** In 2016, Nicolaou et al. reported the second total synthesis of trioxacarcin A (**143**), together with trioxacarcin D (**163**) and C (**164**, Scheme 2.22) [47–49]. The synthesis began with a Hauser-Kraus annulation between iodocyanophthalide **154** and cyclic enone **145** under the promotion of *t*BuOLi, providing tricyclic **155** in 70% yield. Subsequent Stille coupling of aryl iodide **155** with vinyl stannane **156**, followed by asymmetric epoxidation, led to epoxide **157**. A DABCO-mediated Baylis–Hillman reaction of aldehyde **157** with enone **158** provided **159**, which underwent a epoxyketone rearrangement under the action of a monodentate Lewis acid (i.e. $BF_3{\cdot}Et_2O$) to furnish **160** bearing the key dioxabicyclo[2.2.1]heptane motif with the desired configurations. Notably, the use of a bidentate Lewis acid, such as $SnCl_4$, led to the formation of the opposite diastereomers. After installation of the epoxide moiety, oxidation at C4, and deprotection, **160** was successfully converted into the suitably protected trioxacarcin aglycone **149**. The trioxacarcinose B residue was introduced by glycosylation of **149** with glycosyl *ortho*-alkynylbenzoate **161** in the presence of $PPh_3AuOTf$, affording the desired **151** in 91% yield with excellent α-selectivity

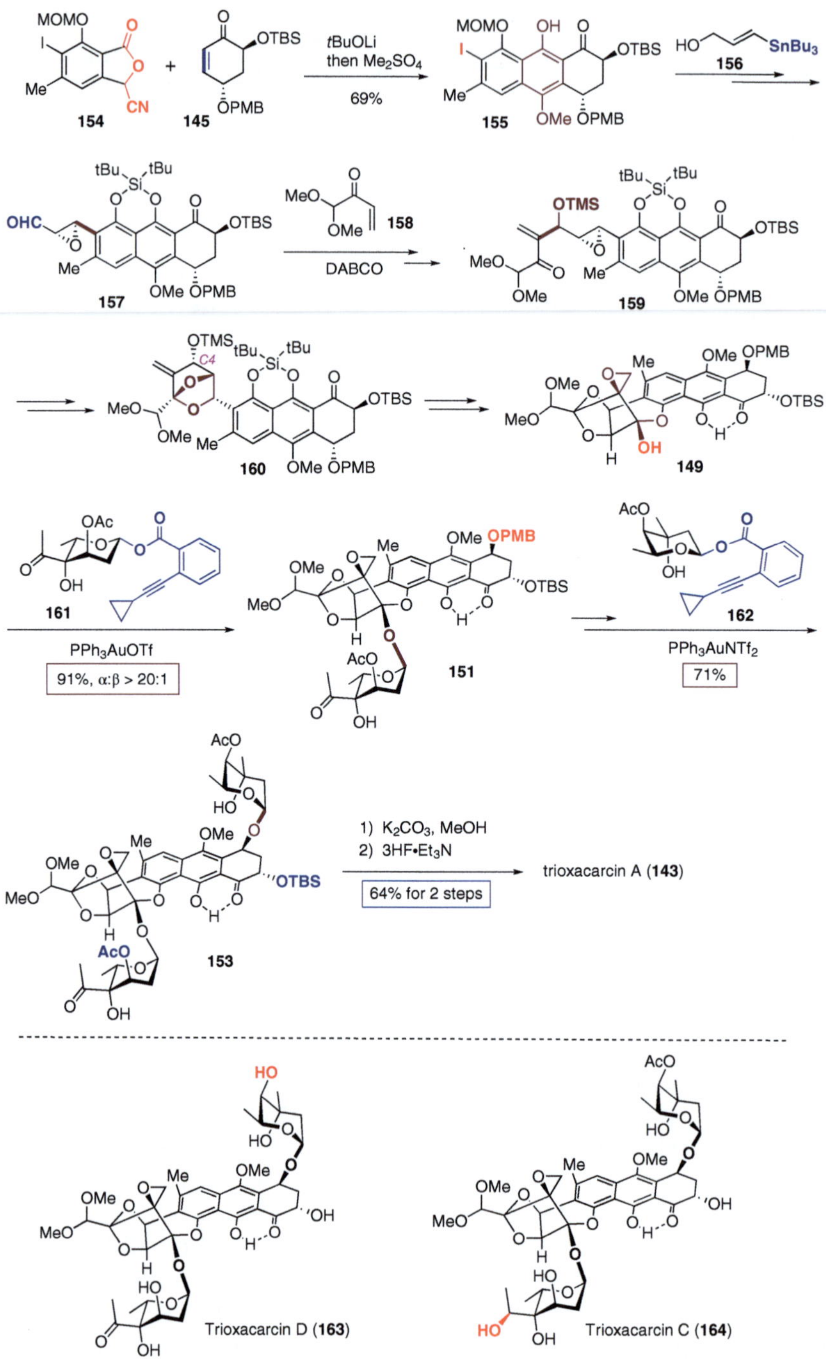


**Scheme 2.22** Total synthesis of trioxacarcin A (**142**), C (**164**), and D (**163**) by Nicolaou and coworkers.

($\alpha/\beta > 20:1$). After removal of the PMB group, the second glycosylation with glycosyl *ortho*-alkynylbenzoate **162** under the catalysis of $PPh_3AuNTf_2$ provided α,α-bisglycoside **153** in 71% yield. Finally, selective removal of the acetyl group on the trioxacarcinose B residue and the TBS group accomplished the synthesis of trioxacarcin A (**143**). By the same token, trioxacarcin D (**163**) and C (**164**) were synthesized.

## Daunorubicin

Daunorubicin (**165**), a congener of the clinically important anticancer drug doxorubicin (adriamycin, **1**), was originally discovered by Brockmann et al. in 1963 (Figure 2.7a) [50]. These compounds comprise a 7,8,9,10-tetrahydro-5,12-naphthacenequinone framework and an α-deoxy(amino)sugar, namely daunosamine, attached at the saturated D ring. In 1974, Acton and coworkers reported the first synthesis of daunorubicin (**165**), which also represents a formal synthesis of doxorubicin (adriamycin, **1**) (Figure 2.7b) [51]. Recently, Neefjes, Codée, and coworkers reported the synthesis of a series of doxorubicin/aclarubicin hybrids, employing the gold(I)-catalyzed glycosylation with glycosyl *ortho*-alkynylbenzoate donors as the key step (Figure 2.7c) [52].

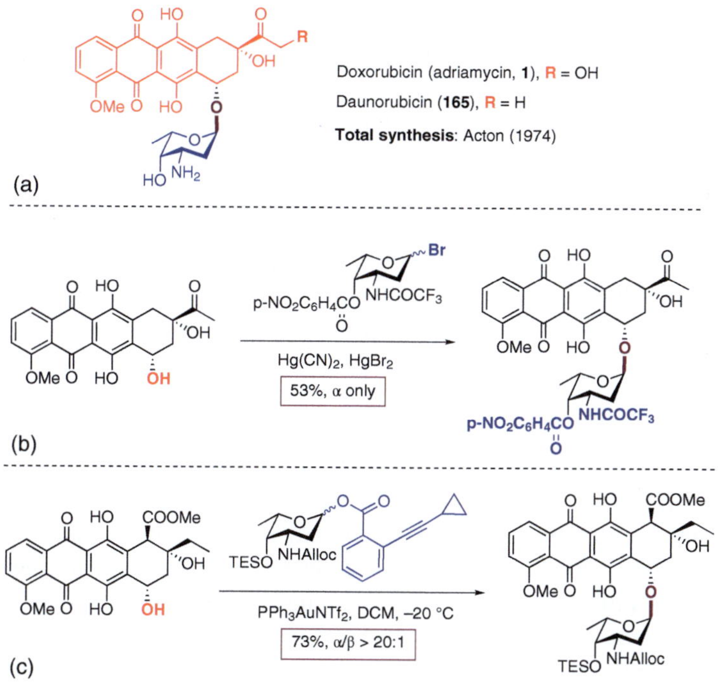


**Figure 2.7** Daunorubicin (**165**) and the glycosylation steps in the synthesis. (a) Structure of daunorubicin. (b) Key glycosylation step in the synthesis of daunorubicin by Acton and coworkers. (c) Key glycosylation step in the synthesis of daunorubicin by Neefjes, Codée, and coworkers.

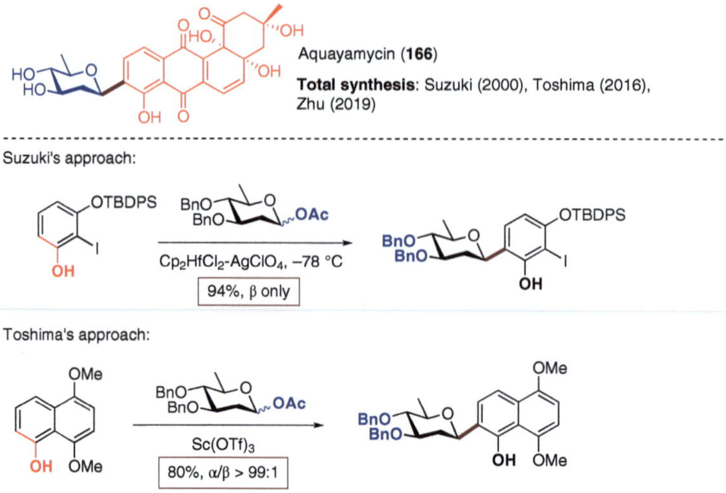


**Figure 2.8** Aquayamycin (**166**) and the *C*-glycosylation steps in the total synthesis.

## Aquayamycin

Aquayamycin (**166**) was isolated in 1967 by Umezawa and coworkers from the culture broth of *Streptomyces misawanensis* (Figure 2.8) [53]. It shows potent activities against gram-positive bacteria and tumor cells. Structurally, it features a *C*-glycosylated benz[*a*]anthraquinone skeleton and two hydroxyl groups at the ring junction (C-4a and C-12b). The total synthesis of aquayamycin (**166**) has been reported by the groups of Suzuki and Toshima in 2002 [54–57] and 2016 [58], respectively. Both syntheses utilize *C*-glycosylation of phenols to construct the *C*-glycoside linkage at an early stage. In 2019, Zhu and coworkers reported the total synthesis of aquayamycin (**166**) using a convergent synthetic strategy [59], which was previously used for the synthesis of derhodinosylurdamycin A (Figure 2.10).

## Vineomycin $A_1$

Vineomycin $A_1$ (**167**) is an angucycline antibiotic, which was first isolated by Ōmora et al. from the culture broth of *Streptomyces matensis* subsp. *vineus* in 1977 (Figure 2.9) [60]. This compound has demonstrated various biological activities, including antibacterial, enzyme inhibitory, and antitumor activities. In 2019, Takahashi, Toshima, and coworkers reported the first synthesis of vineomycin $A_1$ (**167**) through direct glycosylation of a naturally occurring aquayamycin derivative with a disaccharide sulfide in the presence of NIS and TfOH [61].

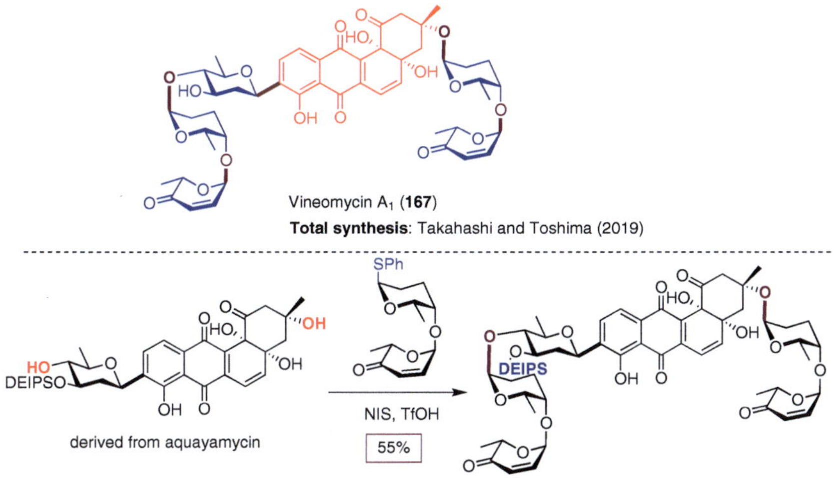


**Figure 2.9** Vineomycin $A_1$ (**167**) and the key *O*-glycosylation step in the total synthesis.

## Derhodinosylurdamycin A

Derhodinosylurdamycin A (**168**) was isolated as a major urdamycin metabolite from *Streptomyces fradiae* Tü2717 mutant lacking UrdGT1a in 1989 (Figure 2.10) [62]. It shows significant activity against L1210 leukemia cell lines ($IC_{50}$ = 0.75 μg $ml^{-1}$), comparable to urdamycinA ($IC_{50}$ = 0.55 μg $ml^{-1}$). Structurally, derhodinosylurdamycin A (**168**) is a longer congener of aquayamycin (**166**), containing two additional deoxysugars at the aryl C–D-olivoside residue. In 2015, Zhu and coworkers reported the total synthesis of derhodinosylurdamycin A (**168**), employing a late-stage installation of the disaccharide residue [63].

## Jadomycins

Jadomycins are produced when *Streptomyces venezuelae* is fed under stress conditions with amino acids (including non-natural amino acids) as the nitrogen source [64]. These compounds constitute a unique family of angucycline antibiotics featuring an 8*H*-benzo[*b*]phenanthridine skeleton fused with an amino acid unit and an α-L-digitoxose unit at C12-OH. Jadomycins show potent antitumor activities, attributable to their ability to inhibit the Aurora-B kinase and induce DNA cleavage. In 2013, Yu and coworkers reported the total synthesis of jadomycin B (**169**), S (**171**), T (**172**), and ILEVS1080 (**170**), wherein the *O*-glycosylation was only achieved under Mistunobu conditions with a lactol donor (Figure 2.11) [65]. Previous efforts from the O'Doherty group led to the synthesis of a carbasugar analog of jadomycin B [66]. In 2022, Akagi and coworkers reported the total synthesis of jadomycin B (**169**) by employing a similar Mistunobu glycosylation step [67].

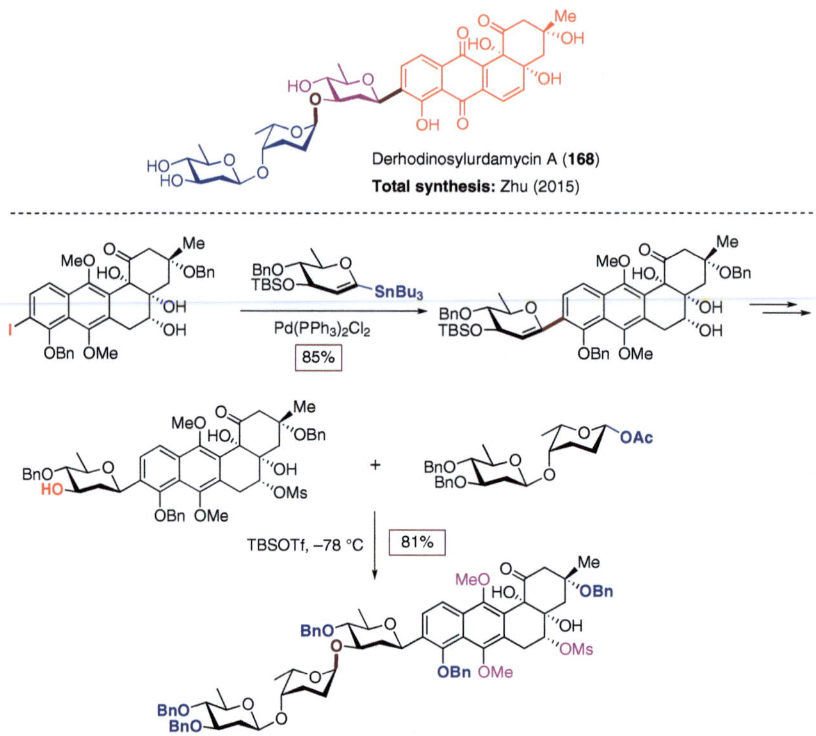


**Figure 2.10** Derhodinosylurdamycin A (**168**) and the glycosylation steps in the total synthesis.

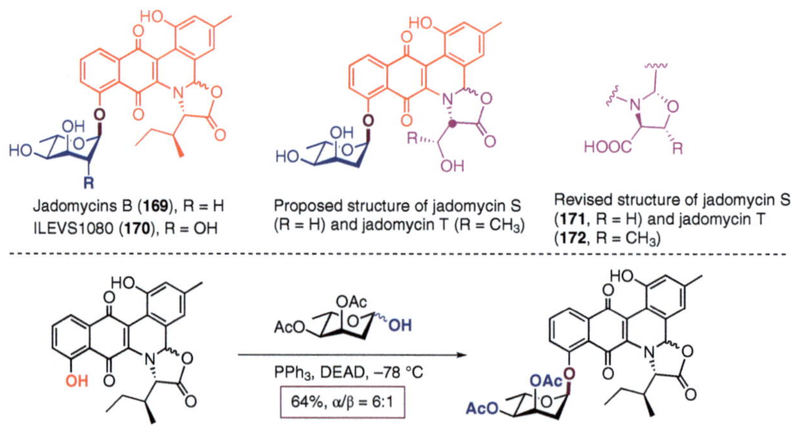


**Figure 2.11** Jadomycins and the glycosylation step in the total synthesis.

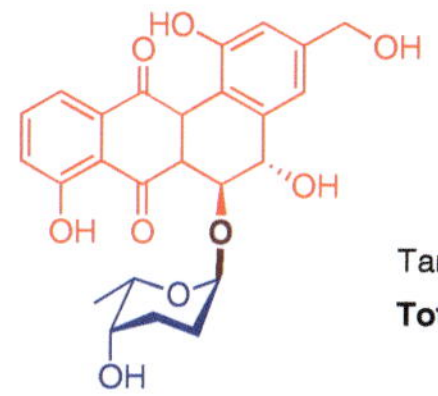

Tan-1085 (**173**)
**Total synthesis**: Suzuki (2004, 2009), Shi (2019)

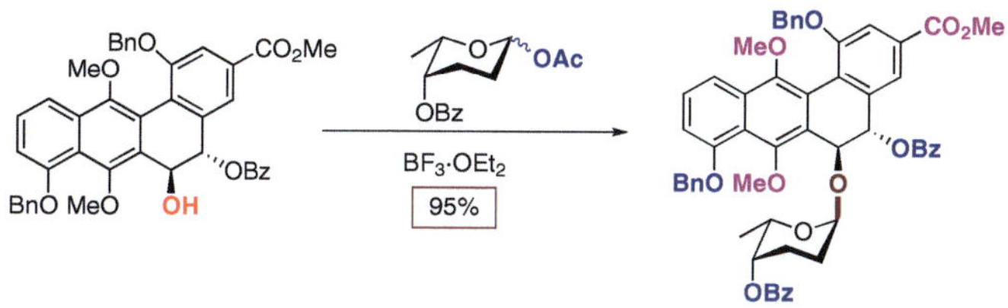


**Figure 2.12** Tan-1085 (**173**) and the glycosylation step in the total synthesis.

## Tan-1085

TAN-1085 (**173**) is produced by *Streptomyces* sp. S-11106 and exhibits potent antitumor activities (Figure 2.12) [68]. Structurally, it consists of a unique *trans*-9,10-dihydrophenanthrene-9,10-diol scaffold, with the C6-OH being glycosylated with L-rhodinose through α-glycosidic bond. In 2004, Suzuki and coworkers reported the first total synthesis and stereochemical assignment of TAN-1085 (**173**). The vicinal diol in the aglycone was synthesized in racemic form, and the subsequent glycosylation led to a pair of the diastereomers. In 2009, the same group reported the second-generation total synthesis, which involved an asymmetric synthesis of the vicinal diol through an axial-to-central chirality transfer strategy [69]. In 2019, Shi and coworkers reported their asymmetric total synthesis of TAN-1085 (**173**), in which a Pd-catalyzed atroposelective C–H olefination was employed for the construction of the axially chiral biaryl scaffold, which was then converted into the chiral diol via axial-to-central chirality transfer [70].

## Benanomicins and Pradimicins

The benanomicin/pradimicin antibiotics (BpAs) were isolated from the culture of *Actinomyces* species in 1988 (Figure 2.13) [71]. Structurally, they feature a benzo[*a*]naphthacene core with various amino acid and disaccharide appendages. These compounds exhibit potent antifungal and anti-HIV activities, attributable to $Ca^{2+}$-mediated specific binding to the mannose-rich oligosaccharides presented on fungi or virus surfaces, making them "lectin-mimics." In 2005, Suzuki and coworkers reported the total synthesis of benanomicn B (**174**) [72], then the synthesis of benanomicin A (**176**) and pradimicin A (**175**) in 2007 [73]. The glycosylation was realized at a late stage with a disaccharide fluoride as donor under the promotion of $Cp_2HfCl_2$/AgOTf.

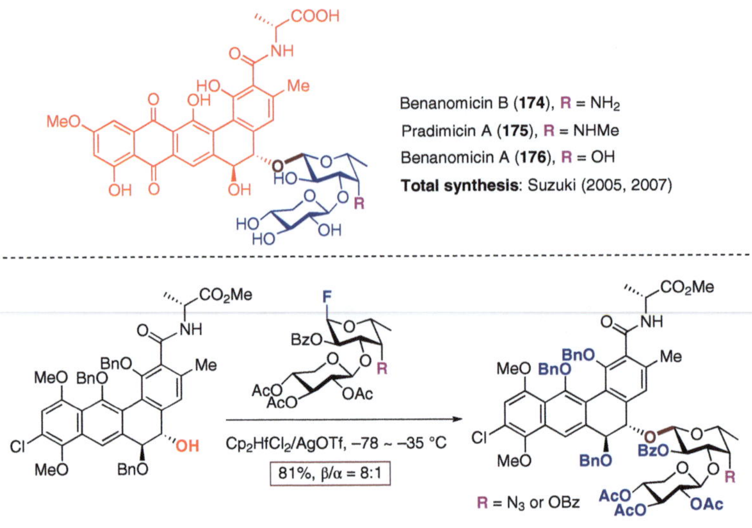


**Figure 2.13** Benanomicins B (**174**) and A (**176**) and pradimicin A (**175**) and the glycosylation step in the total synthesis.

## Pluramycins

Kidamycin (**177**), a C-aryl bisglycoside isolated from *Streptomyces phaeoerticillatus*, is light and acid sensitive and easily transformed into isokidamycin (**178**, the β-anomer) upon treatment with acid (Figure 2.14) [74]. Saptomycin B (**179**) was isolated from *Streptomyces* sp. HP530 from a soil sample collected in Chiba, Japan in 1993 [75]. All these compounds belong to the pluramycin family of antibiotics, which share a common bis-*C*-glycoside chromophore with a distinctive six-carbon-atom side chain. Saptomycin H (**180**) is another congener in this family, featuring a side chain bearing an epoxide with unknown stereochemistry and a *C*-glycoside of 4-*O*-acetyl-*N*,*N*-dimethyl-L-vancosamine. They display a broad range of antibacterial, antifungal, and anticancer properties. In 2010, Martin and coworkers reported the total synthesis of isokidamycin (**178**), featuring the use of a silicon tether as a disposable regiocontrol element for an intramolecular Diels-Alder reaction between a substituted naphthyne and a glycosyl furan, followed by an *O* → *C*-glycoside rearrangement [76, 77]. In 2014, Suzuki and coworkers reported the total synthesis and structure assignment of saptomycin B (**179**), featuring a regioselective installation of the two amino sugars (L-vancosamine and D-angolosamine) onto the tricycle scaffold and an efficient construction of the tetracyclic skeleton by an aldol reaction followed by pyranone formation [78, 79]. In 2022, Suzuki and coworkers reported the first total synthesis of saptomycin H (**180**) using the common mono-glycosylated intermediate, and the unidentified absolute stereochemistry of the oxiranyl side chain has been determined as 14*R*, 16*S* [80].

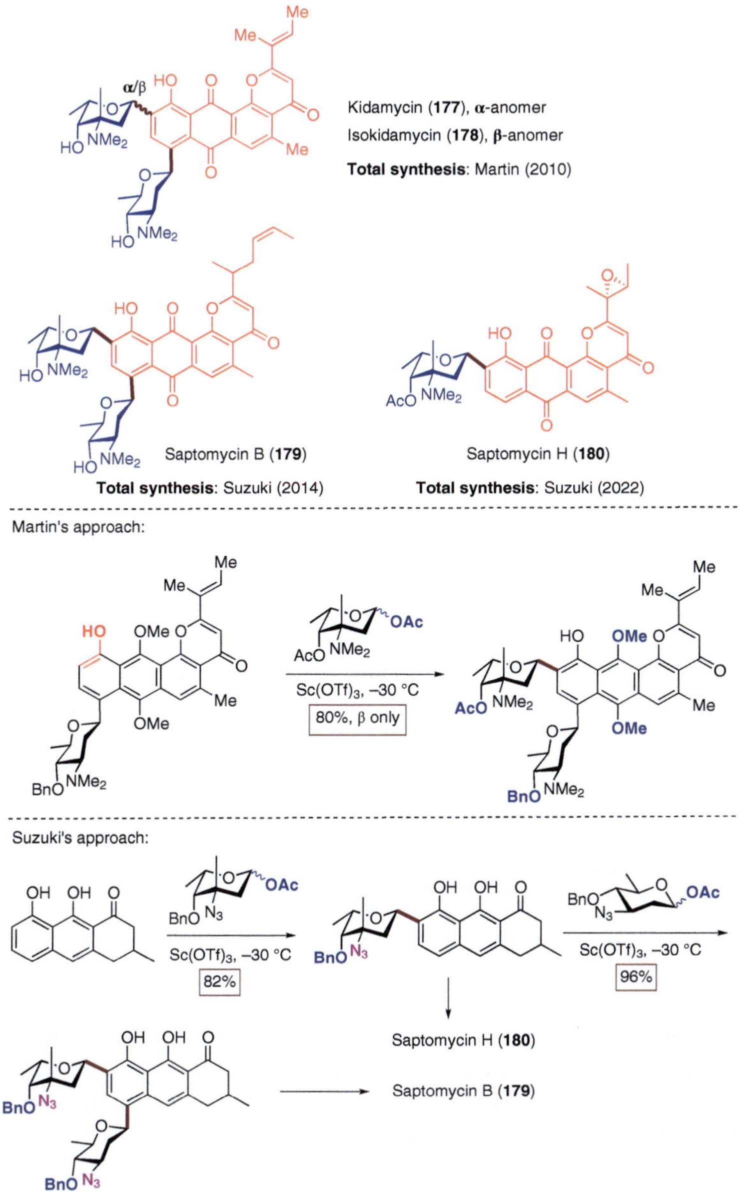


**Figure 2.14** Kidamycin (**177**), isokidamycin (**178**), saptomycin B (**179**), saptomycin H (**180**) and the *C*-glycosylation steps in their total synthesis.

## Marmycins

Marmycin A (**181**) and B (**182**) were isolated from a marine *Streptomyces*-related actinomycete in 2007 (Figure 2.15) [81]. These compounds have an unprecedented structural feature in which the aminosugar is fused with the aglycone via a

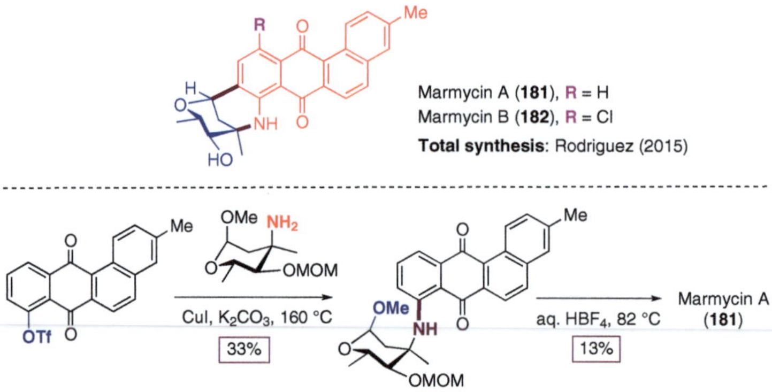


**Figure 2.15** Marmycin A (**181**) and B (**182**) and the glycosylation step in the total synthesis.

*C*-glycosidic bond and an aryl C—N bond. Marmycin A (**181**) was reported to exhibit significant cytotoxicity against cancer cell lines, particularly with an $EC_{50}$ value of 60.5 nM against human colon carcinoma cell line HCT-116. In 2015, Rodriguez and coworkers reported the total synthesis of marmycin A (**181**) and B (**182**), wherein the sugar was installed through a copper-catalyzed Ullmann coupling, followed by a low-yielding intramolecular Friedel–Crafts *C*-glycosylation [82].

## Cassialoin

Cassialoin (**183**) was isolated from the heartwood of *Cassia garrettiana* CRAIB [83] and the roots of *Rheum emodi* Wall [84], which have been used in Ayurvedic and Asian folk medicines. It is a representative member of the anthrone *C*-glycosides, featuring a *C*-glycoside at the hydroxy C10. Cassialoin (**183**) exhibits inhibitory effects on tumor growth and metastasis in the abdomen, reduces the expression of CD31 (an angiogenesis marker) in tumors, and increases the numbers of CD8+ T and natural killer cells in the small intestine and spleen of colon 26-bearing mice. In 2008, Suzuki and coworkers reported the total synthesis and structure validation of cassialion (**183**), in which the sugar residue was introduced via the addition of a glycal anion onto a C10-ketone derivative (Figure 2.16) [85].

## FD-594

FD-594 (**184**), a highly oxygenated polycyclic xanthone-type natural product, was isolated from *Streptomyces* sp. TA-0256 by Kakinuma and coworkers in 1998 (Figure 2.17) [86]. Its hexacyclic framework comprises an isochromanone fragment (A–B ring), a *trans*-9,10-dihydrophenanthrene-9,10-diol fragment (B–C–D ring), a xanthone fragment (D–E–F ring), and a β-linked trisaccharide unit including D-oleandrose and D-olivose. FD-594 exhibits antitumor and antibacterial properties

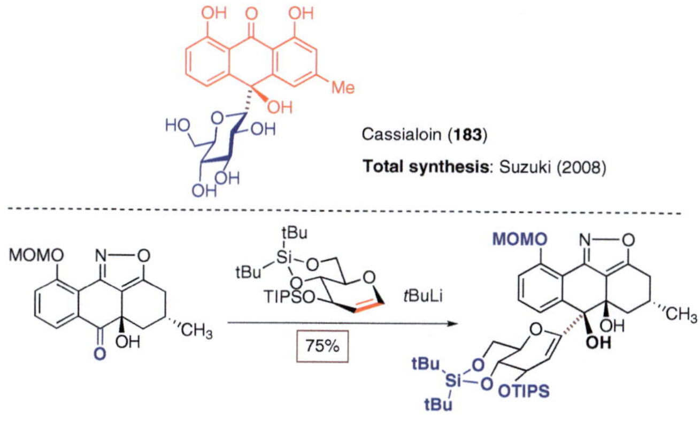


**Figure 2.16** Cassialoin (**183**) and the *C*-glycosylation step in the total synthesis.

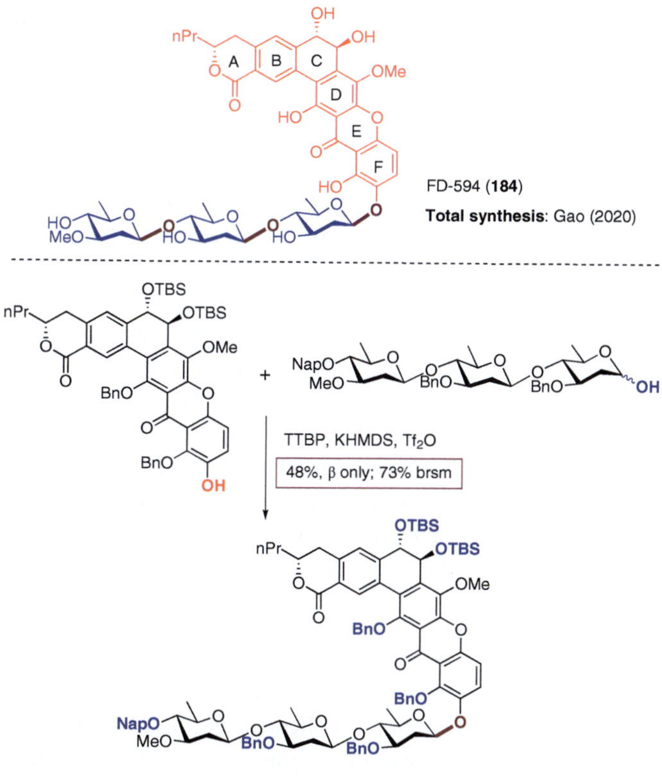


**Figure 2.17** FD-594 (**184**) and the key glycosylation step in the total synthesis.

against P388 and HeLa tumor cells ($IC_{50}$ = 0.25 and 0.10 μg ml$^{-1}$, respectively) and gram-positive bacteria such as *Staphylococcus aureus* 209P-J ($IC_{50}$ = 0.10 μg ml$^{-1}$). In 2020, Gao and coworkers reported the total synthesis of FD-594 (**184**), featuring a late-stage dehydrative glycosylation to install the trisaccharide fragment [87].

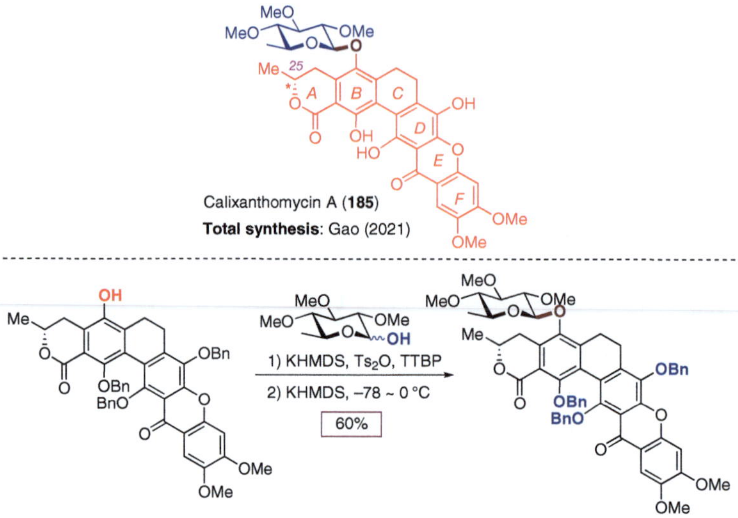


**Figure 2.18** Calixanthomycin A (**185**) and the key glycosylation step in the total synthesis.

## Calixanthomycin A

Calixanthomycin A (**185**) was isolated from the cultures of *Staphylococcus albus* BAC-AB1692/916/170 by Brady and coworkers in 2014 (Figure 2.18) [88], which displayed potent antiproliferative activity against HCT-116 cells ($IC_{50}$ = 0.43 nM). Structurally, calixanthomycin A (**185**) features a highly oxygenated angular hexacyclic skeleton comprising a lactone A ring, a xanthone D–E–F ring, and a β-linked monosaccharide, but the absolute configuration of the lactone (C25) and monosaccharide was unestablished. In 2021, Gao and coworkers reported the first asymmetric total synthesis and structural determination of calixanthomycin A (**185**) through a modular strategy in LLS 15 steps [89]. The β-monosaccharide was introduced with high diastereoselectivity at a late stage through an $S_N2$-type reaction of the in situ generated glycosyl tosylate species under basic conditions.

## Lactonamycin

Lactonamycin (**186**) was isolated in 1996 by Matsumoto et al. from a culture broth of *Streptomyces rishiriensis* MJ773-88K4 collected at Yokohama, Japan (Figure 2.19) [90]. Its potent antimicrobial activities against gram-positive bacteria, including methicillin-resistant *Staphylococcus aureus* (MRSA) and vancomycin-resistant *Enterococcus* (VRE), as well as its cytotoxicity against various tumor cell lines make it an attractive lead compound for drug discovery. Structurally, lactonamycins feature a hexacyclic core incorporating a densely oxygenated hydrofuran hydrofuranone (EF) ring system, which is glycosylated by L-rhodinose (lactonamycin **186**) or digitoxose (lactonamycin Z **187**). However, the absolute configuration and stereochemical relationship between the aglycone and sugar

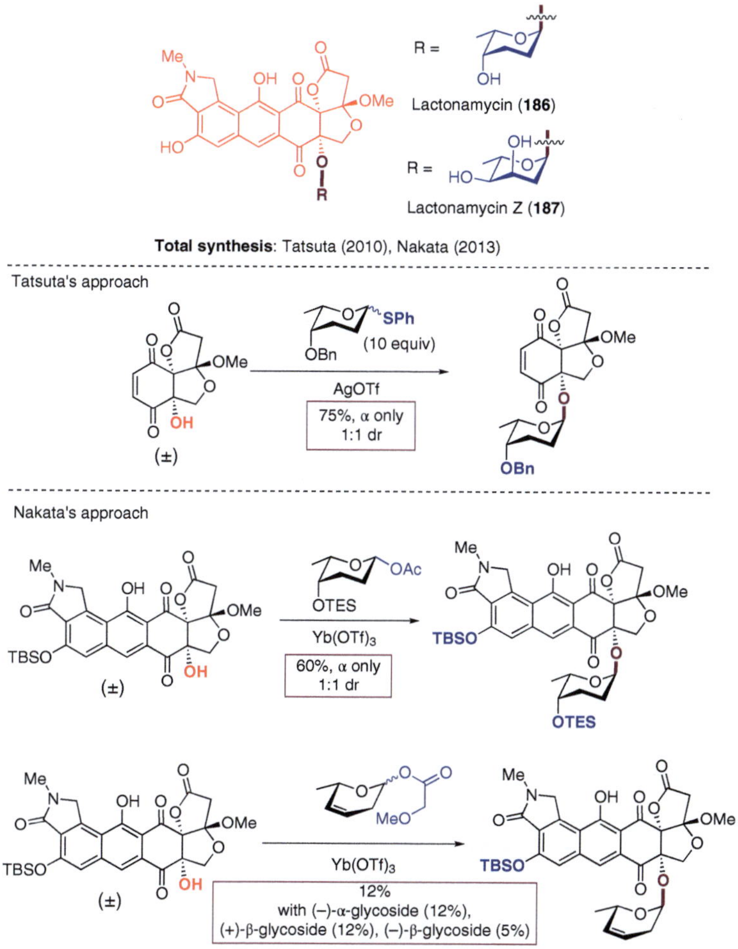


**Figure 2.19** Lactonamycin (**186**) and lactonamycin Z (**187**) and the key glycosylation steps in the total synthesis.

moieties have not been determined. In 2010, Tatsuta et al. reported the first total synthesis of lactonamycin (**186**) using an early-stage glycosylation strategy on the racemic EF ring scaffold [91]. In 2013, Nakata and coworkers reported the total synthesis of lactonamycin (**186**) and lactonamycin Z (**187**) through a late-stage glycosylation strategy on the racemic aglycone [92].

## Lomaiviticin A(–)

Lomaiviticin A (**188**) was isolated from the fermentation of strain *LL-37I366* (also named *Micromonosporalomaivitiensis*) in a seawater medium in the presence of HP20 resin by He et al. in 2001 (Figure 2.20) [93]. It possesses half-maximal inhibitory potencies ($IC_{50}$s) in the nanomolar to picomolar range against cultured

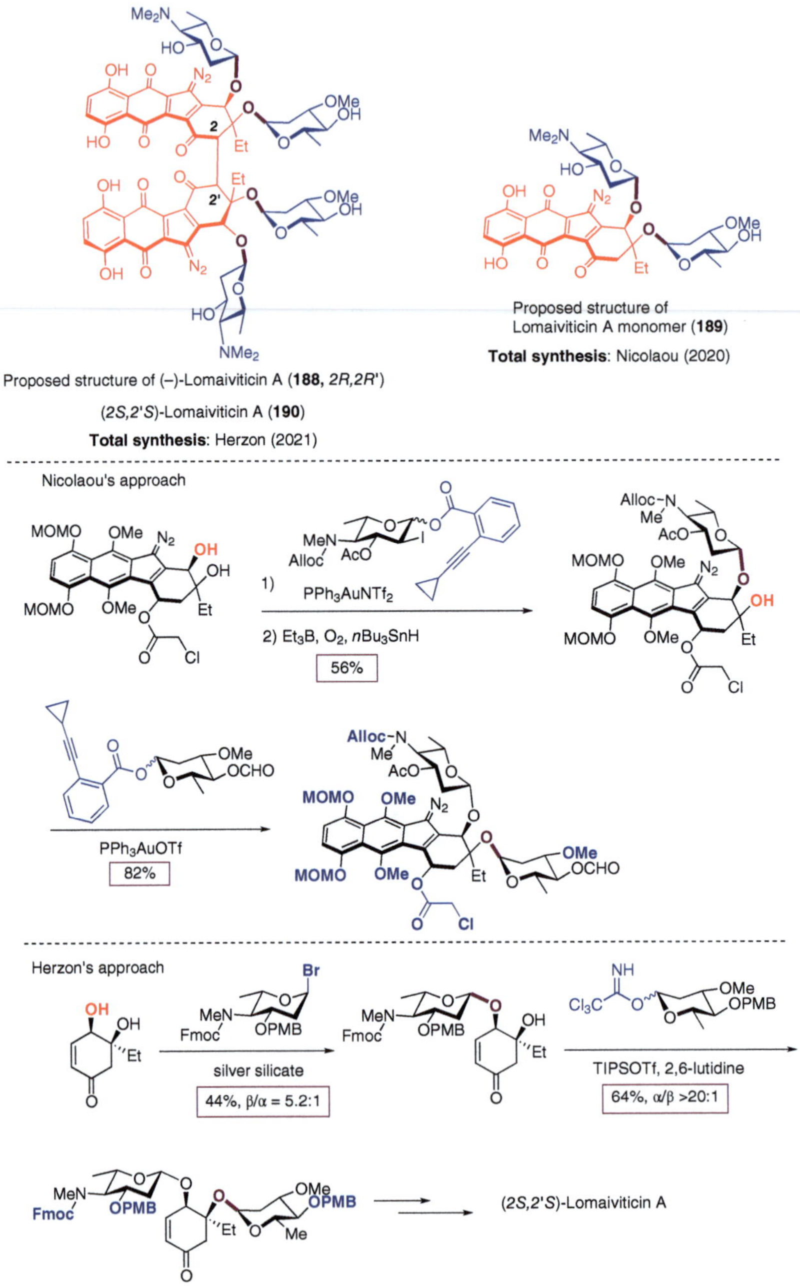


**Figure 2.20** Lomaiviticin A (**188**), lomaiviticin A monomer (**189**), (2*S*,2′*S*)-lomaivitivin A (**190**) and the key glycosylation steps in the syntheses.

human cancer cell lines due to the induction of double-strand breaks (DSBs) in DNA via the addition of DNA with its diazo residues, with the aminosugar residues stabilizing the complex by an electrostatic interaction and preorganizayion of the molecule in its binding conformation. Structurally, (−)-lomaiviticin A (**188**) is a dimeric diazobenzofluorene glycosides bearing two unusual glycosidic bonds that are susceptible to hydrolysis under mildly acidic conditions. Moreover, stereospecific glycosylation in both steps is particularly challenging due to the absence of C2 substituents in either sugar. In 2020, Nicolaou et al. reported the total synthesis of lomaiviticin A monomer (**189**) through a late-stage glycosylation strategy employing the gold(I)-catalyzed glycosylation with glycosyl *ortho*-alkynylbenzoate donor as the key step [94]. In 2021, Herzon and coworkers reported the total synthesis of (2*S*,2′*S*)-lomaivitivin A (**190**), an unnatural diastereomer of (−)-lomaiviticin A (**188**), through an early-stage glycosylation and oxidative coupling of the monomeric diazofluorene [95, 96]. The cytotoxicity and genotoxicity studies suggested that this stereoisomer of lomaiviticin A is not biologically active.

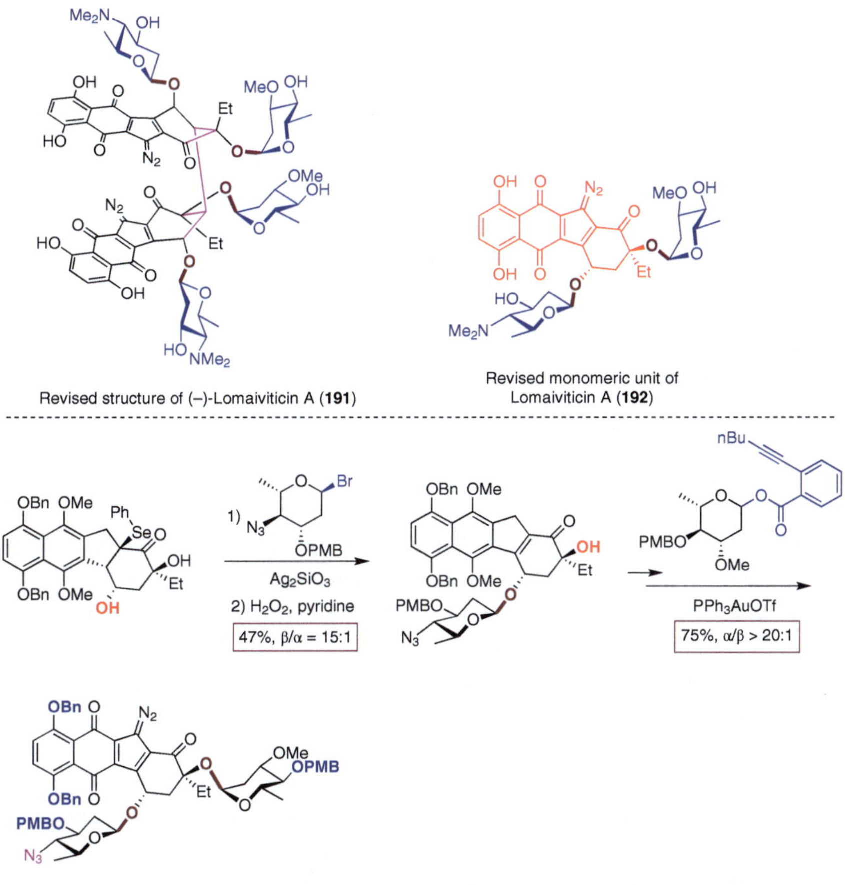


**Figure 2.21** Revised structures of (−)-lomaiviticin A (**191**) and lomaiviticin A monomer (**192**) and the key glycosylation step in the syntheses of lomaiviticin A monomer by Herzon and coworkers.

In fact, determining the structure of lomaiviticins has been challenging, with only 6 of the 19 carbon atoms in the monomeric aglycone unit proton-attached, making NMR spectroscopic analysis difficult. Previous structure assignments proposed the presence of a 4,5-dihydroxycyclohex-2-ene-1-one ring with the bridging bond located at C6. However, this proposal has not been confirmed, as no lomaiviticin has been prepared by total synthesis or successfully analyzed via X-ray crystallography. In 2021, microED studies by the Nelson and Herzon groups revealed that (−)-lomaiviticin C contains a 4,6-dihydroxycyclohex-2-ene-1-one residue with the bridging carbon–carbon bond located at C5. Due to the demonstrated interconversion of lomaiviticins A, B, and C, the structures of (−)-lomaiviticin A (**191**) and the monomer (**192**) were revised (Figure 2.21) [97]. In 2022, the Herzon group reported the synthesis of the revised structure of the monomeric unit of (−)-lomaiviticin A, which involved the installation of a β-linked 2,4,6-trideoxy-4-aminoglycoside using a modified Koenigs-Knorr glycosylation, and the introduction of an α-linked 2,6-dideoxyglycoside through the gold-catalyzed glycosylation with a glycosyl *o*-alkynylbenzoate donor [98].

## References


1 Hertweck, C. (2009). *Angew. Chem. Int. Ed.* 48: 4688–4716.
2 Zhou, H., Li, Y., and Tang, Y. (2010). *Nat. Prod. Rep.* 27: 839–868.
3 Lown, J.W. (1993). *Chem. Soc. Rev.* 22: 165–176.
4 Rohr, J. and Thiericke, R. (1992). *Nat. Prod. Rep.* 9: 103–137.
5 Krohn, K. and Rohr, J. (1997). *Bioorganic Chemistry Deoxysugars, Polyketides and Related Classes: Synthesis, Biosynthesis, Enzymes* (ed. J. Rohr), 127–195. Heidelberg: Springer Berlin.
6 Kharel, M.K., Pahari, P., Shepherd, M.D. et al. (2012). *Nat. Prod. Rep.* 29: 264–325.
7 Henkel, T., Rohr, J., Beale, J.M., and Schwenen, L. (1990). *J. Antibiot.* 43: 492–503.
8 Yang, X., Fu, B., and Yu, B. (2011). *J. Am. Chem. Soc.* 133: 12433–12435.
9 Guo, Y. and Sulikowski, G.A. (1998). *J. Am. Chem. Soc.* 120: 1392–1397.
10 Roush, W.R. and Bennett, C.E. (2000). *J. Am. Chem. Soc.* 122: 6124–6125.
11 Yu, B. and Wang, P. (2002). *Org. Lett.* 4: 1919–1922.
12 Tanaka, H., Yamaguchi, S., Yoshizawa, A. et al. (2010). *Chem. Asian J.* 5: 1407–1424.
13 Zhou, M. and O'Doherty, G.A. (2008). *Org. Lett.* 10: 2283–2286.
14 Baryal, K.N. and Zhu, J. (2015). *Org. Lett.* 17: 4530–4533.
15 McDonald, F.E. and Reddy, K.S. (2001). *J. Organomet. Chem.* 617–618: 444–452.
16 Roush, W.R. and Neitz, R.J. (2004). *J. Org. Chem.* 69: 4906–4912.
17 Yang, X., Wang, P., and Yu, B. (2013). *Chem. Rec.* 13: 70–84.
18 Yang, X. and Yu, B. (2016). *Synthesis* 48: 1693–1699.
19 Lai, Y.-H., Mondal, S., Su, H.-T. et al. (2021). *RSC Adv.* 11: 9426–9432.

20 Brazhnikova, M.G., Krugliak, E.B., Kovsharova, I.N. et al. (1962). *Antibiotiki* 7: 39–44.

21 Lombó, F., Menéndez, N., Salas, J.A., and Méndez, C. (2006). *Appl. Microbiol. Biotechnol.* 73: 1–14.

22 Thiem, J. and Gerken, M. (1985). *J. Org. Chem.* 50: 954–958.

23 Thiem, J. and Schöttmer, B. (1987). *Angew. Chem. Int. Ed. Engl.* 26: 555–557.

24 Toshima, K., Nozaki, Y., Mukaiyama, S. et al. (1995). *J. Am. Chem. Soc.* 117: 3717–3727.

25 Roush, W.R., Lin, X., and Straub, J.A. (1991). *J. Org. Chem.* 56: 1649–1655.

26 Sebesta, D.P. and Roush, W.R. (1992). *J. Org. Chem.* 57: 4799–4802.

27 Roush, W.R. and Lin, X.-F. (1995). *J. Am. Chem. Soc.* 117: 2236–2250.

28 Roush, W.R., Briner, K., Kesler, B.S. et al. (1996). *J. Org. Chem.* 61: 6098–6099.

29 Roush, W.R., Hartz, R.A., and Gustin, D.J. (1999). *J. Am. Chem. Soc.* 121: 1990–1991.

30 Lima, V.Q.G., Albert, C.A., and Lima, O.G. (1964). *An. Acad. Bras. Cienc.* 36: 317.

31 Suzuki, K., Sulikowski, G.A., Friesen, R.W., and Danishefsky, S.J. (1990). *J. Am. Chem. Soc.* 112: 8895–8902.

32 Raghavan, S. and Kahne, D. (1993). *J. Am. Chem. Soc.* 115: 1580–1581.

33 Gildersleeve, J., Smith, A., Sakurai, K. et al. (1999). *J. Am. Chem. Soc.* 121: 6176–6182.

34 Ōmora, S., Tanaka, H., Oiwa, R. et al. (1977). *J. Antibiot.* 30: 906–916.

35 Kawashima, A., Kishimura, Y., Tamai, M., and Hanada, K. (1989). *Chem. Pharm. Bull.* 37: 3429–3431.

36 Danishefsky, S., Uang, B.J., and Quallich, G. (1985). *J. Am. Chem. Soc.* 107: 1285–1293.

37 Bolitt, V., Mioskowski, C., Kollah, R.O. et al. (1991). *J. Am. Chem. Soc.* 113: 6320–6321.

38 Tius, M.A., Gomez-Galeno, J., Gu, X.Q., and Zaidi, J.H. (1991). *J. Am. Chem. Soc.* 113: 5775–5783.

39 Matsumoto, T., Katsuki, M., Jona, H., and Suzuki, K. (1991). *J. Am. Chem. Soc.* 113: 6982–6992.

40 Chen, Q., Zhong, Y., and O'Doherty, G.A. (2013). *Chem. Commun.* 49: 6806–6808.

41 Chen, C.L., Sparks, S.M., and Martin, S.F. (2006). *J. Am. Chem. Soc.* 128: 13696–13697.

42 Yu, X. and O'Doherty, G.A. (2008). *Org. Lett.* 10: 4529–4532.

43 Kusumi, S., Tomono, S., Okuzawa, S. et al. (2013). *J. Am. Chem. Soc.* 135: 15909.

44 Tomita, F., Tamaoki, T., Morimoto, M., and Fujimoto, K. (1981). *J. Antibiot.* 34: 1519–1524.

45 Švenda, J., Hill, N., and Myers, A.G. (2011). *Proc. Natl. Acad. Sci. U.S.A.* 108: 6709.

46 Magauer, T., Smaltz, D.J., and Myers, A.G. (2013). *Nat. Chem.* 5: 886.

47 Nicolaou, K.C., Cai, Q., Qin, B. et al. (2015). *Angew. Chem. Int. Ed.* 54: 3074–3078.

48 Nicolaou, K.C., Cai, Q., Sun, H. et al. (2016). *J. Am. Chem. Soc.* 138: 3118–3124.

49 Nicolaou, K.C., Chen, P., Zhu, S. et al. (2017). *J. Am. Chem. Soc.* 139: 15467–15478.

50 Brockmann, H.J. (1963). *Chem. Ber.* 96: 1771–1778.

51 Acton, E.M., Fujiwara, A.N., and Henry, D.W. (1974). *J. Med. Chem.* 17: 659–660.

52 Wander, D.P.A., van der Zanden, S.Y., van der Marel, G.A. et al. (2020). *J. Med. Chem.* 63: 12814–12829.

53 Sezaki, M., Kondo, S., Maeda, K. et al. (1970). *Tetrahedron* 26: 5171–5190.

54 Takashi, M., Tsukasa, S., Hiroki, Y., and Keisuke, S. (1995). *Chem. Lett.* 24: 677–678.

55 Matsumoto, T., Yamaguchi, H., Hamura, T. et al. (2000). *Tetrahedron Lett.* 41: 8383–8387.

56 Matsumoto, T., Yamaguchi, H., Tanabe, M. et al. (2000). *Tetrahedron Lett.* 41: 8393–8396.

57 Yamaguchi, H., Konegawa, T., Tanabe, M. et al. (2000). *Tetrahedron Lett.* 41: 8389–8392.

58 Kusumi, S., Nakayama, H., Kobayashi, T. et al. (2016). *Chem. Eur. J.* 22: 18733–18736.

59 Acharya, P.P., Khatri, H.R., Janda, S., and Zhu, J. (2019). *Org. Biomol. Chem.* 17: 2691–2704.

60 Omura, S., Tanaka, H., Oiwa, R. et al. (1977). *J. Antibiot.* 30: 908–916.

61 Matsumoto, Y., Kuriki, H., Kitamura, T. et al. (2019). *J. Org. Chem.* 84: 14724–14732.

62 Henkel, T., Ciesiolka, T., Rohr, J., and Zeeck, A. (1989). *J. Antibiot.* 42: 299.

63 Khatri, H.R., Nguyen, H., Dunaway, J.K., and Zhu, J. (2015). *Chem. Eur. J.* 21: 13553–13557.

64 Ayer, S.W., McInnes, A.G., Thibault, P. et al. (1991). *Tetrahedron Lett.* 32: 6301–6304.

65 Yang, X. and Yu, B. (2013). *Chem. Eur. J.* 19: 8431–8434.

66 Shan, M., Sharif, E.U., and O'Doherty, G.A. (2010). *Angew. Chem. Int. Ed.* 49: 9492–9495.

67 Akagi, Y., Mori, Y., Sato, Y. et al. (2022). *Tetrahedron Lett.* 101: 153919.

68 Kanamaru, T., Nozaki, Y., and Muroi, M. (1991). *Chem. Abstr.* 115: 47759n.

69 Mori, K., Ohmori, K., and Suzuki, K. (2009). *Angew. Chem. Int. Ed.* 48: 5633–5637.

70 Fan, J., Yao, Q.-J., Liu, Y.-H. et al. (2019). *Org. Lett.* 21: 3352–3356.

71 Takeuchi, T., Hara, T., Naganawa, H. et al. (1988). *J. Antibiot.* 41: 807–811.

72 Ohmori, K., Tamiya, M., Kitamura, M. et al. (2005). *Angew. Chem. Int. Ed.* 44: 3871–3874.

73 Tamiya, M., Ohmori, K., Kitamura, M. et al. (2007). *Chem. Eur. J.* 13: 9791–9823.

74 Kanda, N. (1971). *J. Antibiot.* 24: 599–606.

75 Abe, N., Nakakita, Y., Nakamura, T. et al. (1993). *J. Antibiot.* 46: 1530–1535.

76 O'Keefe, B.M., Mans, D.M., Kaelin, D.E., and Martin, S.F. (2010). *J. Am. Chem. Soc.* 132: 15528–15530.

77 O'Keefe, B.M., Mans, D.M., Kaelin, D.E., and Martin, S.F. (2011). *Tetrahedron* 67: 6524–6538.

**78** Kitamura, K., Ando, Y., Matsumoto, T., and Suzuki, K. (2014). *Angew. Chem. Int. Ed.* 53: 1258–1261.

**79** Kitamura, K., Maezawa, Y., Ando, Y. et al. (2014). *Angew. Chem. Int. Ed.* 53: 1262–1265.

**80** Shimura, J., Ando, Y., Ohmori, K., and Suzuki, K. (2022). *Org. Lett.* 24: 1439–1443.

**81** Martin, G.D.A., Tan, L.T., Jensen, P.R. et al. (2007). *J. Nat. Prod.* 70: 1406–1409.

**82** Cañeque, T., Gomes, F., Mai, T.T. et al. (2015). *Nat. Chem.* 7: 744–751.

**83** Hata, K., Baba, K., and Kozawa, M. (1978). *Chem. Pharm. Bull.* 26: 3792–3797.

**84** Krenn, L., Pradhan, R., Presser, A. et al. (2004). *Chem. Pharm. Bull.* 52: 391–393.

**85** Koyama, Y., Yamaguchi, R., and Suzuki, K. (2008). *Angew. Chem. Int. Ed.* 47: 1084–1087.

**86** Qiao, Y.F., Okazaki, T., Ando, T. et al. (1998). *J. Antibiot.* 51: 282–287.

**87** Xie, T., Zheng, C., Chen, K. et al. (2020). *Angew. Chem. Int. Ed.* 59: 4360–4364.

**88** Kang, H.-S. and Brady, S.F. (2014). *J. Am. Chem. Soc.* 136: 18111–18119.

**89** Chen, K., Xie, T., Shen, Y. et al. (2021). *Org. Lett.* 23: 1769–1774.

**90** Matsumoto, N., Tsuchida, T., Maruyama, M. et al. (1996). *J. Antibiot.* 49: 953.

**91** Tatsuta, K., Tanaka, H., Tsukagoshi, H. et al. (2010). *Tetrahedron Lett.* 51: 5546–5549.

**92** Adachi, S., Watanabe, K., Iwata, Y. et al. (2013). *Angew. Chem. Int. Ed.* 52: 2087–2091.

**93** He, H., Ding, W.-D., Bernan, V.S. et al. (2001). *J. Am. Chem. Soc.* 123: 5362–5363.

**94** Nicolaou, K.C., Chen, Q., Li, R. et al. (2020). *J. Am. Chem. Soc.* 142: 20201–20207.

**95** Kaneko, M., Li, Z., Burk, M. et al. (2021). *J. Am. Chem. Soc.* 143: 1126–1132.

**96** Rose, J.A., Mahapatra, S., Li, X. et al. (2020). *Chem. Sci.* 11: 7462–7467.

**97** Kim, L.J., Xue, M., Li, X. et al. (2021). *J. Am. Chem. Soc.* 143: 6578–6585.

**98** Xu, Z., DiBello, M., Wang, Z. et al. (2022). *J. Am. Chem. Soc.* 144: 16199–16205.

# 3

# Enediyne Glycosides

Enediyne natural products are characterized by a unique (*Z*)-1,5-diyn-3-ene motif in their structures, with many of them featuring conjugation with unusual branched, amino-, or thio-containing sugar moieties. These antibiotics gained attention in the 1980s when Lee et al. reported the structures of calichemicins (**1**) derived from the fermentation produced by *Micromonospora echinospora* ssp. calichensis [1, 2]. Almost at the same time, Golik et al. reported the characterization of esperamicins (**2**) obtained from the cultures of *Actinomadura verrucosospora* (Figure 3.1) [3, 4]. It is noteworthy that another type of enediyne natural product, neocarzinostatin chromophore (**3**), which possesses a distinct nine-numbered bicyclo[7.3.0]dodecadiynene scaffold, was originally isolated from *Streptomyces carzinostaticus* var. F-41 (ATCC15944) in 1965, whose structure was later determined in 1985 [5, 6]. To date, more than 40 enediynes have been identified from various marine and terrestrial sources [7–9].

Enediynes demonstrate potent antitumor and antimicrobial properties, with esperamicins and calicheamicins as the most active ones. Their biological activities stem from their ability to induce sequence-specific double-strand breaks, damaging DNA. Figure 3.2 presents a concise depiction of the mechanism of action for calicheamicins. The trisulfide moiety cleavage, triggered by the addition of an endogenous nucleophile, results in the formation of a thiyl species, which subsequently attacks the enone moiety, leading to the formation of a tricyclic intermediate. This transformation sets the stage for an enediyne → diyl conversion, with the diyl acting as the active DNA-cleaving agent through hydrogen abstraction from the proximal 2-deoxyribose residues. While the key elements to the biological activity lie within the enediyne core, the oligosaccharide domain displays important roles in delivery and recognition. Notably, it exhibits highly selective binding to TCCT, TCTC, and TTTT sequences within duplex DNA.

Significant efforts have been devoted to the synthesis and medicinal studies of various enediyne derivatives. However, only five complex natural enediyne glycosides have been successfully synthesized to date. These notable achievements include the total synthesis of calicheamicin $\gamma_1{}^I$ (Nicolaou et al. and Danishefsky et al.), namenamicin (Nicolaou et al), shishijimicin (Nicolaou et al.), neocarzinostatin chromophore (Myers et al.), and maduropeptin chromophore (Hirama et al.).

*Carbohydrate Chemistry in the Total Synthesis of Naturally Occurring Glycosides*,
First Edition. Biao Yu and Xiaoyu Yang.
© 2024 WILEY-VCH GmbH. Published 2024 by WILEY-VCH GmbH.

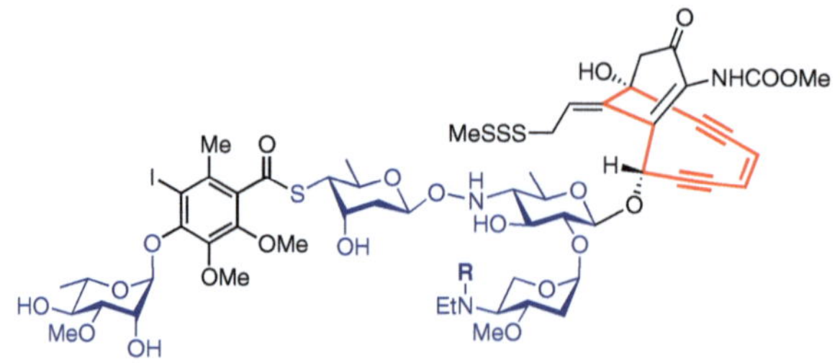

Calicheamicin $\gamma_1^{I}$ (**1**), R = H
N-Acetylcalicheamicin $\gamma_1^{I}$, R = Ac

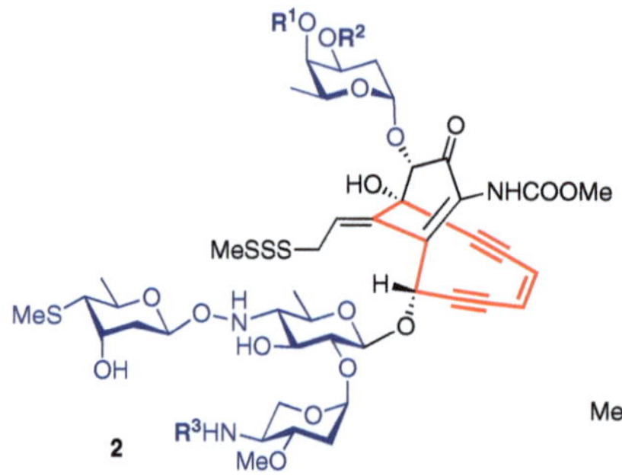

Esperamicin $A_1$, $R^1$ = H, $R^2$ = **Ar**, $R^3$ = iPr
Esperamicin $A_{1b}$, $R^1$ = **Ar**, $R^2$ = H, $R^3$ = iPr
Esperamicin $A_2$, $R^1$ = H, $R^2$ = **Ar**, $R^3$ = Et

**Ar** =

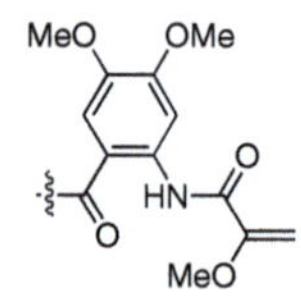

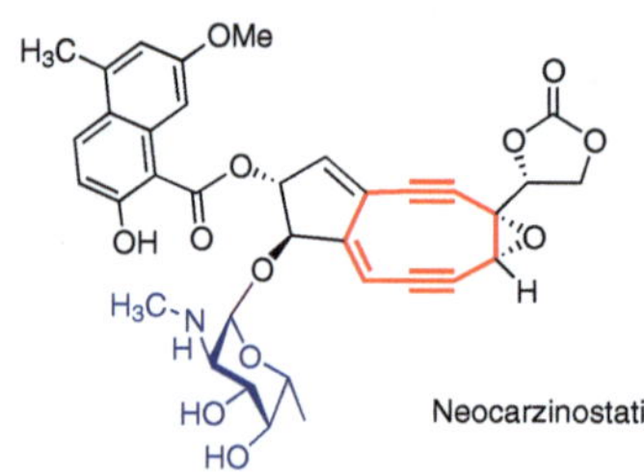

Neocarzinostatin chromophore (**3**)

**Figure 3.1** The representative enediyne glycosides: neocarzinostatin chromophore, calicheamicins, and esperamicins.

## Calicheamicin $\gamma_1^{I}$

During the early stage of calicheamicin discovery, it was found that these compounds exhibited remarkable activity in the biochemical prophage induction assay, even at concentrations below 1 pg/ml. Notably, they demonstrated potent activity against both gram-positive and gram-negative bacteria. Moreover, calicheamicins exhibited extraordinary potency against murine tumors such as P338 and L1210 leukemia, as well as solid neoplasms including colon 26 and B-16 melanoma, at low doses of 0.15–5 μg/kg. Calicheamicin $\gamma_1^{I}$ (**1**), as the prototypical member of the calicheamicin family, represents a formidable synthetic target owing to

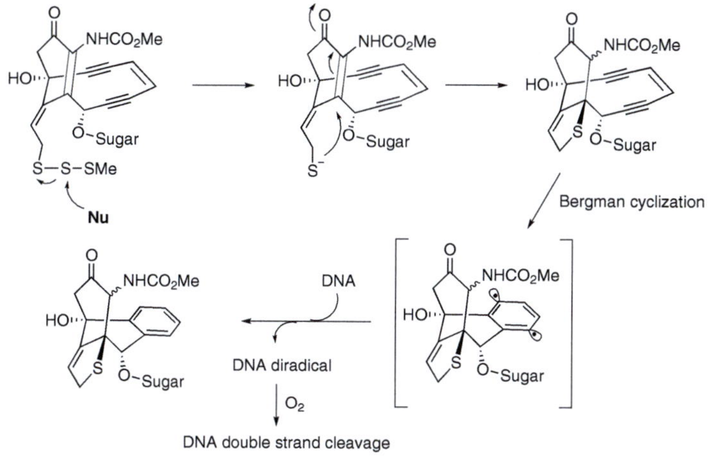


**Figure 3.2** The mechanism of DNA cleavage by calicheamicins.

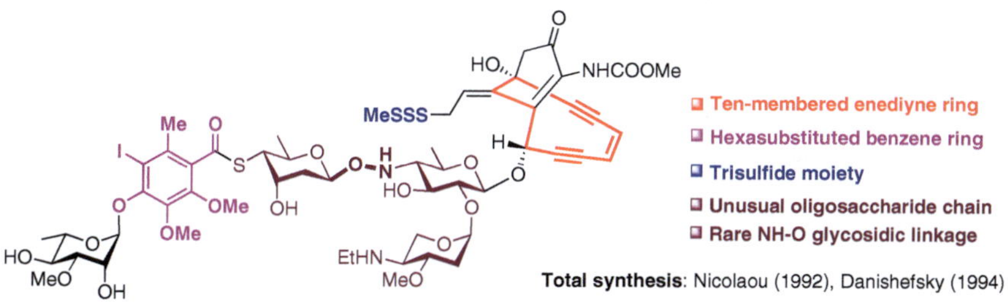


**Figure 3.3** Calicheamicin $\gamma_1{}^I$ (**1**) and its structural features.

its exceptional structural characteristics. These distinctive features include a 10-membered enediyne ring, a hexasubstituted benzene ring, a trisulfide moiety, an uncommon oligosaccharide chain, and a rare NH–O glycosidic linkage (Figure 3.3). The total synthesis of calicheamicin $\gamma 1^I$, accomplished by Nicolaou et al. in 1993 and Danishefsky et al. in 1994, stands as a landmark achievement in the field of organic synthesis.

**Total synthesis by the Nicolaou group**. In 1993, the Nicolaou group reported the first total synthesis of calicheamicin $\gamma_1{}^I$ [10–12]. The overall synthetic strategy involved the independent syntheses of the oligosaccharide fragment and the aglycone fragment, with a subsequent coupling of the appropriate forms of the two fragments in the later stage of the synthesis. The synthesis of the tetrasaccharide fragment **14** is illustrated in Scheme 3.1. Glycosylation of D-fucose-derived compound **4** and L-serine-derived **5** proceeded smoothly under standard conditions ($AgClO_4$, $SnCl_2$) for glycosylation with glycosyl fluoride, resulting in the formation of disaccharide **6** in 86% yield as a 4.5 : 1 mixture of the α/β-anomers, with the desired thermodynamically favored α-anomer predominating. Following the removal of the cyclic carbonate and regioselective oxidation, the resulting

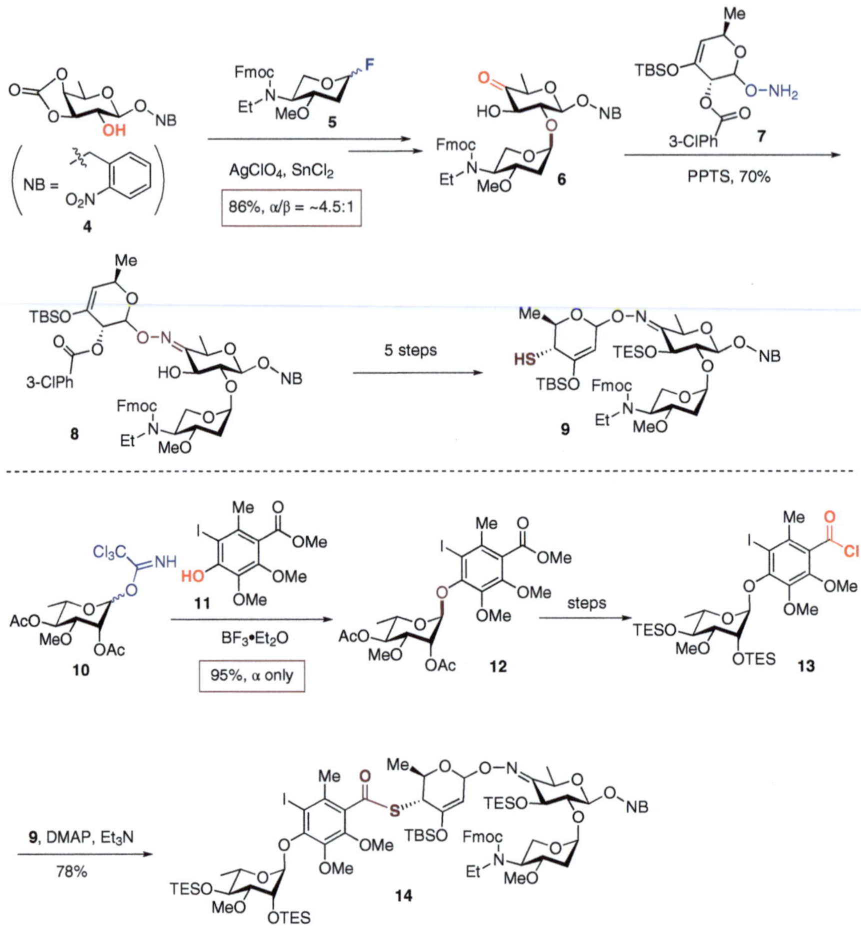

Scheme 3.1 Synthesis of tetrasaccharide building block **14**.

disaccharide hydroxyl ketone **6** was successfully coupled with sugar hydroxylamine **7** using a catalytic amount of pyridinium *p*-toluenesulfonate (PPTS), affording trisaccharide **8** in 70% yield as a single geometrical isomer. Further manipulation of protecting groups, along with the introduction of the thiol group, led to the formation of multigram quantities of trisaccharide **9** for subsequent assembly. On the other hand, coupling of L-rhamnosyl trichloroacetimidate **10** with phenol **11** in the presence of $BF_3{\cdot}Et_2O$ gave the desired α-glycoside **12** in an excellent 95% yield. After appropriate transformations of the *O*-protecting groups and other functional groups, condensation of the resulting acid chloride **13** with trisaccharide thiol **9** provided the desired tetrasaccharide **14** in 78% yield.

Tetrasaccharide **14**, featuring a selectively removable 2-nitrobenzyl (NB) group to protect the anomeric position, was transformed into glycosyl imidate donor **15** with five steps (Scheme 3.2). The final glycosylation between aglycone derivative **16** and imidate **15** in the presence of $BF_3{\cdot}Et_2O$ led to the desired coupling product **17** in

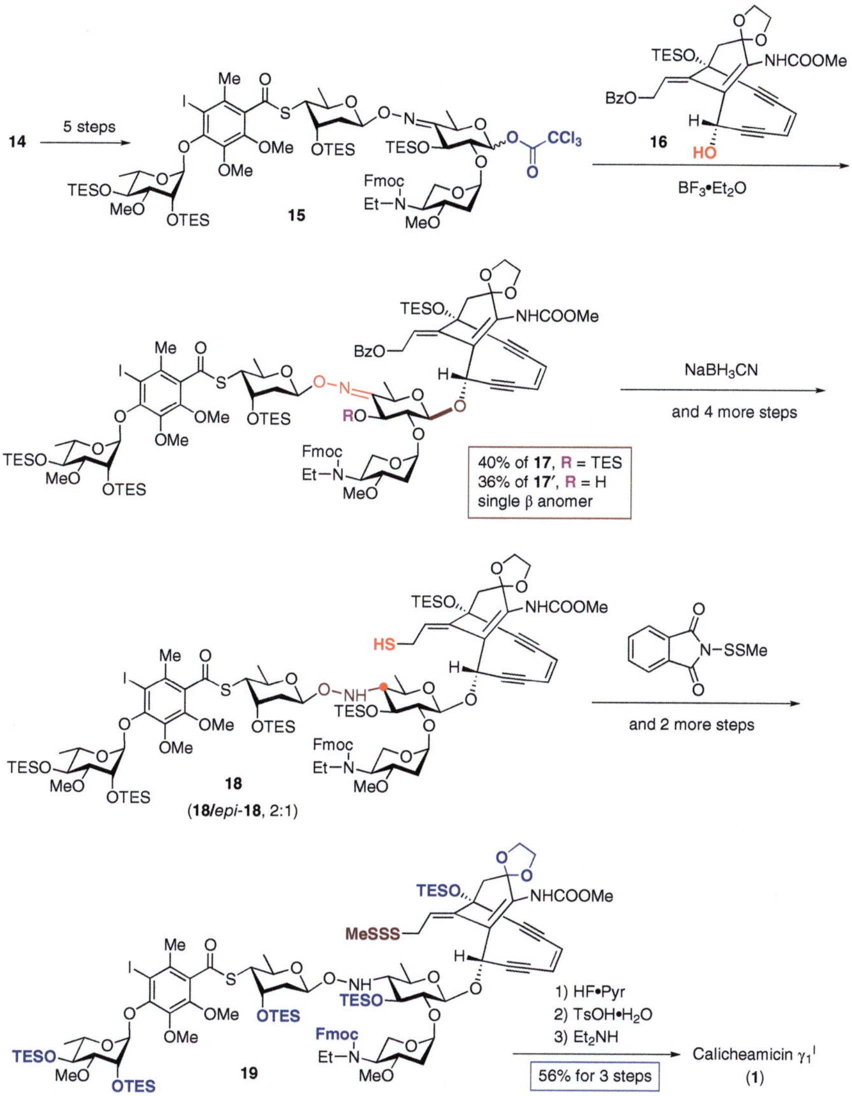


**Scheme 3.2** Coupling of the oligosaccharide with the aglycone and completion of the total synthesis of calicheamicin $\gamma_1{}^I$ (**1**) by Nicolaou and coworkers.

40% yield as a single β-anomer, accompanied by its monodesilylated derivative **17′** in 36% yield. After the introduction of the first sulfur atom, the oxime bond underwent reduction with $NaCNBH_3/BF_3{\cdot}Et_2O$, resulting in a mixture of C4-α-isomer and C4-β-isomer (**18**, α/β = ~2 : 1). Following separation of these two epimers through column chromatography, the trisulfide moiety was installed via treatment of the C4-β-isomer with excess *N*-(methyldithio)phthalimide (PhthNSSMe), yielding **19** in 57% yield. The total synthesis of calicheamicin $\gamma_1{}^I$ (**1**) was finally completed by sequential removal of the triethylsilyl (TES), acetal, and Fmoc groups with HF·pyr, TsOH, and $Et_2NH$, respectively (56% yield for three steps).

**Total synthesis by the Danishefsky group**. After years of dedicated research [13–16], the Danishefsky group successfully accomplished the total synthesis of calicheamicin $\gamma_1^I$ in 1994 (Scheme 3.3) [17]. The regioselective coupling of monosaccharide diol **20** with glycal **21** in the presence of $I^+ClO_4$-(*sym*-collidine)$_2$ provided the C2-*O*-glycoylation product **22** in a moderate yield. Subsequent removal of the iodide and treatment with triflic anhydride gave the disaccharide OTf derivative **23**. Meanwhile, coupling of phenol **25** with L-rhamnosyl trichloroacetimidate **24** in the presence of $BF_3 \cdot Et_2O$ afforded aryl α-glycoside **26** in 89% yield. Following steps of

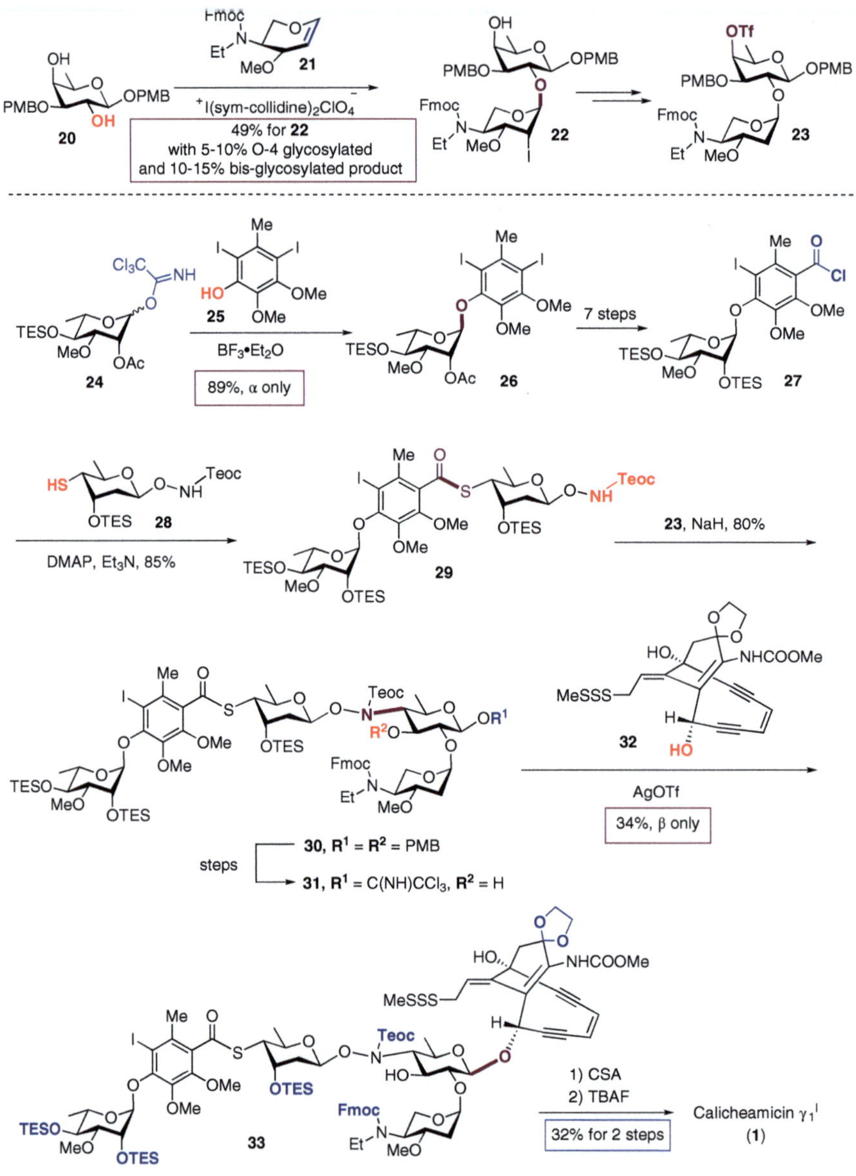


**Scheme 3.3** Total synthesis of calicheamicin $\gamma_1^I$ (**1**) by Danishefsky and coworkers.

protecting group manipulation and Pd-catalyzed selective methoxycarbonylation, the desired acyl chloride **27** was obtained. Condensation of sugar thiol **28** with acyl chloride **27**, mediated by 4-*N,N*-dimethylaminopyridine (DMAP) and $Et_3N$, generated aryldisaccharide **29** in 85% yield. Subsequently, alkylation of the nitrogen atom of the Teoc carbamate in **29** with triflate **23**, under basic conditions, provided the desired N–O linked tetrasaccharide **30** in 80% yield. Aryltetrasaccharide **30** underwent further protecting group manipulations and transformations, yielding the desired trichloroacetimidate donor **31**. The late-stage glycosylation of trisulfide aglycone **32** with donor **31**, utilizing AgOTf as the catalyst, provided the desired β-glycoside **33** in 38% yield without detection of the corresponding α-anomer. Upon deketalization with camphorsulfonic acid (CSA), the *O*-TES, *N*-Fmoc, and *N*-Teoc groups in the carbohydrate domain were deprotected with a single step of tetra-*n*-butylammonium fluoride (TBAF) treatment (32% yield for two steps), thus completing the total synthesis of calicheamicin $\gamma_1^I$ (**1**).

**Synthesis of the calicheamicin oligosaccharide by the Kahne group.** In 1994, the Kahne group reported a concise and highly stereoselective synthesis of the oligosaccharide fragment of calicheamicin using the sulfoxide glycosylation reaction (Scheme 3.4) [18]. The glycosylation of *N*-hydroxyurethane **35** with glycosyl sulfoxide **34** in the presence of $Tf_2O$ provided the *O*-glycosylated product

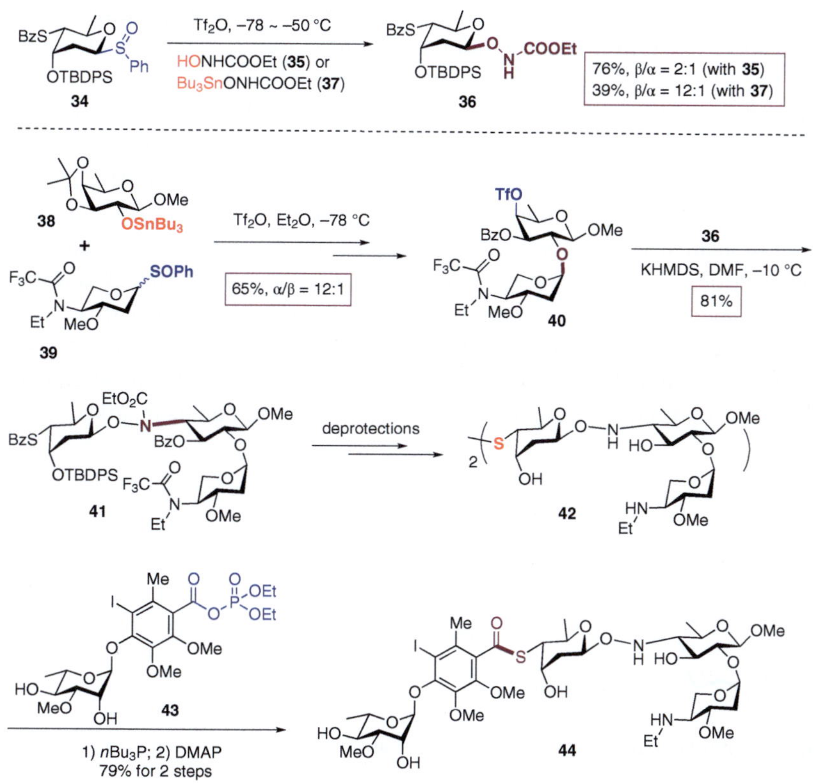


**Scheme 3.4** Concise synthesis of calicheamicin oligosaccharide **44** by Kahne and coworkers.

**36** in 76% yield with moderate β-selectivity (β/α = 2 : 1). The observed β-selectivity was attributed to the bulky silyl group installed at the C3-OH position, which favored the nucleophilic attack from the β face. Interestingly, utilizing *O*-stannyl-*N*-hydroxyurethane **37** as the nucleophile in this reaction resulted in **36** in moderate yield (39%) but high β-selectivity (β/α = 12 : 1). On the other hand, coupling of glycosyl sulfoxide **39** with stannyl alkoxide derivative of fucose **38**, using $Tf_2O$ at −78 °C, afforded the desired disaccharide in 65% yield with high α-selectivity (α/β = 12 : 1); subsequent functional group transformations (three steps) led to **40**. An $S_N2$ displacement of the axial triflate at disaccharide **40** with glycosyl hydroxylamine derivative **36**, under basic conditions, provided the desired *N*–*O* linked trisaccharide **41** in 81% yield. After removal of the protecting groups on the hydroxyl groups and thiol group, selective thiol acylation of **42** with the mixed carboxylic phosphoric anhydride of aryl rhamnose derivative **43** provided the calicheamicin oligosaccharide **44** in good yield.

## Namenamicin

Namenamicin (**45**) was isolated from the orange-colored ascidian *Polysyncraton lithostrotum*, collected off the coast of the Fijian island Namenalala, in 1996 (Figure 3.4) [19]. However, it is likely that namenamicin is produced by a micromonospora bacterium living symbiotically within the ascidian. Namenamicin exhibited potent *in vitro* cytotoxicity, with a mean $IC_{50}$ = 3.5 nM. *In vivo* assay demonstrated its ability to increase the life span of mice bearing P388 leukemia (ILS) by 40% at a dosage of 3 μg/kg. Moreover, namenamicin displayed significant antimicrobial properties, such as minimum inhibitory concentration (MIC) values of 0.001 μM against *Staphylococcus aureus*, 0.03 μM against *Enterococcus faecium*, and 0.004 μM against *Ustilago maydis*. The same compound was later isolated by the Fusetani group, emphasizing its exceptional potency against cancer cell lines ($IC_{50}$ = 13 pM against 3Y1; $IC_{50}$ = 34 pM against HeLa; $IC_{50}$ = 3.3 pM against P388). In comparison with calicheamicins, namenamicin contains the same enediyne warhead; however, the attached carbohydrate moiety displays significant differences, particularly with a unique branched and *S*-methyl substituted sugar linked to the

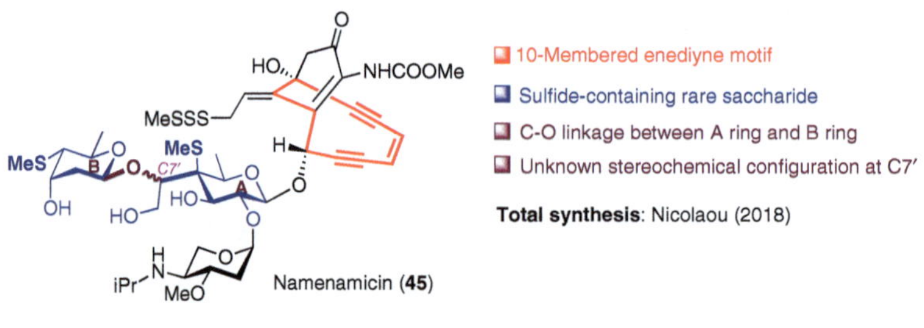


**Figure 3.4** Namenamicin (**45**) and its structural features.

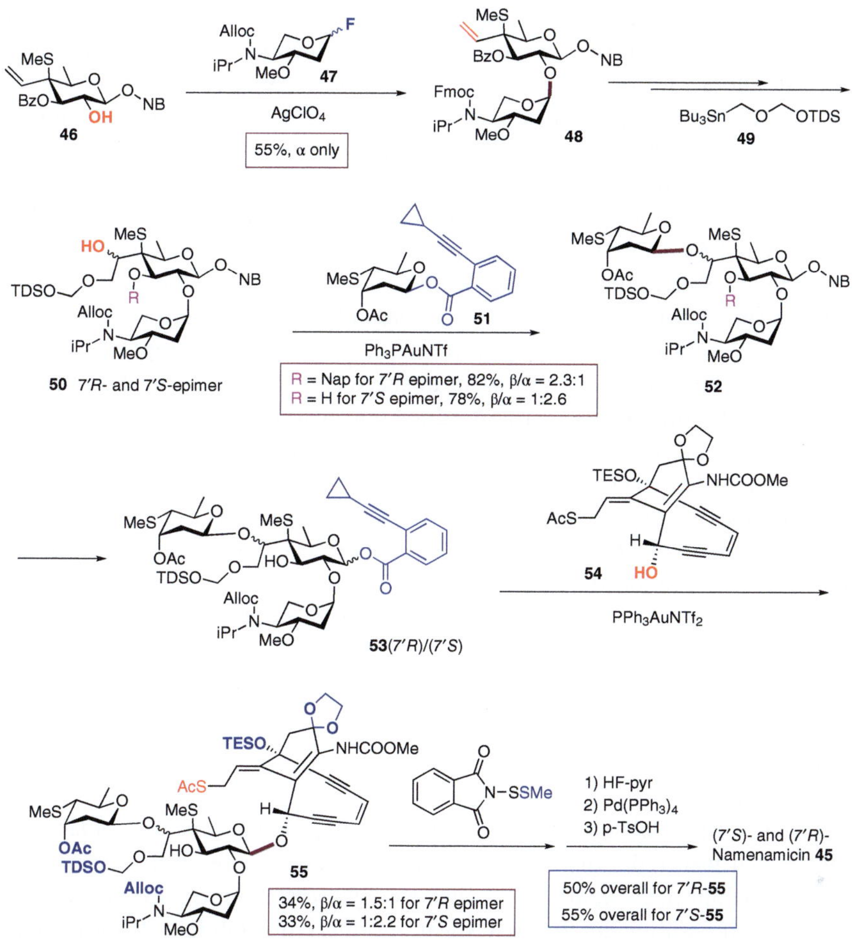

Scheme 3.5 Total synthesis of namenamicin (45) and assignment of the stereochemistry at C7′ by Nicolaou and coworkers.

aglycone. It is worth noting that the configuration at C7′ could not be assigned by analytical methods.

**Total synthesis by the Nicolaou group**. In 2018, the Nicolaou group reported the total synthesis of two C7′-epimers of namenamicin, thereby establishing the complete structure of the natural product (Scheme 3.5) [20]. Glycosylation of the C2–OH of sugar **46** with glycosyl fluoride **47**, under the promotion of $AgClO_4$, provided disaccharide **48** in 55% yield with exclusive α-selectivity. The introduction of an aldehyde group was followed by adding organometallic species derived from stannane **49**, resulting in the formation of both epimers of alcohol **50**, which could be separated by column chromatography. Interestingly, coupling of the 7′(*R*)-epimer of **50** with glycosyl *ortho*-alkynylbenzoate **51**, in the presence of $Ph_3PAuOTf$, furnished the desired trisaccharide **52**(7′*R*) in 82% yield, albeit with a moderate selectivity of β/α = ~ 2.3 : 1. In contrast, attempts to couple 7′(*S*)-epimer with glycosyl donor **51** failed to afford the coupled β-anomer, most likely due to the steric hindrance by the neighboring Nap protecting group. Therefore, the Nap group was removed to give the corresponding hydroxy disaccharide, whose glycosylation with donor **51** successfully provided the desired trisaccharide **52**(7′*S*) in 78% yield with a β/α selectivity of 1 : 2.6. Further transformations were carried out to convert trisaccharide **52**(7′*R*) and **52**(7′*S*) into the corresponding glycosyl *o*-alkynylbenzoate donor **53**, respectively. Subsequent coupling of these donors (**53**) with the endiyne fragment **54**, in the presence of $Ph_3PAuNTf_2$, afforded the advanced glycoside **55** (34% yield, β/α = 3 : 2 for **53**(7′*R*); 33% yield, β/α = 1 : 2.2 for **53**(7′*S*)). Attempts to improve the yields and the stereoselectivity were unsuccessful, likely due to the intricate nature of the coupling partners. To introduce the trisulfide motif, the thiol group was treated with PhthNSSMe, followed by sequential treatment with HF·pyridine (desilylation and deformylation), $Pd(PPh_3)_4$ (removal of Alloc), and *p*-TsOH (cleavage of ketal). These transformations resulted in the formation of (7′*R*)-namenamicin and (7′*S*)-namenamicin **45** in 50% and 55% yield, respectively. Comparison of the $^1H$ and $^{13}C$ NMR data of these two epimers with those of the natural product led to the assignment of the (7′*S*)-epimer **45** as the structure of the naturally occurring namenamicin.

## Shishijimicin A

Shishijimicins were isolated from the thin encrusting orange ascidian *Didemnum proliferum* and featured a distinct 10-membered endiyne core motif (Figure 3.5). What sets shishijimicins apart from other enediynes is the presence of a unique carboline motif residing in the first sugar residue. Shishijimicin A (**56**) stands out as the most potent congener of the shishijimicin family, exhibiting remarkable $IC_{50}$ values of 2.0, 1.8, and 0.47 pM against 3Y1, HeLa, and P388 cell lines, respectively.

The Nicolaou group accomplished the first total synthesis of shishijimicin A (**56**) in 2015 [21], followed by an improved streamlined total synthesis in 2018

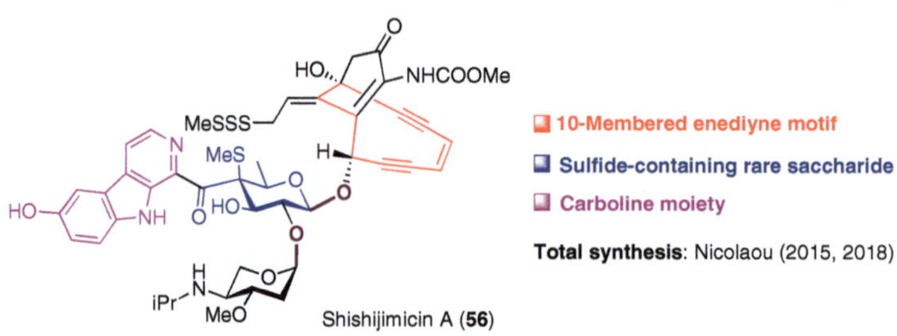

Shishijimicin A (**56**)

**Figure 3.5** Shishijimicin A (**56**) and its structural features.

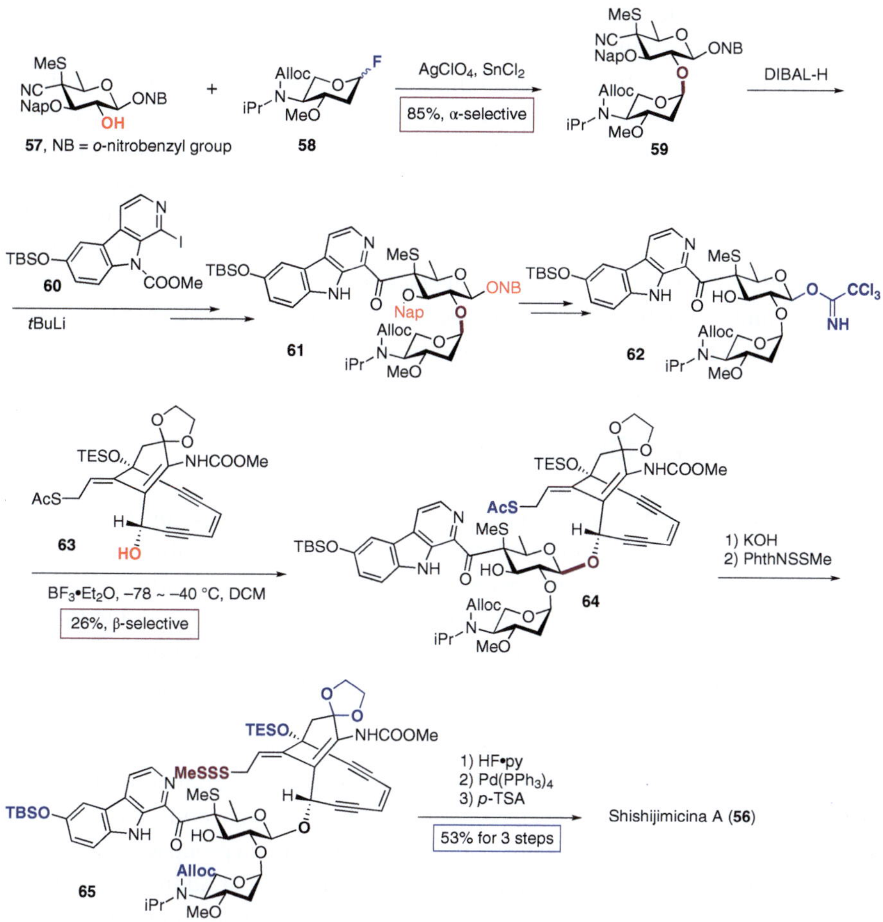

**Scheme 3.6** Total synthesis of shishijimicin A (**56**) by Nicolaou and coworkers.

(Scheme 3.6) [22]. The synthesis began with the coupling of glycosyl acceptor **57**, bearing a C4-CN group, with glycosyl fluoride **58** in the presence of $AgClO_4$ and $SnCl_2$, which provided the desired disaccharide **59** in an excellent yield of 85%. Subsequent conversion of the CN group to aldehyde was followed by the addition of carboline anion and subsequent oxidation, leading to the formation of disaccharide **61**. Several steps were required to transform **61** into disaccharide trichloroacetimidate **62**, which was then coupled with the enediyne aglycone **63** in the presence of $BF_3{\cdot}Et_2O$, yielding the coupled glycoside **64** in 26% yield. Attempts to improve the yield of this step by using alternative glycosyl donors or different reaction conditions were unsuccessful, which was attributed to the obstructive influence of the C-4 axial SMe group hindering the formation of the desired β-glycoside. The thiol group was released under basic conditions, and was then converted into trisulfide **65** upon treatment with PhthNSSMe. Finally, the silyl, Alloc, and acetal groups were removed sequentially (53% for three steps), culminating in the completion of the total synthesis of shishijimicin A (**56**).

With sufficient quantities of shishijimicin A rendered by the total synthesis, the binding and scission of double-stranded DNA by shishijimicin A were investigated [23]. The results provide evidence in support of the proposed hypothesis that DNA strand cleavage occurs due to the formation of 1,4-benzenoid diradicals, which are generated through Bergman cycloaromatization of the enediyne core of shishijimicin A upon activation by thiols. Moreover, with their ultimate goal of developing targeted cancer therapies, Nicolaou and coworkers reported the design, synthesis, and evaluation of four antibody-drug conjugates (ADCs) [24], with the shishijimicin A payload being conjugated through their phenolic moiety.

## Neocarzinostatin Chromophore

Neocarzinostatin was first isolated from *S. carzinostaticus* var. F-41 by Rikimaru and coworkers in 1965 [5]. Its chromophore, known as neocarzinostatin chromophore (**3**), features a unique nine-membered epoxy diyne scaffold (Figure 3.6). Similar to other enediyne antibiotics, it exhibits potent cytotoxicity via inducing DNA cleavage. The biochemical process is initiated by nucleophilic addition of a thiol to C12, followed by a rapid Bergeman rearrangement that generates a carbon-centered biradical.

In 1998, Myers and coworkers reported the first total synthesis of (+)-neocarzinostatin chromophore (**3**, Scheme 3.7) [25, 26]. The synthesis commenced with the addition of epoxy diyne **66** to cyclopentenone **67** under basic conditions, providing **68** in 75% yield. Subsequent transformations of the functional group led to the formation of alkynyl aldehyde **69**, which underwent intramolecular addition under basic conditions to afford the desired nine-membered diyne **70**. Sequential condensation of **70** with naphthoic acid **71** and further transformations resulted in the formation of alcohol **72**. Glycosylation of **72** with glycosyl trichloroacetimidate **73**, in the presence of $BF_3{\cdot}Et_2O$ at −30 °C, afforded the desired glycoside **74** in 51% yield with an excellent α-selectivity. The excellent α-selectivity is attributed to the directing effect of the 2-amino group, being facilitated by hydrogen bonding with the alcoholic acceptor. Finally, removal of the TES groups upon treatment with HF·pyr afforded (+)-neocarzinostatin chromophore (**3**) in 49% yield.

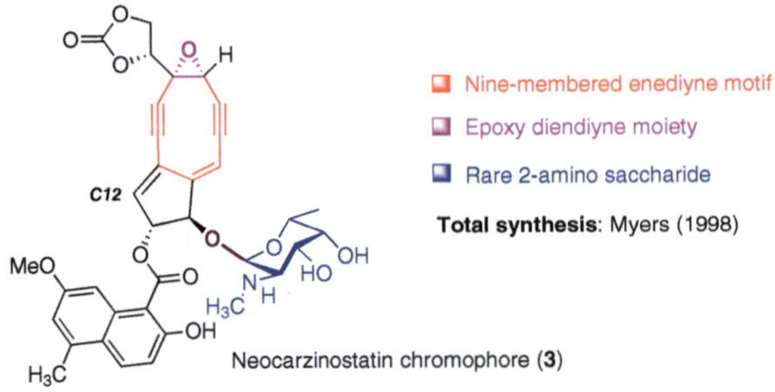


**Figure 3.6** (+)-Neocarzinostatin chromophore (**3**) and its structural features.

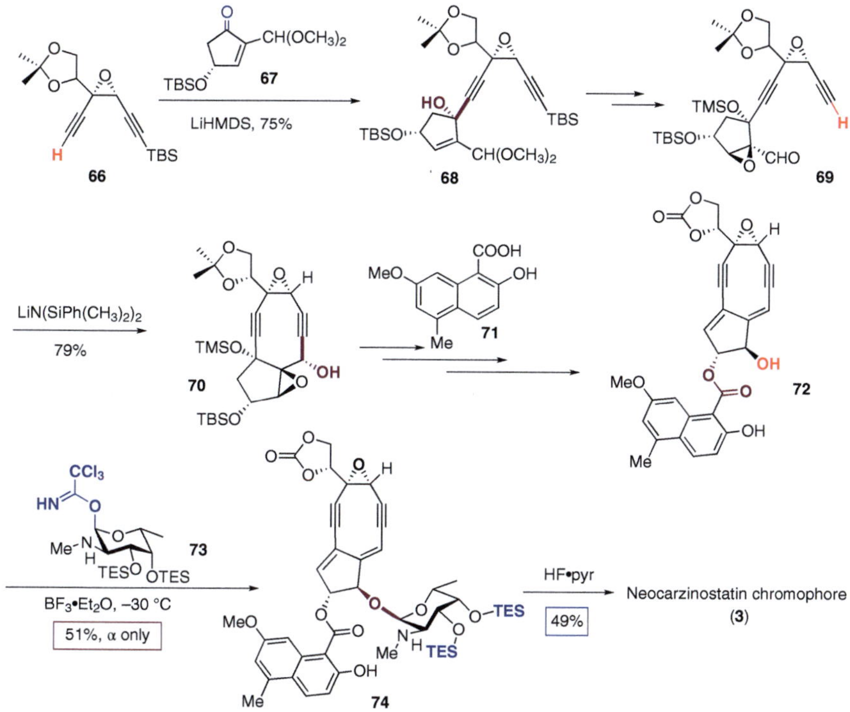


**Scheme 3.7** Total synthesis of (+)-neocarzinostatin chromophore (**3**) by Myers and coworkers.

## Maduropeptin Chromophore

Maduropeptin, a novel member among the family of chromoprotein antibiotics, was first isolated from the broth filtrates of *Actinomadura madurae* by scientists at Bristol-Myers Squibb in 1991 [27]. The chromophore (**75**) showcases a highly strained architecture, which comprises a bicyclo[7.3.0]enediyne scaffold and a 15-membered ansa-macrolactam with atropisomerism. Additionally, a rare saccharide is attached to the aglycone through a tertiary hydroxyl group (Figure 3.7). Although compound **75** represents the methanol adduct of a structurally undefined native chromophore, it retains DNA cleavage property observed in enediyne antibiotics.

In 2009, Hirama and coworkers reported the total synthesis and structural revision of (−)-maduropeptin chromophore (Scheme 3.8) [28, 29]. Under basic conditions, intramolecular addition of the alkynyl residue to the aldehyde in **77** provided the nine-membered cyclic diyne **78** with 82% yield. Subsequent functional group transformations led to **79**, which underwent reduction of the azido group using $PPh_3$, followed by lactamization to provide bicyclic compound **80**. A few steps were required to convert **80** into the desired aglycone derivative **81**. Glycosylation of **81** with glycosyl trichloroacetimidate **82**, in the presence of TMSOTf, led to the formation of the desired β-glycoside **83** in 40% yield without detecting the anomeric isomer. Finally,

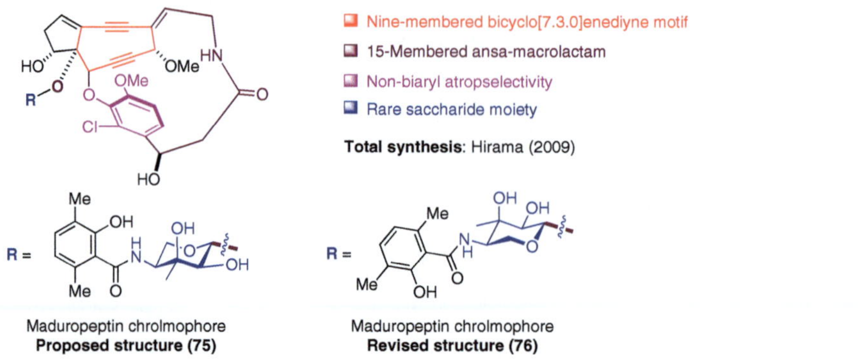


**Figure 3.7** The proposed structure of maduropeptin chrolmophore (**75**) and the revised structure (**76**).

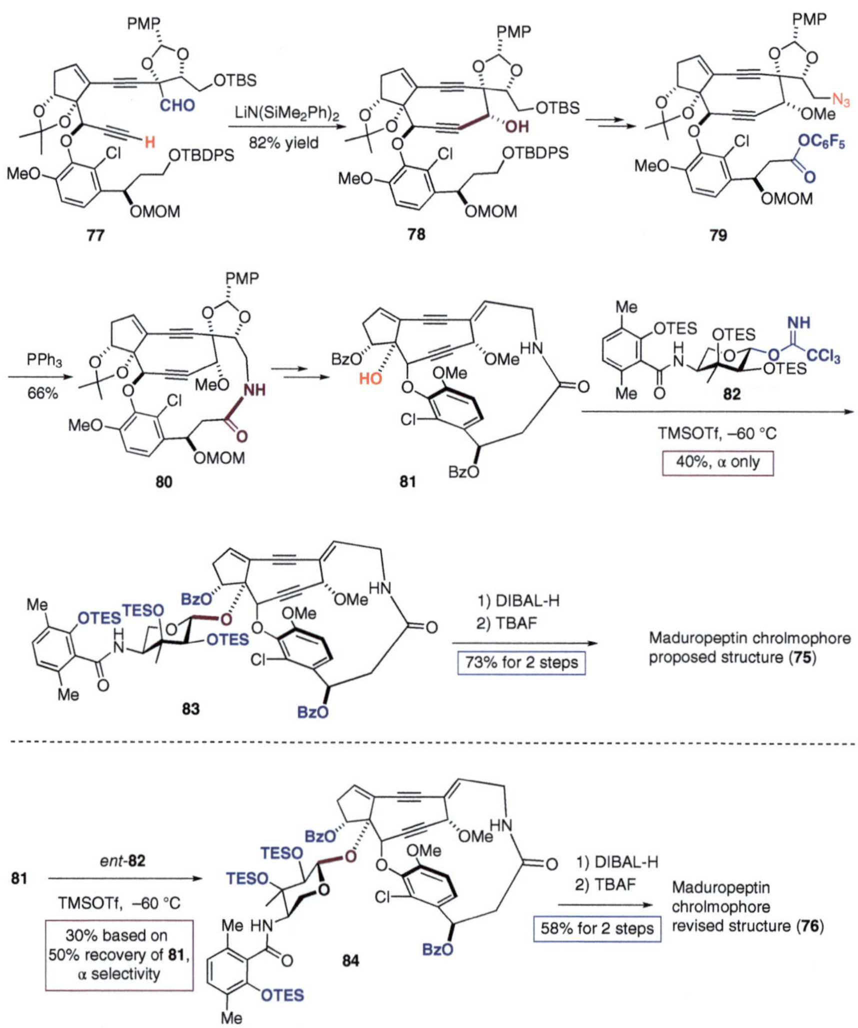


**Scheme 3.8** Total synthesis and structural revision of (−)-maduropeptin chromophore by Hirama and coworkers.

sequential removal of the Bz and TES groups (73% yield for two steps) completed the total synthesis of **75**. However, the NMR data of the synthetic compound did not match with the data obtained from the natural product. As a result, the authors proposed a revised structure of maduropeptin chromophore (**76**), which incorporated an enantiomeric sugar residue. Glycosylation of aglycone **81** with the enantiomer of glycosyl donor **82** (*ent*-**82**) under similar conditions provided the coupled glycoside **84** in 30% yield (based on 50% recovery of **81**). Removal of the Bz and TES protecting groups afforded **76** (58% for two steps), whose analytical data agreed with those of the natural product. Thus, Hirama and coworkers accomplished the total synthesis and structural revision of (−)-maduropeptin chromophore.

## References


**1** Lee, M.D., Dunne, T.S., Chang, C.C. et al. (1987). *J. Am. Chem. Soc.* 109: 3466–3468.

**2** Lee, M.D., Dunne, T.S., Siegel, M.M. et al. (1987). *J. Am. Chem. Soc.* 109: 3464–3466.

**3** Golik, J., Clardy, J., Dubay, G. et al. (1987). *J. Am. Chem. Soc.* 109: 3461–3462.

**4** Golik, J., Dubay, G., Groenewold, G. et al. (1987). *J. Am. Chem. Soc.* 109: 3462–3464.

**5** Ishida, N., Miyazaki, K., Kumagai, K., and Rikimaru, M. (1965). *J. Antibiot.* 18: 68.

**6** Edo, K., Mizugaki, M., Koide, Y. et al. (1985). *Tetrahedron Lett.* 26: 331–334.

**7** Nicolaou, K.C. and Dai, W.-M. (1991). *Angew. Chem. Int. Ed. Engl.* 30: 1387–1416.

**8** Smith, A.L. and Nicolaou, K.C. (1996). *J. Med. Chem.* 39: 2103–2117.

**9** Joshi, M.C. and Rawat, D.S. (2012). *Chem. Biodivers.* 9: 459–498.

**10** Groneberg, R.D., Miyazaki, T., Stylianides, N.A. et al. (1993). *J. Am. Chem. Soc.* 115: 7593–7611.

**11** Nicolaou, K.C., Hummel, C.W., Nakada, M. et al. (1993). *J. Am. Chem. Soc.* 115: 7625–7635.

**12** Smith, A.L., Pitsinos, E.N., Hwang, C.K. et al. (1993). *J. Am. Chem. Soc.* 115: 7612–7624.

**13** Cabal, M.P., Coleman, R.S., and Danishefsky, S.J. (1990). *J. Am. Chem. Soc.* 112: 3253–3255.

**14** Halcomb, R.L., Wittman, M.D., Olson, S.H. et al. (1991). *J. Am. Chem. Soc.* 113: 5080–5082.

**15** Halcomb, R.L., Boyer, S.H., and Danishefsky, S.J. (1992). *Angew. Chem. Int. Ed. Engl.* 31: 338–340.

**16** Hitchcock, S.A., Chu-Moyer, M.Y., Boyer, S.H. et al. (1995). *J. Am. Chem. Soc.* 117: 5750–5756.

**17** Hitchcock, S.A., Boyer, S.H., Chu-Moyer, M.Y. et al. (1994). *Angew. Chem. Int. Ed. Engl.* 33: 858–862.

**18** Kim, S.H., Augeri, D., Yang, D., and Kahne, D. (1994). *J. Am. Chem. Soc.* 116: 1766–1775.

**19** McDonald, L.A., Capson, T.L., Krishnamurthy, G. et al. (1996). *J. Am. Chem. Soc.* 118: 10898–10899.

**20** Nicolaou, K.C., Li, R., Lu, Z. et al. (2018). *J. Am. Chem. Soc.* 140: 8091–8095.

**21** Nicolaou, K.C., Lu, Z., Li, R. et al. (2015). *J. Am. Chem. Soc.* 137: 8716–8719.

**22** Nicolaou, K.C., Li, R., Lu, Z. et al. (2018). *J. Am. Chem. Soc.* 140: 12120–12136.

**23** Zhang, H., Li, R., Ba, S. et al. (2019). *J. Am. Chem. Soc.* 141: 7842–7852.

**24** Nicolaou, K.C., Li, R., Chen, Q. et al. (2020). *J. Am. Chem. Soc.* 142: 12890–12899.

**25** Myers, A.G., Hammond, M., Wu, Y. et al. (1996). *J. Am. Chem. Soc.* 118: 10006–10007.

**26** Myers, A.G., Liang, J., Hammond, M. et al. (1998). *J. Am. Chem. Soc.* 120: 5319–5320.

**27** Hanada, M., Ohkuma, H., Yonemoto, T. et al. (1991). *J. Antibiot.* 44: 403–414.

**28** Komano, K., Shimamura, S., Inoue, M., and Hirama, M. (2007). *J. Am. Chem. Soc.* 129: 14184–14186.

**29** Komano, K., Shimamura, S., Norizuki, Y. et al. (2009). *J. Am. Chem. Soc.* 131: 12072–12073.

# 4
# Flavonoid Glycosides

Flavonoids refer to polyphenolic compounds that exhibit a C6–C3–C6 skeleton. These secondary metabolites are widely distributed throughout the plant kingdom. Currently, more than 5000 flavonoid compounds have been identified, and they are categorized into several sub-groups based on structural differences of the oxidation state of the C-ring (Figure 4.1). Natural flavonoids typically have functionalized rings bearing OH, OMe, *iso*-prenyl, and glycosyl groups, while the OH and glycosyl residues can be further functionalized by acyl, glycosyl, or other residues [1, 2].

Flavonoid glycosides are compounds that the flavonoid aglycones bounded with various saccharide moieties. Thus far, more than 2500 flavonoid glycosides have been characterized and this number is still increasing rapidly. These ubiquitous metabolites play a variety of important roles in the growth and development of plants, especially as interspecies signaling molecules. As common ingredients of human diets, flavonoid glycosides have shown many beneficial effects on human health that have attracted great attention [3]. In fact, a number of homogeneous flavonoid glycosides have demonstrated a wide spectrum of pharmacological activities, such as antimicrobial, anticancer, radical-scavenging, and neuroprotective activity. Flavonoid glycosides can be divided into three categories based on the bond connecting the flavonoid and saccharide subunits, i.e. flavonoid *O*-glycosides, *C*-glycosides, and mixed *O*,*C*-glycosides (Figure 4.2).

Since the first synthesis of anthocyanins by Robinson and coworkers in 1926 [4], numerous efforts have been devoted to the synthesis of flavonoid glycosides and great achievements have been made [5–10]. The glycosylation reactions involved in the synthesis of flavonoid glycosides can present specific challenges. Typically, two major factors contribute to such difficulties: (i) phenolic hydroxyl groups have lower nucleophilicity compared to their aliphatic alcohol counterparts, and (ii) electron-withdrawing substituents and intramolecular hydrogen bonding can further diminish the nucleophilic activity of phenolic flavonoid hydroxyl groups. In addition, many flavonoid derivatives exhibit poor solubility in conventional glycosylation solvents and are susceptible to degradation under both acidic and basic conditions.

Strategies for the synthesis of flavonoid glycosides can be grouped into three main types (Figure 4.3). The first strategy involves direct glycosylation of the elaborated flavonoid derivative with sugar donors, which is primarily used for the total

*Carbohydrate Chemistry in the Total Synthesis of Naturally Occurring Glycosides*,
First Edition. Biao Yu and Xiaoyu Yang.
© 2024 WILEY-VCH GmbH. Published 2024 by WILEY-VCH GmbH.

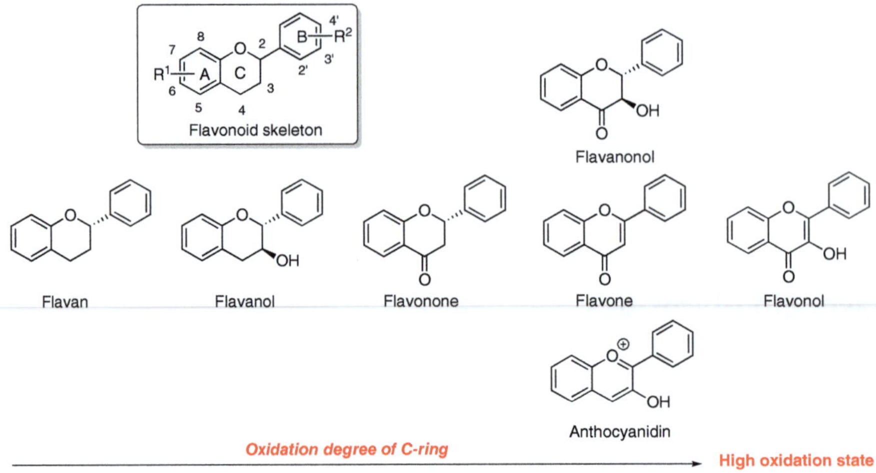


**Figure 4.1** Various skeletons of flavonoids and their oxidation state of C-ring.

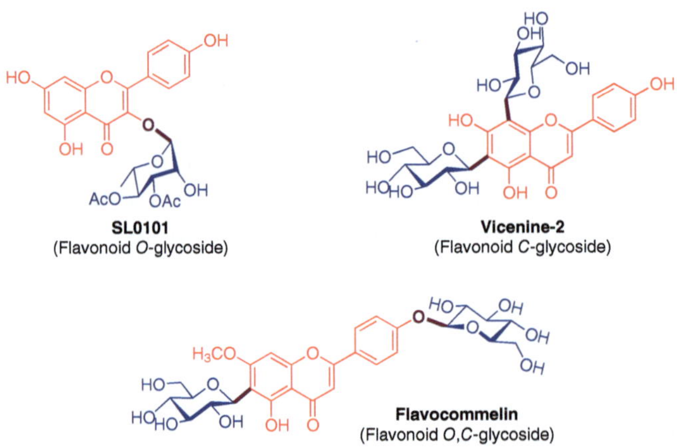


**Figure 4.2** Selected examples of flavonoid *O*-glycoside, *C*-glycosides, and mixed *O,C*-glycoside.

synthesis of flavonoid *O*-glycosides. This approach is appropriate for flavonoids that are abundant in plants or commercially available. The second strategy involves elaboration of the flavonoid moiety (especially the construction of the C-ring) after construction of the glycosidic linkage. This strategy works well for the synthesis of flavonoid *C*-glycosides because of the stability of the *C*-glycosidic bond and the harsh conditions used in *C*-glycosylation reactions. The third strategy involves manipulation of the oxidation state of the product after glycosidic bond formation. Unlike the second strategy, this approach does not include any skeleton-building steps. This tactic is suitable for the synthesis of cynadine-like products, where purification of this type of flavonoid glycosides is arduous.

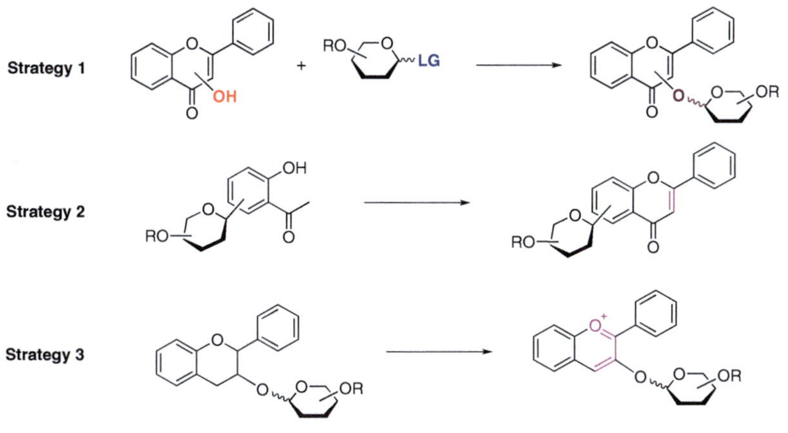


**Figure 4.3** General strategies used in the synthesis of flavonoid glycosides.

## Flavonol 3-*O*-glycosides

### Quercetin 3-sophorotrioside

Flavonol 3-*O*-glycosides are the most commonly occurring flavonoid *O*-glycosides, displaying a diverse range of biological activities. Considerable effort has been dedicated to the synthesis and derivatization of this type of natural glycosides.

Quercetin 3-sophorotrioside (**1**), a flavonol triglycoside (3-*O*-β-D-glucopyranosyl-(1 → 2)-β-D-glucopyranosyl-(1 → 2)-β-D-glucopyranoside), was isolated from the young seedpods of *Pisum sativum* in 1995 [11]. This compound shows protective effects in liver injury induced in mice with D-galactosamine and lipopolysaccharide. In 2004, Linhardt and coworkers reported the first synthesis of quercetin 3-sophorotrioside (**1**, Scheme 4.1) [12]. 7,4′-Di-*O*-benzyl quercetin (**2**) was chosen as the glycosylation aglycone, which was obtained via three steps of protecting group manipulation from naturally occurring quercetin. Glycosylation of **2** with glucosyl bromide **3** in a mixed solvent of aqueous $K_2CO_3/CHCl_3$ in the presence of tetrabutylammonium bromide (TBAB) at 50 °C produced the C-3 glycosylated compound **4** in 90% yield. Notably, glycosylation of **2** with glucopyranosyl trichloroacetimidate or thioglycoside under typical conditions failed to give the desired product. Benzylation of the 5,3′-hydroxyl groups and removal of the acetyl group gave the desired acceptor, which was coupled with glucosyl trichloroacetimidate donor **5** under the catalysis of AgOTf, affording quercetin disaccharide **6** in excellent yield (87%). After deprotection of the acetyl group, coupling with trichloroacetimidate donor **5** afforded quercetin trisaccharide **7** in 90% yield. Finally, deprotection of the benzyl group by hydrogenation catalyzed by 20% $Pd(OH)_2$ on charcoal and removal of acetyl groups using NaOMe/MeOH accomplished the synthesis of quercetin 3-sophorotrioside (**1**) in 81% yield. It is worth mentioning that convergent glycosylation of aglycone **2** with a trisaccharide bromide donor or coupling of **4** with a disaccharide imidate donor failed to afford the desired product, indicating the sensitivity of the glycosylation reactions to the stereo environment.

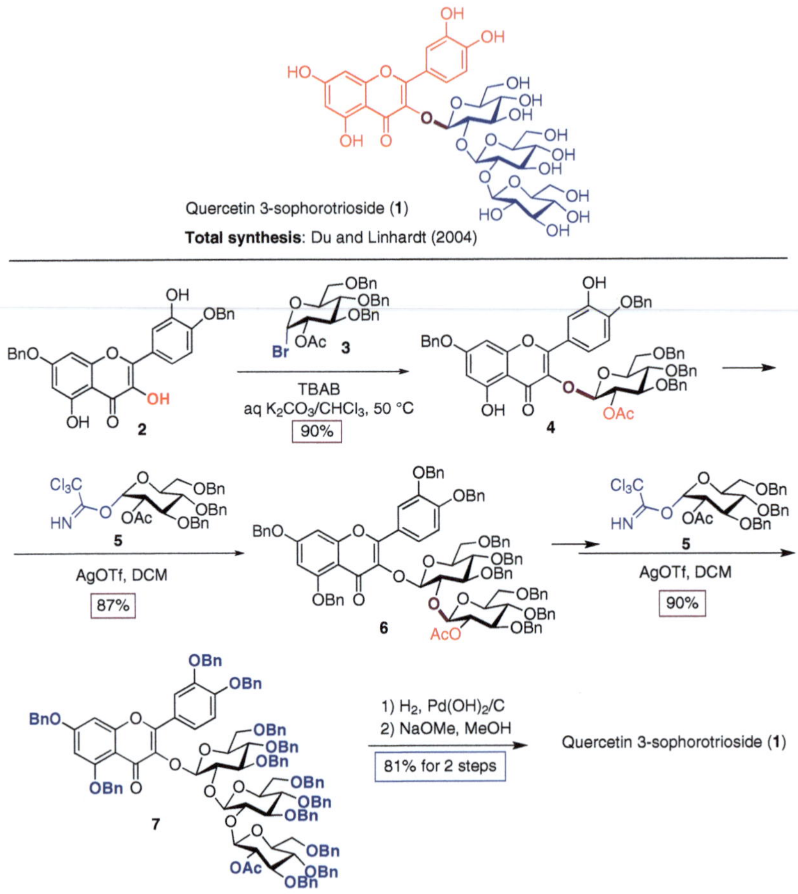

**Scheme 4.1** Synthesis of quercetin 3-sophorotrioside (**1**) by Du, Linhardt, and coworkers

## SL0101

SL0101 (**8**) is a member of the acylated kaempferol L-rhamnosides, which was isolated from a South American dogbane plant named *Forsteronia refracta* [13, 14]. It could inhibit Rsk1 and Rsk2 to a greater extent than Rsk3, despite the fact that Rsk2 and Rsk3 are 80% homologous at the level of primary sequence. Furthermore, in contrast to its potent inhibition of Rsk2 ($IC_{50} \sim 89$ nM), SL0101 was not found to inhibit upstream kinases such as MEK, Raf, and PKC. In 2005, Hecht and coworkers reported a short and efficient approach for the synthesis of this flavonoid glycoside (Scheme 4.2) [15]. Treatment of naringenin (**9**) with benzyl bromide and excess $K_2CO_3$ resulted in simultaneous β-elimination and benzyl protection to give the chalcone, which was then treated with catalytic $I_2$ in DMSO at 140 °C to give the desired flavone **10**. Introduction of a 3-OH group was achieved using dimethyldioxirane (DMDO), followed by opening of the epoxide with catalytic *p*-toluenesulfonic acid to afford flavonol **11**. Glycosylation of **11** with L-rhamnosyl bromide **12** in the presence of $Ag_2O$ provided the desired perbenzylated SL0101 (**13**), exclusively as the α-anomer in 60% yield. Other commonly used glycosylation methods such as

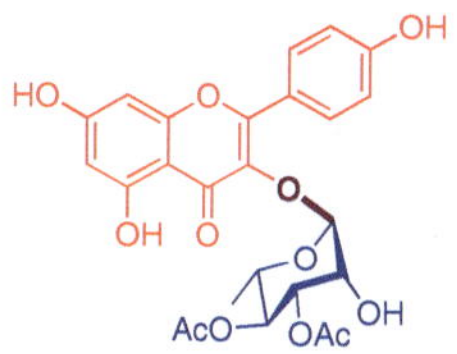


SL0101 (**8**)

**Synthesis**: Hecht (2005), O'Doherty (2006)

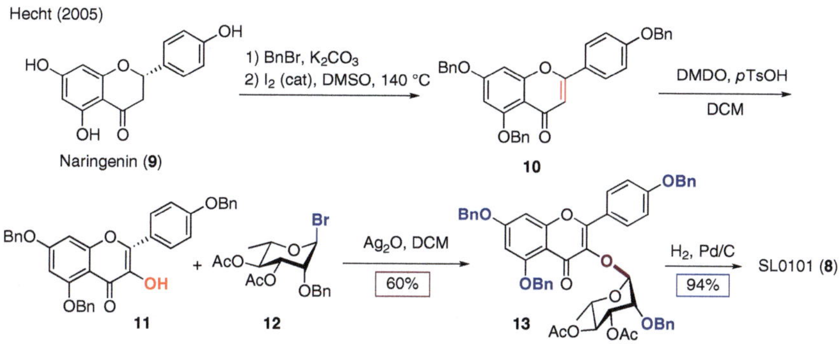


**Scheme 4.2** Synthesis of SL0101 (**8**) by Hecht and coworkers.

benzyltriethylamine bromide and dilute aqueous KOH (phase-transfer-catalyzed [PTC] conditions) failed to provide the desired product. Finally, global debenzylation using Pearlman's catalyst ($Pd(OH)_2/C$) in the presence of $H_2$ led to SL0101 (**8**).

In 2006, O'Doherty and coworkers reported the asymmetric syntheses of SL0101 (**8**) and its analogs via a novel palladium-catalyzed glycosylation method (Scheme 4.3) [16]. Asymmetric reduction of acylfuran **14** under the Noyori reduction conditions gave alcohol **15**, which upon rearrangement by treatment with NBS and protection with $(Boc)_2O$ gave pyranone **16**. Glycosylation was carried out with known flavonol **11** and L-pyranone **16** under palladium catalysis to afford the expected glycoside **17** in 85% yield with complete α-selectivity. $NaBH_4$ reduction of enone **17** and acylation gave the allylic acetate with excellent diastereoselectivity (dr > 20 : 1), which was subjected to dihydroxylation with $OsO_4$ and NMO to give rhamno-diol **18** in good yield. Selective introduction of the C3′-*O*-Ac group afforded **19**, which was hydrogenated over a Pd/C catalyst to accomplish the synthesis of SL0101 (**8**) in 87% yield. In contrast to the previous synthesis, this route provided easy access to the D-sugar enantiomer of SL0101, allowing investigation of the role of the sugar in its structure–activity relationship (SAR). Indeed, the O'Doherty group applied the palladium-catalyzed glycosylation/postglycosylation approach to the synthesis of a number of SL0101 analogs, some of which have shown improved RSK1/2 inhibitor activity and longer biological half-life *in vivo* [17–19].

## Flavone and Isoflavone 7-*O*-Glycosides

### A-76202

A-76202 (**23**), namely 4′,8-dihydroxyisoflavon-7-yl-α-D-arabinofuranoside, was isolated from *Rhodococcus* sp. SANK61694. It is a potent inhibitor of endoplasmic

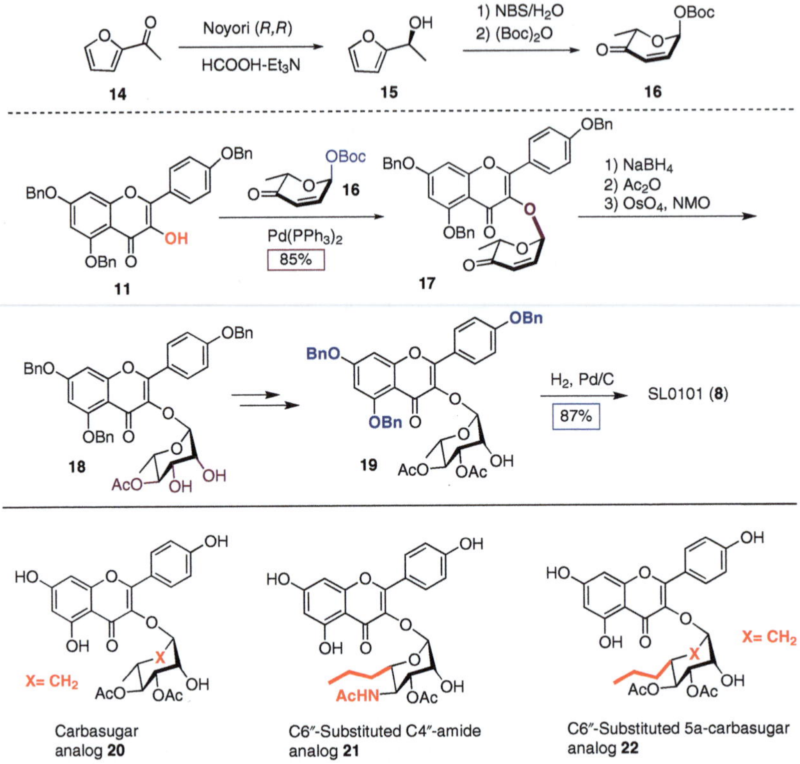

**Scheme 4.3** Synthesis of SL0101 (**8**) and analogs (**20–22**) by O'Doherty and coworkers.

reticulum α-glucosidases I and II, which are involved in the processing of secretory, cell membrane and virus surface glycoproteins. In 1999, Shiozaki and coworkers reported the first synthesis of A-76202 (**23**, Scheme 4.4) [20]. The 4′ and 8 positions of the 4′,7,8-trihydroxyisoflavone were selectively protected by treatment with allyl bromide and DBU to give **24** in moderate yield. Treatment of **24** with *n*BuLi generated the corresponding lithium salt, which was reacted with 2,3,5-tri-*O*-benzoyl-α-D-arabinofuranosyl bromide **25** under reflux conditions to give the expected α-anomer **26** in 26% yield (53% yield based on recycling of **24**), without isolation of the undesired β-anomer. Deprotection of the allyl groups with $RhCl_3$ hydrate followed by saponification of the three benzoyl esters with NaOH accomplished the synthesis of A-76202 (**23**) in 59% yield (two steps). It is worth mentioning that other glycosyl donors, such as glycosyl fluoride **27**, 1,2-anhydro-glycosyl donor **28**, and 1,5-anhydro-glycosyl donor **29**, could not yield the desired glycosylated products. In 2001, the same group reported the synthesis of several derivatives of A-76202 containing the 4′,8-dihydroxyisoflavon-7-yl hexopyranoside moiety and their activities against rat liver α-glucosidase [21].

In 2003, Yu and coworkers reported a concise synthetic route toward flavonoid 7-*O*-glycosides, including A-76202 (**23**, Scheme 4.5) [22]. Their synthesis was highlighted by a highly regioselective removal of the 7-*O*-acyl group of the peracylated

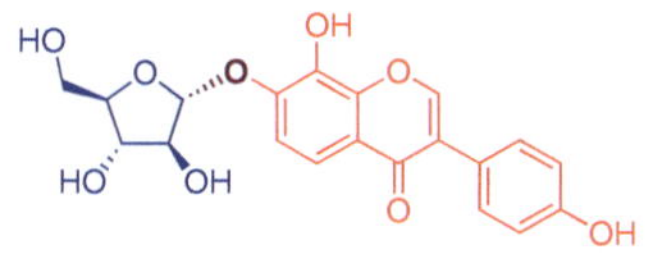

A-76202 **(23)**
**Synthesis**: Shiozaki (1999), Yu (2003)

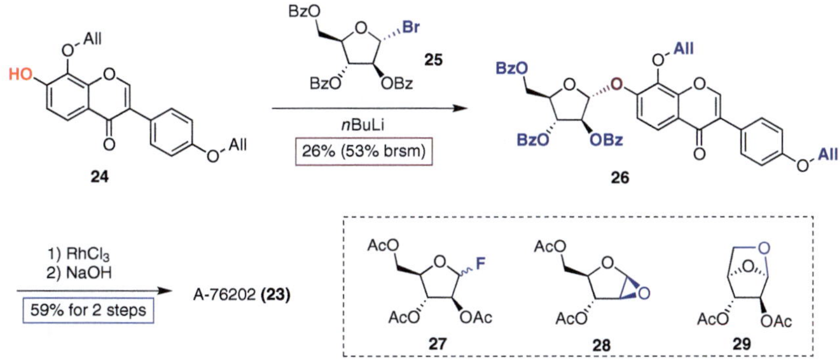

**Scheme 4.4** Synthesis of A-76202 (**23**) by Shiozaki and coworkers.

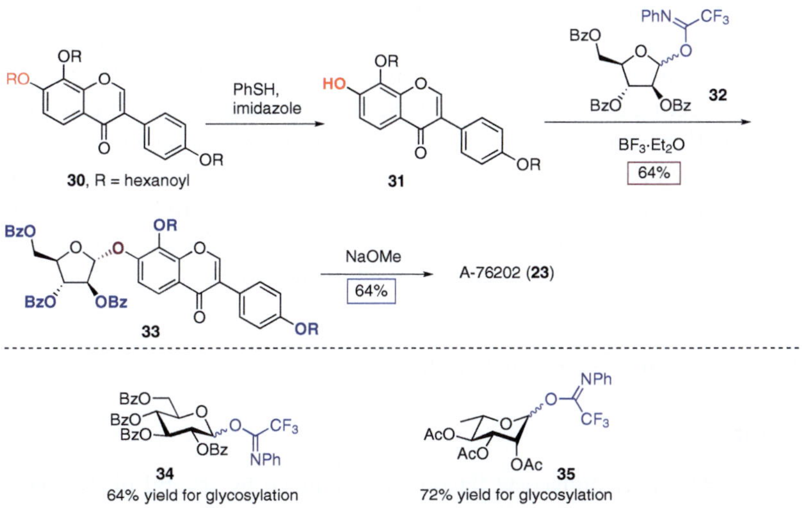

**Scheme 4.5** Synthesis of flavonoid and isoflavonoid 7-*O*-glycosides by Yu and coworkers.

flavones, isoflavones, and flavonols (PhSH, imidazole, NMP), and an effective glycosylation with glycosyl *N*-phenyltrifluoroacetimidate donors. For the synthesis of A-76202, treatment of the per-hexanoyl substituted **30** with PhSH and imidazole afforded C7-OH product **31** in high yield and regioselectivity. Glycosylation of **31** with arabinofuranosyl *N*-phenyltrifluoroacetimidate **32** in the presence of $BF_3 \cdot Et_2O$ afforded the desired α-anomer **33** in 64% yield. Careful removal of the acyl protecting groups under basic conditions led to A-76202 (**23**) in 64% yield. Notably, other six-membered glycosyl *N*-phenyltrifluoroacetimidate donors (e.g. **34** and **35**) could also be coupled with **31** with high efficiency and stereoselectivity.

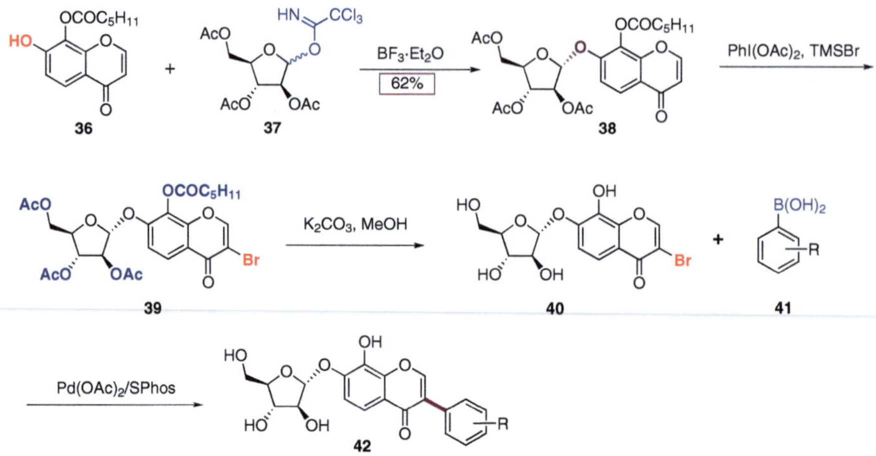


**Scheme 4.6** Divergent synthesis of B-ring congeners of A-76202 by Yu and coworkers.

In 2008, Yu and coworkers developed a highly divergent approach to the synthesis of the B-ring varied isoflavone glycosides analogous to A-76202 (Scheme 4.6) [23]. Glycosylation of chromone **36** with α-D-arabinofuranyl imidate **37** promoted by $BF_3$·$OEt_2$ led to α-glycoside **38** in 62% yield. Regioselective bromination of **38** with $PhI(OAc)_2$ and TMSBr gave **39**, which upon treatment with $K_2CO_3$ in MeOH gave access to the protecting group-free glycoside **40**. Finally, Suzuki–Miyaura coupling of bromide **40** with a variety of arylboronic acids **41** provided isoflavone glycosides with a varied B-ring (**42**) in moderate yields. Evaluation of the inhibitory activities of these compounds against the α-glucosidases of rat liver microsome led to the conclusion that variations at the B-ring of the isoflavone 7-*O*-glycosides are largely tolerated to retain the potent α-glucosidase inhibitory activities.

## Flavonol 3-*O*- and 7-*O*-Bisglycosides

### Kaempferitrin

Kaempferitrin (**43**), a 3,7-diglycosyl flavone, is produced by several plant species and has been reported to have a significant hypoglycemic effect in diabetic rats and antioxidant properties comparable to those of quercetin [24]. A report in 2001 indicated that this compound exhibited antimicrobial activity comparable to several antibiotics against gram-positive and gram-negative bacteria [25]. In 2007, Shaw and coworkers reported a short synthesis of kaempferitrin (**43**), involving the first bis-glycosylation of a dihydroxyflavone (Scheme 4.7) [26]. The key cyclization of benzoate ester **44** to 5-hydroxyflavone **45** was furnished by heating **44** in the presence of $K_2CO_3$ in pyridine in moderate yield. Three steps of protecting group manipulation gave C3,C7-bis-OH flavone **46**. After several attempts at the bis-glycosylation reaction, it was found that coupling of **46** with tri-*O*-acetylrhamnosyl bromide **47** under the activation of $Ag_2O$ could give the desired bis-glycosylated flavone **48** in high conversion (>95% by LC/MS) and exclusively as the α-isomer. Finally, removal

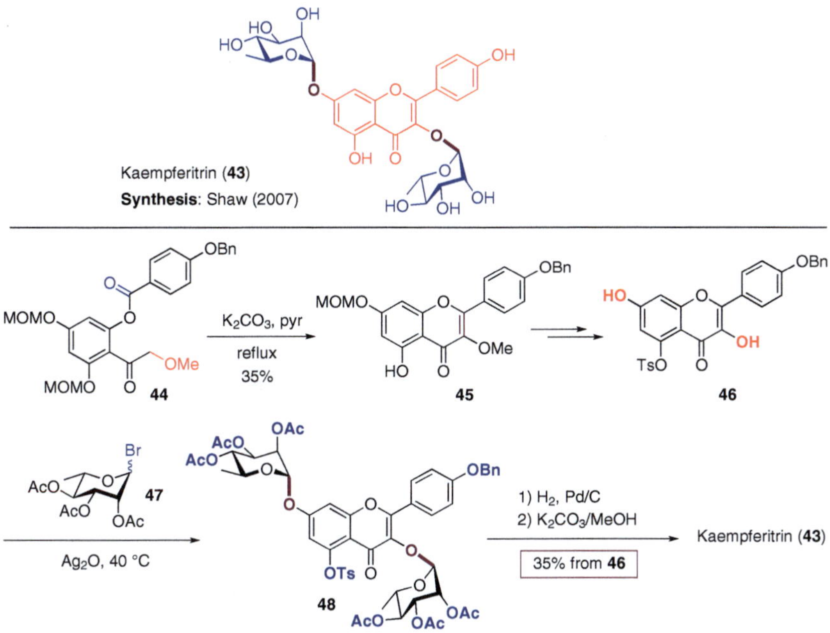


**Scheme 4.7** Synthesis of kaempferitrin (**43**) by Shaw and coworkers.

of the protecting groups under standard hydrogenation conditions followed by treatment with $K_2CO_3$ in MeOH completed the first synthesis of kaempferitrin (**43**) in 35% yield (three steps from **46**).

## Flavonol 5-*O*-Glycosides

### Camellianin B

Much effort has been devoted to the synthesis of flavonoid 3-*O*, 7-*O*, 3,7-di-*O*-glycosides, but the synthesis of flavonol 5-*O*-glycosides has been largely overlooked, despite the fact that the number of this subclass of flavonoid glycosides is rapidly increasing and some of them have been shown to possess interesting bioactivities. Camellianin B (originally proposed as **49**), a typical flavonoid 5-*O*-glycoside, was initially isolated from *Camellia sinensis* L. in 1987. Although camellianin B has been known for almost three decades, no bioactivity study has been conducted due to its extremely low content in natural sources. Structurally, apigenin, the aglycone of camellianin B, lacks the 3-hydroxyl group, which dramatically reduces the reactivity of 5-OH due to the increased strength of intramolecular hydrogen bonding. In 2016, Sun and coworkers reported the synthetic investigation of camellianin B (**49**, Scheme 4.8) [27]. The challenging glycosylation with the aglycone was achieved by coupling with glycosyl *ortho*-alkynyl benzoate donor **52** and suitably protected aglycone **51** under the promotion of a catalytic amount of Au(I) complex, giving the desired β-anomer **53** in a reasonable yield of 61%. Debenzylidenation with TFA

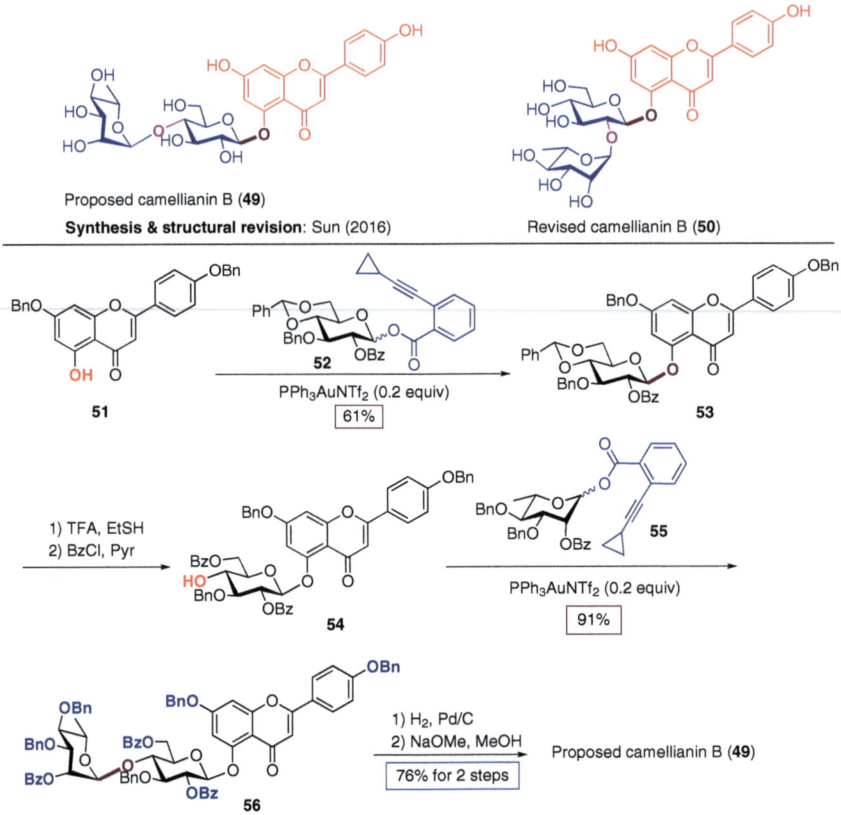


**Scheme 4.8** Synthesis of camellianin B and structural revision by Sun and coworkers.

and EtSH followed by regioselective benzoylation at the C-6′ furnished C4′-OH acceptor **54**. Further gold(I)-catalyzed glycosylation between **54** and rhamnosyl *ortho*-alkynylbenzoate **55** gave access to the desired α-anomer **56** in excellent yield (91%), where the rhamnosyl donor was armed (with *O*-3″ and *O*-4″ protected by benzyl groups). Finally, successive palladium-mediated hydrogenolysis and base-mediated saponification of **56** provided the proposed camellianin B (**49**). However, apparent discrepancies in the spectroscopic data between those of the synthetic sample and those reported in the literature prompted the authors to synthesize analogs as well as peracetylated derivatives, eventually leading to the structural revision of camellianins B (**50**) through 1D and 2D NMR analyses.

## Flavone 6-*C*-Glycosides

### Chafurosides

A novel flavone *C*-glycoside, chafuroside A (**57**), isolated from the extraction of Oolong tea, displays potent inhibitory activity against DNFB (2,4-dinitrofluorobenzene) induced contact hypersensitivity in mice, at a concentration of

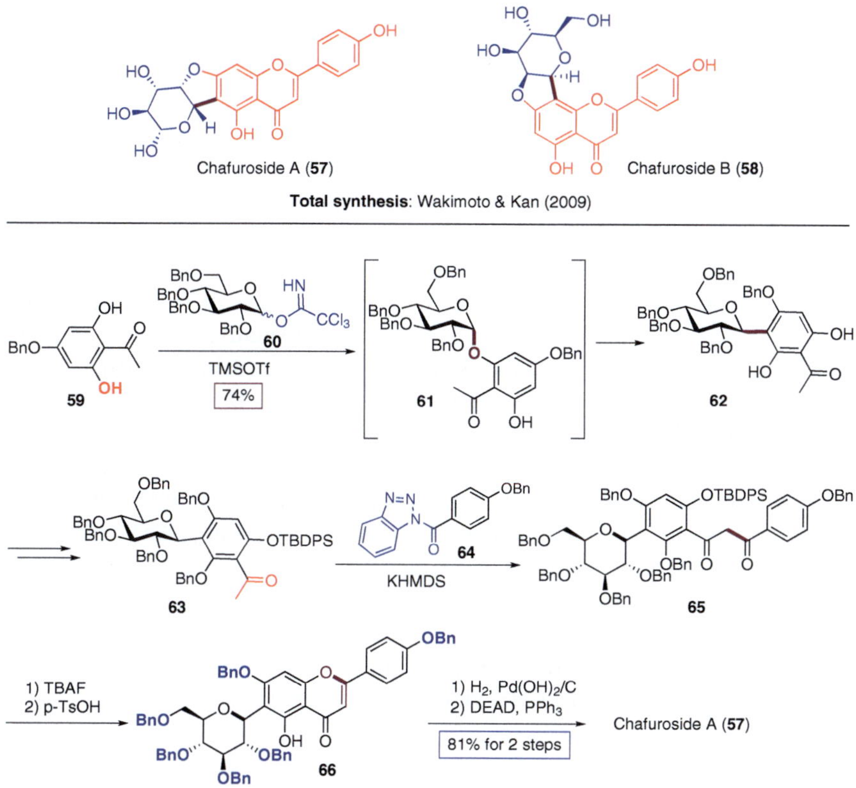


**Scheme 4.9** Synthesis of chafuroside A (**57**) by Kan and coworkers.

1.0–10 μg/kg. In 2009, Kan and coworkers reported a concise synthesis of chafurosides A (**57**) and its regioisomer chafurosides B (**58**) (Scheme 4.9) [28]. The synthesis of chafuroside A (**57**) began with the glycosylation of acetophenone derivative **59** with tetra-*O*-benzyl-D-glucosyl imidate (**60**) in the presence of TMSOTf, which underwent the *O*-glycosidation and successive *O*-*C* rearrangement (Fries rearrangement) to give aryl-*C*-glucoside **62** in 74% yield. Successive protection of the phenolic OH with TBDPS and Bn group gave key intermediate **63**, which upon treatment with KHMDS and 1-acylbenzotriazole **64** provided the desired acylation product β-diketone **65**. After removal of the TBDPS group with TBAF, sequential cyclization and dehydration were performed by heating in toluene with *p*-TsOH to afford selectively protected isovitexin derivative **66**. Finally, after deprotection of the benzyl group under standard hydrogenation conditions, the critical dihydrobenzofuran ring was formed by Mitsunobu reaction to give access to chafuroside A (**57**) in 81% yield (for two steps). Interestingly, intermolecular reactions and/or polymerized products were not observed. Chafuroside B (**58**) was synthesized by a similar synthetic route in five steps.

# Cyanidin 3-*O*-Glycosides

## Cyanidin 3-*O*-β-D-glucoside

Anthocyanins are pigments widely found in flowers, leaves, fruits, and the roots of higher plants, ranging in colors from red through purple to blue. Nowadays, anthocyanins are attracting attention not only as food colorants, but also for nutritional and medicinal reasons, as well as potential application in solar cell devices. In 2006, Kondo, Yoshida, and coworkers reported the first chemical synthesis of cyanidin 3-*O*-β-D-glucoside (**67**), which is one of the most popular anthocyanidin monoglucosides, using a biomimetic oxidative reaction of a leucoanthocyanidin compound via a flav-3-en-3-ol derivative (Scheme 4.10) [29]. The syntheses began with the glycosylation of 5,7,3′,4′-tetra-*O*-benzylcatechin (**68**), which was prepared from (+)-catechin according to a known procedure. Glucosylation of the 3-hydroxyl group of **68** with peracetylglucosyl trichloroacetimidate **69** in the presence of catalytic amounts of TMSOTf afforded the desired β-glucoside **70** in 71% yield. After switching of all the benzyl groups to TBS group, DDQ oxidation of **71** gave 3,4-*cis*-leucoanthocyanin **72** as a single isomer in 74% yield. Elimination of 4-OH was performed by treatment of **72** with MsCl and *i*-$Pr_2NEt$ to give **73** in 82% yield. Finally, deprotection of the *O*-acetyl groups of **73** with NaOMe followed by removal of the TBS groups and oxidation in one pot under acidic conditions gave cyanidin 3-*O*-β-D-glucoside **67** in 51% yield (for two steps).

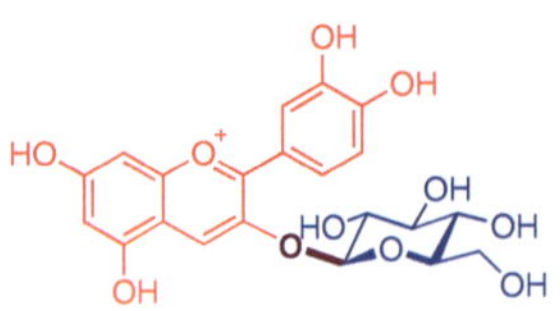

Cyanidin 3-*O*-β-D-glucoside (**67**)
**Synthesis**: Kondo and Yoshida (2006)

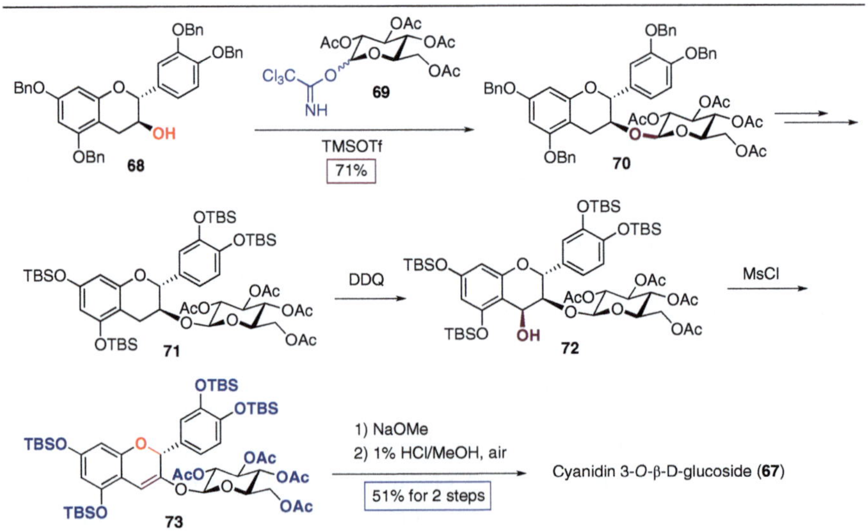


**Scheme 4.10** Synthesis of cyanidin 3-*O*-β-D-glucoside (**73**) by Kondo and coworkers.

## Other Examples

### Quercetin 3-*O*-β-D-Glucuronide

Quercetin 3-*O*-β-D-glucuronide (**74**) was isolated from green beans and from *Reynoutria sachialinensis* [30], which was also the most abundant quercetin metabolite of human diet (Figure 4.4). Interestingly, this glycoside displayed more potent biological activities than quercetin. The Rolando group reported two synthetic approaches to quercetin 3-*O*-β-D-glucuronide (**74**) using glucosyl bromide as donor under basic conditions, starting from quercetin and rutin, respectively [31, 32].

### Quercetin 3-*O*-(2″-Galloyl)-α-L-arabinopyranoside

Quercetin 3-*O*-(2″-galloyl)-α-L-arabinopyranoside (**75**), a representative flavonol *O*-glycoside, was isolated from *Lasiobema japonica*, *Eucalyptus rostrata*, and *Acer okamotoanum* (Figure 4.5) [33]. This compound exhibited potent antibacterial activity against *Escherichia coli* B at a concentration of 100 ppm, along with antioxidant properties. Moreover, it has demonstrated inhibitory effects against HIV-1 integrase ($IC_{50}$ = 18.1 mg/ml). In 2002, Yu and coworkers reported the synthesis of **75** in seven steps with 36% overall yield, by employing a PTC glycosylation with glycosyl bromide donor [34].

### Isoquercitrin Coumarate

Quercetin 3-*O*-[6″-*O*-(*trans*-*p*-coumaroyl)]-β-D-glucopyranoside (**76**), also known as isoquercitrin coumarate, was an acylated quercetin glycoside isolated from *Helichrysum kraussii*, *Melastoma candidum*, and *Marrubium cylleneum* (Figure 4.6) [35]. This compound exhibited strong antihypertensive and tyrosinase inhibitory activity, along with antidiabetic activity. In 2011, Chen and coworkers reported the synthesis of isoquercitrin coumarate (**76**) by employing glycosylation with glycosyl bromide under basic conditions [36].

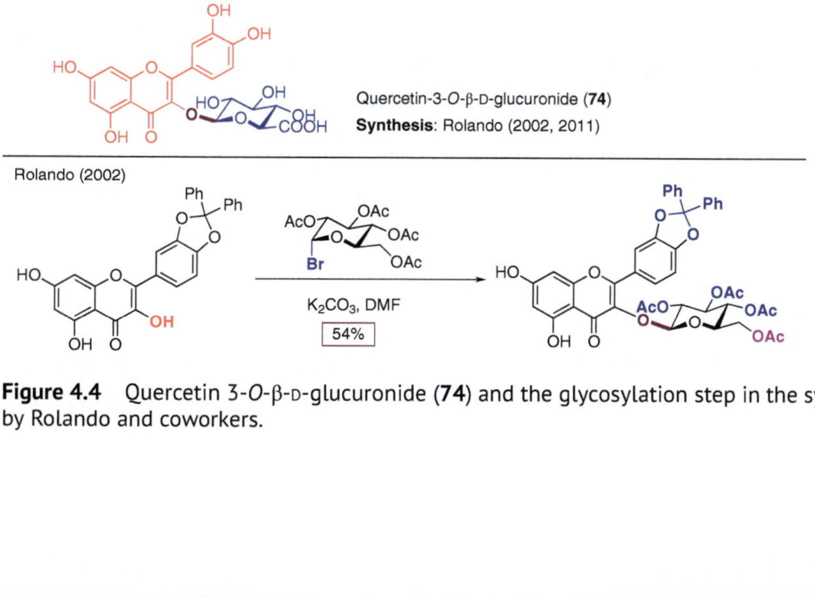


**Figure 4.4** Quercetin 3-*O*-β-D-glucuronide (**74**) and the glycosylation step in the synthesis by Rolando and coworkers.

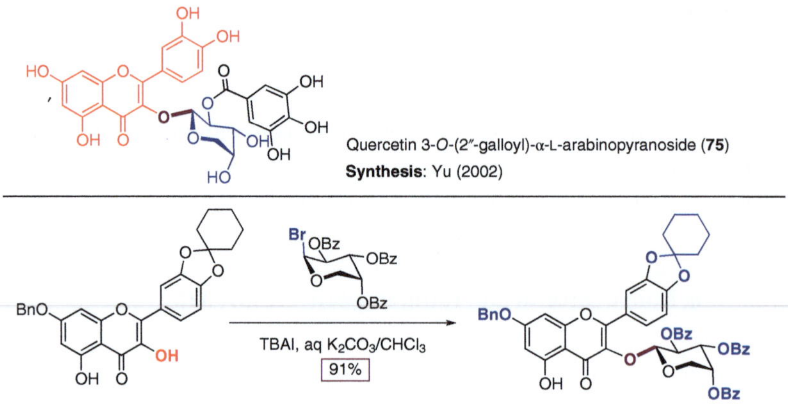


**Figure 4.5** Quercetin 3-*O*-(2″-galloyl)-α-L-arabinopyranoside (**75**) and the glycosylation step in the synthesis by Yu and coworkers.

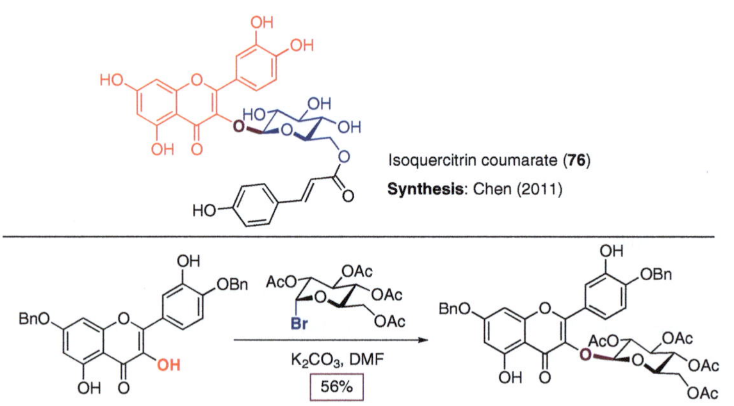


**Figure 4.6** Isoquercitrin coumarate (**76**) and the glycosylation step in the synthesis by Chen and coworkers.

## Kaempferol 3-*O*-(3″,6″-di-*O*-*E*-*p*-coumaroyl)-β-D-Glucopyranoside

Kaempferol 3-*O*-(3″,6″-di-*O*-*E*-*p*-coumaroyl)-β-D-glucopyranoside (**77**) is a representative acylated flavonol 3-*O*-glycoside, which has been isolated from the needles of *Piceaobovata* Ledeb and the leaves of *Stenochlaenapalustris* (Figure 4.7) [37]. This compound can offer significant protection to deep-lying tissues against the harmful effects of UV-B radiation. In 2010, Zhang, Yu, and coworkers reported the first synthesis by employing glycosylation with glucopyranosyl *o*-hexynylbenzoate as a donor under the catalysis of a gold(I) complex [38].

## Platanoside

Kaempferol 3-*O*-[2″,3″-di-*O*-(*E*)-*p*-coumaroyl]-α-L-rhamnopyranoside, also referred as platanoside (**78**), was an acylated flavonol 3-*O*-glycoside isolated from *Platanus*

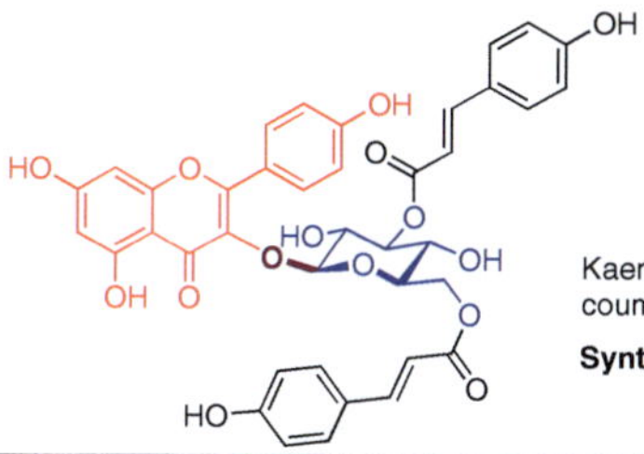

Kaempferol 3-*O*-(3″,6‴-Di-*O*-*E*-*p*-coumaroyl)-β-D-glucopyranoside (**77**)
**Synthesis**: Yu (2010)

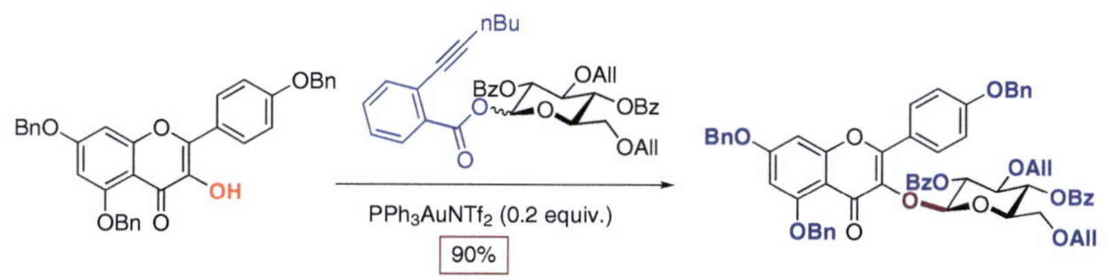

**Figure 4.7** Kaempferol 3-*O*-(3″6″-Di-*O*-*E*-*p*-coumaroyl)-β-D-glucopyranoside (**77**) and the glycosylation step in the synthesis by Yu and coworkers.

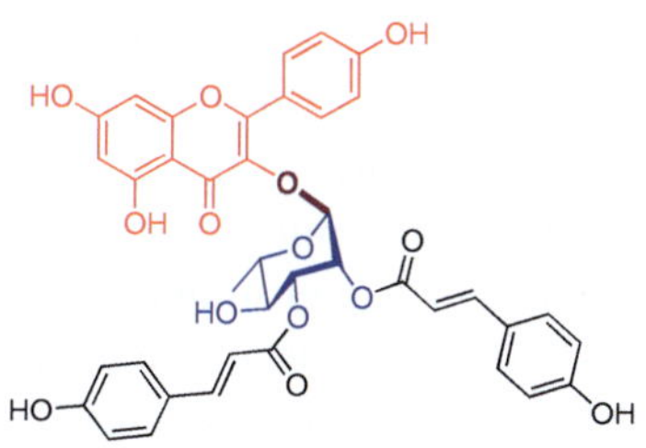

Platanoside (**78**)
**Synthesis**: Yu (2011)

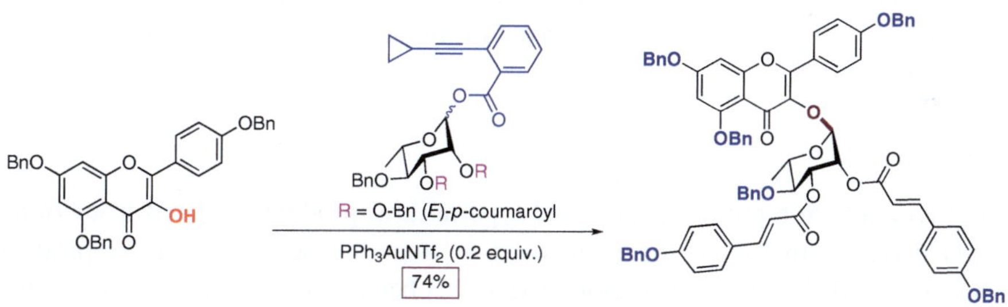

**Figure 4.8** Platanoside (**78**) and the glycosylation step in the synthesis by Yu and coworkers.

*acerifolia*, *Platanus orientalis*, *Platanus occidentalis*, *Pentachondra pumila*, *Faeniculum vulgare*, and *F. dulce* (Figure 4.8) [39]. This compound exhibited selective and potent activity against methicillin-resistant *Staphylococcus aureus* (MRSA), with an $IC_{50}$ of 2.0 mg/ml. In 2011, Sun, Yu and coworkers reported the first synthesis by employing glycosyl *o*-cyclopropylethynylbenzoate as donor and $Ph_3PAuNTf_2$ as a catalyst [40].

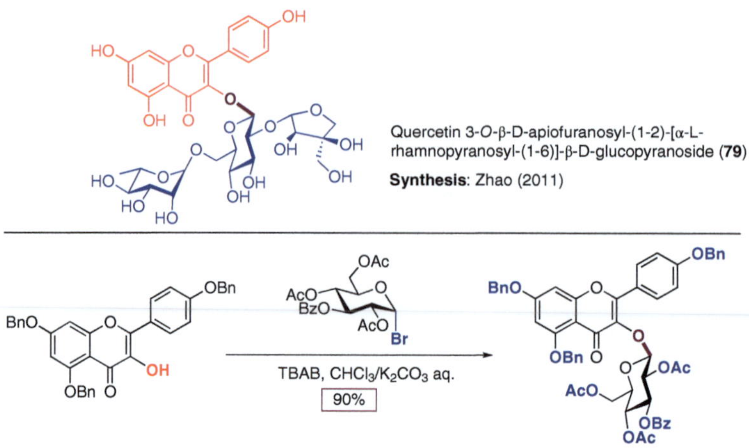


**Figure 4.9** Quercetin trisaccharide **79** and the key glycosylation step in the synthesis by Zhao and coworkers.

## Quercetin-3-*O*-trisaccharide 79

Quercetin-3-*O*-β-D-apiofuranosyl-(1 → 2)-[α-L-rhamnopyranosyl-(1 → 6)]-β-D-glucopyranoside (**79**) is a flavonol *O*-glycoside isolated from the seeds of glandless cotton, whose structure features a unique 2,6-branched trisaccharide (Figure 4.9) [41]. In experimental studies involving forced swimming tests of mice, this compound exhibited significant antidepressant effects. Moreover, it demonstrated the ability to promote the proliferation of hippocampal cells both in chronically stressed mice *in vivo* and cultured hippocampal neural progenitor cells *in vitro*. In 2011, Zhao and coworkers reported a synthesis in 11 linear steps and with an overall yield of 14.6%, with the PTC glycosylation using glycosyl bromide donor to connect the aglycone and saccharide [42].

## Houttuynoid A

Houttuynoids are a group of flavonoid-type natural products isolated from *Houttuynia cordata*, an herbal medicine used to treat the Severe Acute Respiratory Syndrome (SARS) [43]. Houttuynoid A (**80**) has been identified as the most potent member, demonstrating the ability to inhibit HSV-1 multiplication and prevent lesion formation in an HSV-1 infection mouse model (Figure 4.10). Furthermore, this compound exhibited broad antiviral properties against various α-herpesviruses, including HSV-2 and varicella zoster. In 2018, Sun and coworkers reported the first synthesis in nine linear steps, utilizing the glycosylation with glycosyl bromide donor under basic conditions to form the connection between the aglycone and saccharide [44].

## Daidzin, Genistin, Ononin, and Sissotrin

Daidzin (**81**), genistin (**82**), ononin (**83**), and sissotrin (**84**) are isoflavone *O*-glucosides that have been extracted from various plant and food sources (Figure 4.11)

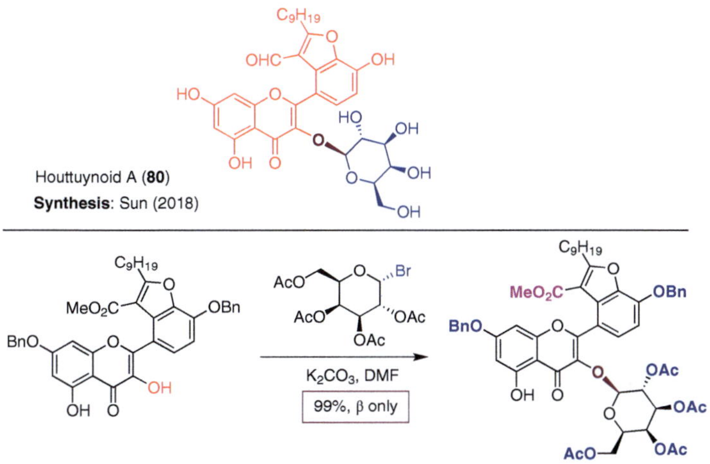


**Figure 4.10** Houttuynoid A (**80**) and the key glycosylation step in the synthesis by Sun and coworkers.

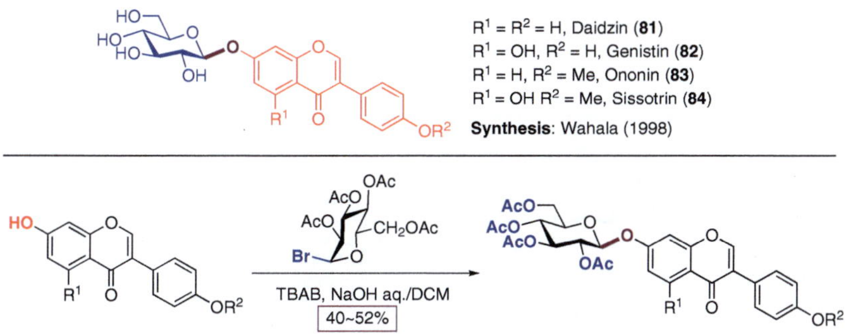


**Figure 4.11** Daidzin (**81**), genistin (**82**), ononin (**83**), sissotrin (**84**), and the key glycosylation step in the syntheses by Wahala and coworkers.

[45]. Isoflavone *O*-glycosides, particularly daidzin (**81**) and genistin (**82**), have garnered significant interest due to findings from numerous biological studies. Notably, daidzin (**81**) exhibited antidipsotropic activity and served as a potent selective inhibitor of human mitochondrial aldehyde dehydrogenase *in vitro*. In 1998, Wahala and coworkers reported a collective synthesis of these compounds by employing glycosyl bromide as donor under the PTC conditions [46].

## Apigenin 7-*O*-Cellobioside

Apigenin 7-*O*-cellobioside (**85**) was isolated from the petals of *Salvia patens* and *Salvia uliginosa* in 1998 (Figure 4.12) [47]. In 2013, Kucukislamoglu and coworkers reported the synthesis in four steps with 76% overall yield, which utilized disaccharide bromide as a donor under the promotion of silver salt [48].

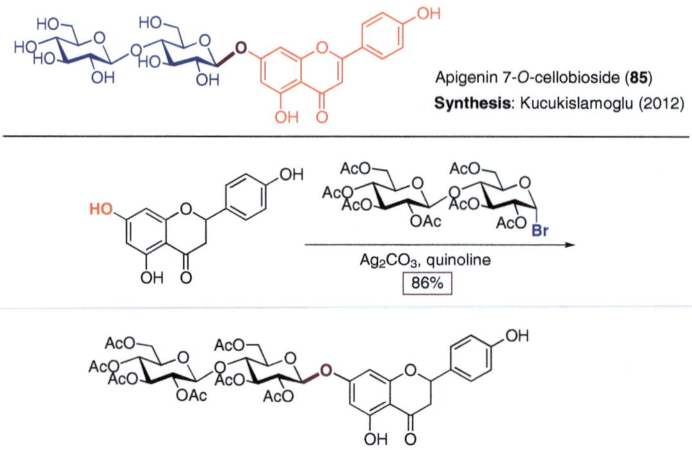


**Figure 4.12** Apigenin 7-*O*-cellobioside (**85**) and the glycosylation step in the synthesis by Kucukislamoglu and coworkers.

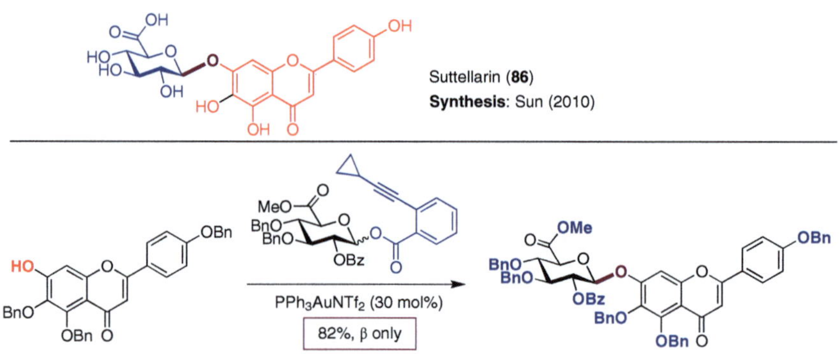


**Figure 4.13** Suttellarin (**86**) and the glycosylation step in the synthesis by Sun and coworkers.

## Suttellarin

Scutellarin (**86**) was isolated from *Erigeron breviscapu*, a well-known traditional Chinese medicinal herb used for the treatment of cerebrovascular diseases (Figure 4.13) [49]. This compound has demonstrated various pharmaceutical activities, including anti-tumoral, neuroprotective, angiogenesis-promoting, and anti-HIV effects. In 2019, Sun and coworkers reported the synthesis of scutellarin (**86**) through direct glycosylation with a glucuronyl *o*-alkynylbenzoate donor in the presence of a gold(I) catalyst [50].

## Calabricoside A

Calabricoside A (**87**), a flavonol triglycoside having the structure of quercetin 3-*O*-[α-L-rhamnopyranosyl-(1 → 2)-α-L-arabinopyranoside]-7-*O*-β-D-glucopyranoside, was

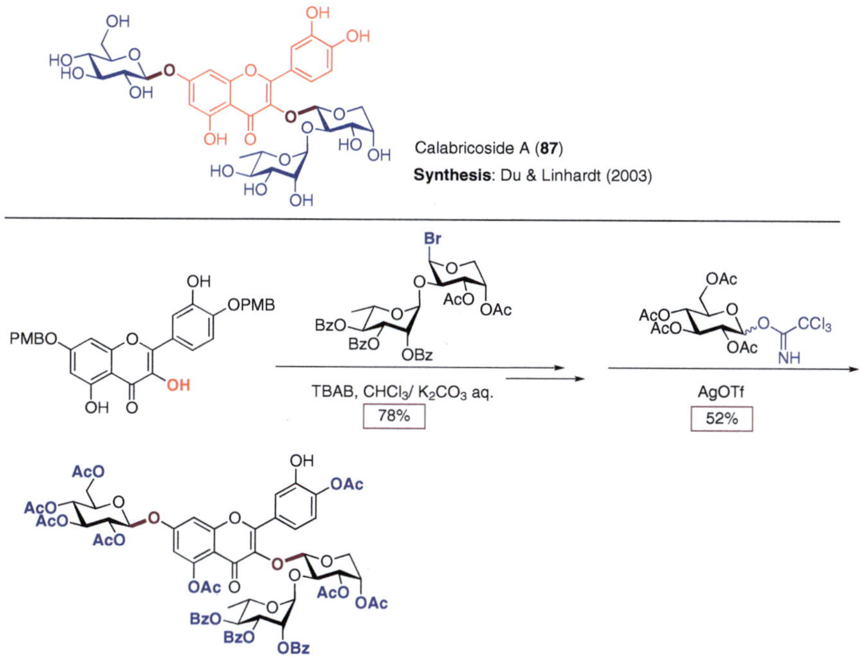


**Figure 4.14** Calabricoside A (**87**) and the glycosylation steps in the synthesis by Du, Linhardt, and coworkers.

isolated from the aerial parts of *Putoria calabrica* (Figure 4.14) [51]. This compound exhibited potent radical-scavenging activity, with an $IC_{50}$ value of 0.25 μM. In 2003, Du, Linhardt, and coworkers reported the first synthesis in 12 linear steps and with a 24.7% overall yield, involving a combination of PTC C-3 glycosylation and AgOTf-promoted homogeneous C-7 glycosylation [52].

## Kaempferol-3-*O*-7-*O*-bisglycoside 88

Kaempferol-3-*O*-β-L-arabinopyranosyl-(1 → 4)-α-L-rhamnopyranoside-7-*O*-α-L-rhamnopyranoside (**88**), a 3,7-triglycosylflavone, was isolated from the aerial parts of *Fagonia taeckholmiana* (Figure 4.15) [53]. The alcohol and water extracts containing this compound exhibited cytotoxic activity against MCF7 human breast tumor cells in culture, with $IC_{50}$ values of 8.72 and 9.80 μg/ml, respectively. In 2011, Liu and coworkers reported the first synthesis in 13 linear steps, achieving an overall yield of 9.2% starting from commercially available kaempferol. The phenol glycoside linkages were efficiently constructed through PTC glycosylation using disaccharide bromide and monosaccharide bromide donors [54].

## Vicenines

Vicenin-3 (**89**) was isolated from *Vitex lucens* (Verb.), *C. sinensis* (Thea.), *Ephedra* sp. (Ephed.), and *Premna integrifolia* (Verb.) (Figure 4.16) [55]. This compound has also

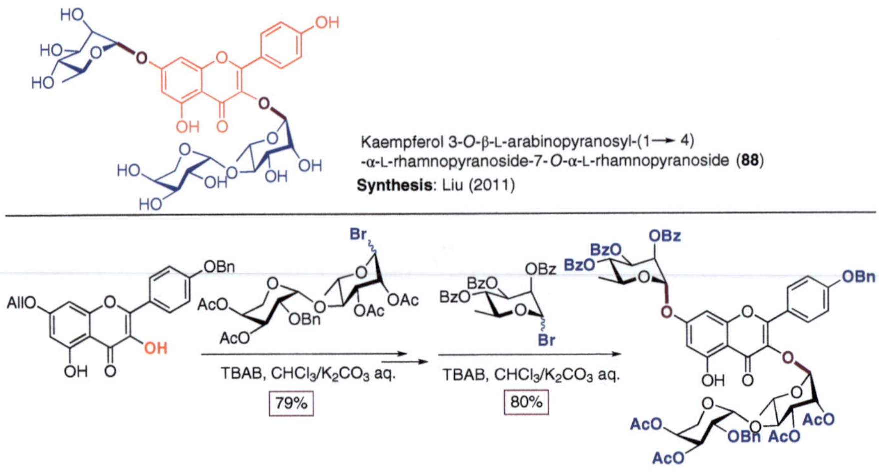


**Figure 4.15** Flavonol bisglycoside **88** and the glycosylation steps in the synthesis by Liu and coworkers.

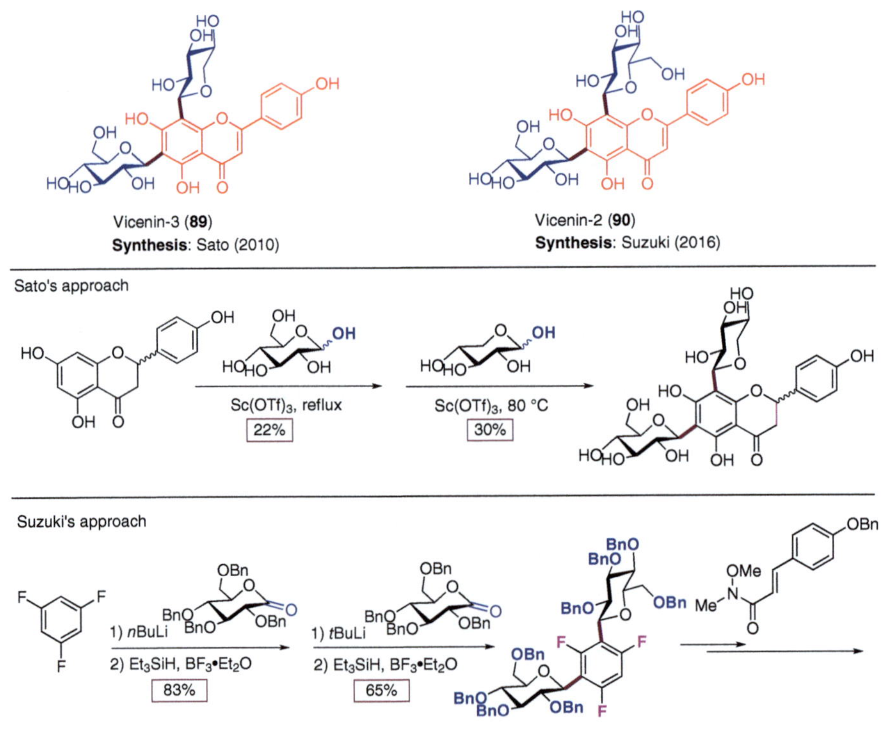


**Figure 4.16** Vicenine-3 (**89**) and vicenin-2 (**90**), and the glycosylation steps in the synthesis by the groups of Sato and Suzuki, respectively.

been identified in the leaves of *Desmodium styracifolium* MERR (Leguminosae), which has a history of use in Chinese folk medicine for treating cholelithiasis, lithiasis, and liver inflammation. Vicenin-2 (**90**), a bis-*C*-glycosyl apigenin, was isolated from *Urtica circularis* and has demonstrated noteworthy biological activities, including anti-inflammatory, anticancer, antidiabetic, and antioxidant effects [56]. In 2010, Sato and coworkers reported the synthesis of vicenin-3 (**89**) from naringenin using a short five-step sequence with an overall yield of 4.4%, which employed two regioselective direct *C*-glycosylation with D-glucose and D-xylose as the key steps [57]. In 2016, Ohmori, Suzuki, and coworkers reported an efficient synthesis of vicenin-2 (**90**) utilizing a novel protocol that entailed the direct bis-β-*C*-glucosylation of 1,3,5-trifluorobenzene, followed by transformation of all fluorine atoms into oxygen functionalities [58].

## Carambolaflavone A

Carambolaflavone A was originally isolated from the leaves of *Averrhoa carambola* in 2005, with its structure initially identified as a flavone 6-C-L-fucoside (**91a**) (Figure 4.17) [59]. This compound exhibited an acute blood glucose-lowering effect through the promotion of glucose-induced insulin secretion and stimulation of glycogen synthesis. In 2018, Sun and coworkers reported the synthesis of carambolaflavone A, involving a scalable synthesis of an appropriately protected 2,4,6-trihydroxyacetophenone and a modified Suzuki *C*-glycosylation method. Through the synthesis, the structure of carambolaflavone A was revised to flavone 6-*C*-D-fucoside (**91b**) [60].

## Flavocommeline

Flavocommeline (**92**), identified as a 7-*O*-methylapigenin 6-*C*-4′-*O*-di-β-D-glucoside, represents one of the constituents of commelinin, a blue supramolecular pigment classified as a metalloanthocyanin (a metal-complex anthocyanin) found in

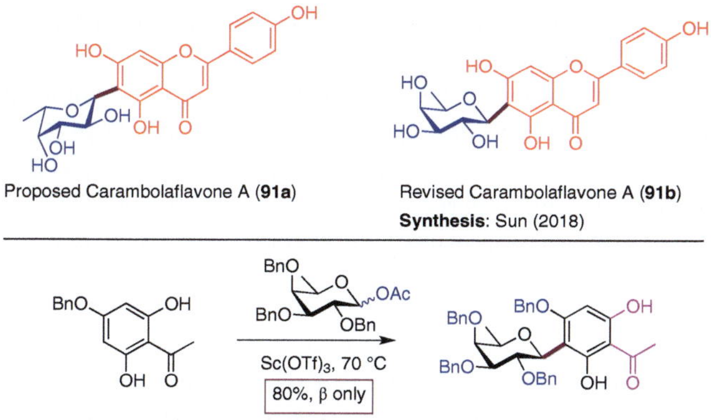


**Figure 4.17** The proposed and revised structure of carambolaflavone A (**91a** and **91b**) and the key glycosylation step in the synthesis by Sun and coworkers.

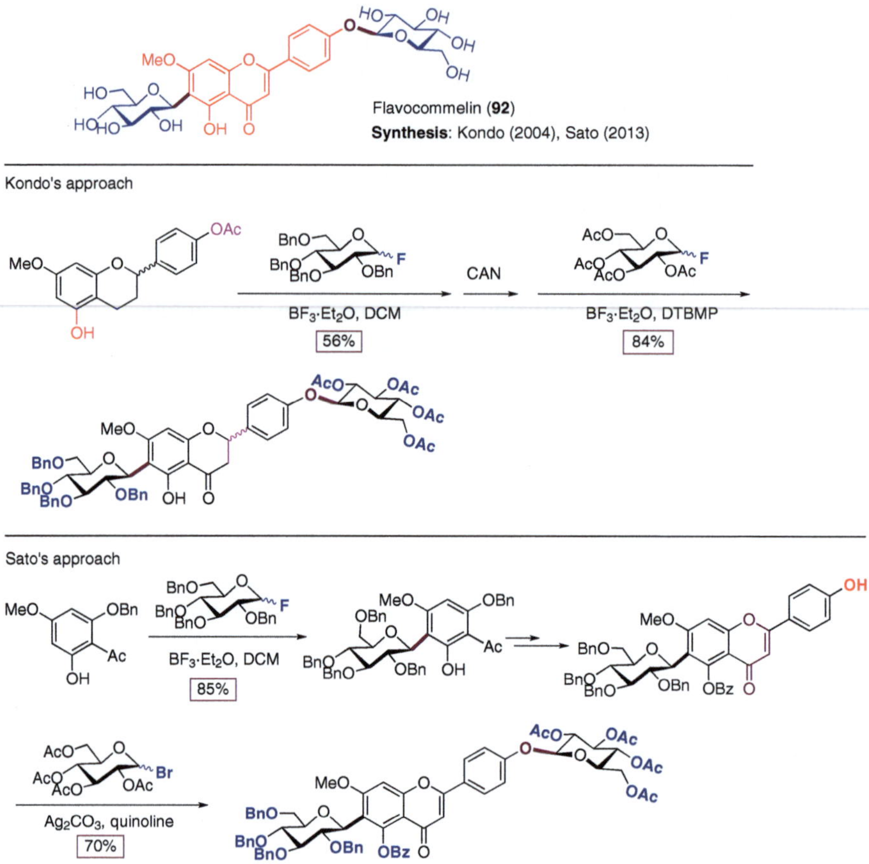


**Figure 4.18** Flavocommeline (**92**) and the glycosylation steps in the syntheses by the Kondo group and the Sato group.

*Commelina communis* (Figure 4.18) [61]. Its structure was elucidated through X-ray crystallographic analysis, revealing a stoichiometric self-assembled supramolecule consisting of six molecules of malonylawobanin, six molecules of flavocommeline (**92**), and two magnesium ions. In 2004, Kondo and coworkers reported the first synthesis of flavocommeline (**92**) via direct 6-*C*-glycosylation of flavan, followed by *O*-glycosylation with glucosyl fluoride donors catalyzed by $BF_3 \cdot Et_2O$ [62]. In 2013, Sato and coworkers reported their synthesis in nine steps, encompassing *C*-glycosylation of a phloroacetophenone derivative with glycosyl fluoride, C-ring formation, and subsequent *O*-glycosylation using a glycosyl bromide donor [63].

## Schaftoside

Schaftoside (**93**), apigenin 6-*C*-β-D-glucopyranosyl-8-*C*-α-L-arabinopyranoside, was isolated from the leaves of sugarcane (*saccharum*) (Figure 4.19) [64]. Its potential benefits include anti-obesity and antidiabetic properties through the inhibition of pancreatic lipase and promotion of glucose-induced insulin secretion, as well as

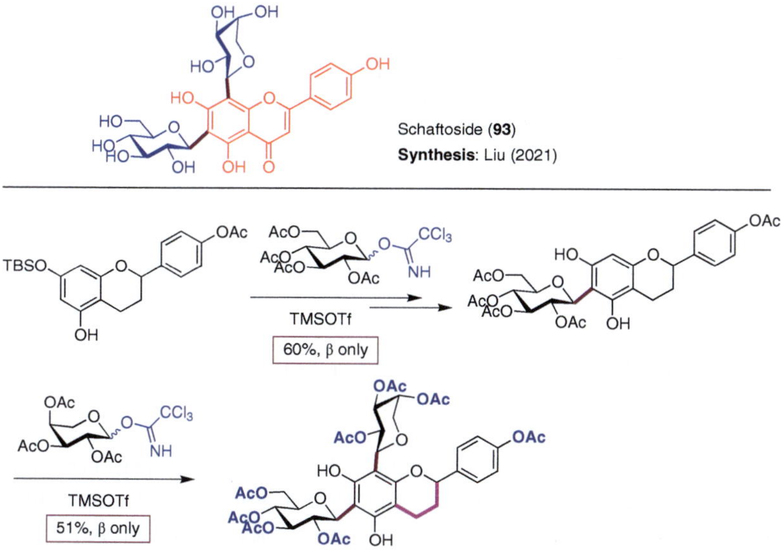


**Figure 4.19** Schaftoside (**93**) and the glycosylation step in the syntheses by Liu and coworkers.

hepatoprotective and antiallergenic effects. However, obtaining this di-*C*-glycosyl flavonoid in high purity from natural sources has proven challenging. In 2021, Liu, Du, and coworkers reported a concise and scalable approach for the synthesis of schaftoside (**93**), involving a double Lewis acid promoted O/C rearrangement to form the *C*-glycoside structure [65].

## Pelargonidin 3-*O*-6″-*O*-acetyl-β-D-glucopyranoside

Pelargonidin 3-*O*-6″-*O*-acetyl-β-D-glucopyranoside (**94**) is an acylated anthocyanin of magenta-colored *Verbena* flowers (Figure 4.20) [66]. This compound exhibited potential antioxidant, hepatoprotective, and inhibitory properties against influenza virus sialidase. In 2007, Kondo and coworkers reported the first synthesis of **94**,

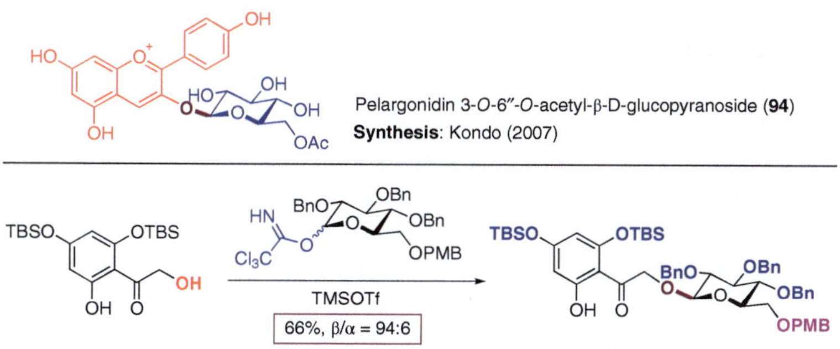


**Figure 4.20** Acylated anthocyanin **94** and the glycosylation step in the synthesis by Kondo and coworkers.

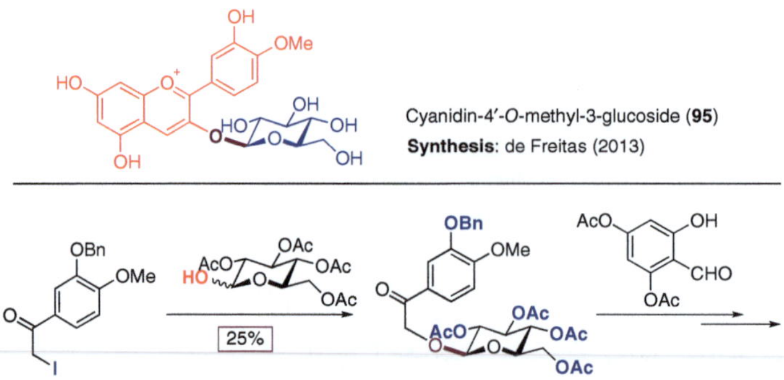


**Figure 4.21** Cyanidin 4′-*O*-methyl-3-glucoside (**95**) and the glycosylation step in the synthesis by de Freitas and coworkers.

involving the synthesis of a crucial protected kaempferol 3-*O*-glucoside intermediate through the Baker–Venkataraman rearrangement of a glucosyloxyacetophenone, followed by Zn–Hg reduction to obtain the acylated anthocyanin structure [67].

### Cyanidin 4′-*O*-methyl-3-*O*-glucoside

Anthocyanins can undergo metabolism in humans to yield various derivatives, including methyl, glucuronyl, and sulfate conjugates. Among these derivatives, cyanidin 4′-*O*-methyl-3-glucoside (**95**) has been detected *in vivo* and holds promise as a valuable standard for biological studies [68]. In 2013, de Freitas and coworkers reported the first synthesis of this compound (**95**) [69]. Their approach involved synthesizing 2,4-diacetoxy-6-hydroxybenzaldehyde and 2-glucopyranosyloxy-acetophenone derivative, coupled with Robinson's acidic aldol condensation (see Figure 4.21).

## References


1 Veitch, N.C. and Grayer, R.J. (2008). *Nat. Prod. Rep.* 25: 555–611.
2 Veitch, N.C. and Grayer, R.J. (2011). *Nat. Prod. Rep.* 28: 1626–1695.
3 Harborne, J.B. and Baxter, H. (1999). *The Handbook of Natural Flavonoids*, 1. Chichester: John Wiley & Sons.
4 Robertson, A. and Robinson, R. (1926). *J. Chem. Soc.* 129: 1713–1720.
5 Sun, J., Laval, S., and Yu, B. (2014). *Synthesis* 46: 1030–1045.
6 Oualid, O. and Artur, M.S.S. (2012). *Curr. Org. Chem.* 16: 859–896.
7 Kin-ichi, O., Kumi, Y., and Tadao, K. (2011). *Curr. Org. Chem.* 15: 2567–2607.
8 Kitamura, K., Ando, Y., Matsumoto, T., and Suzuki, K. (2018). *Chem. Rev.* 118: 1495–1598.
9 Xu, H., Li, Z., Wu, Y. et al. (2019). *Chin. J. Org. Chem.* 39: 1875–1890.
10 Khodzhaieva, R.S., Gladkov, E.S., Kyrychenko, A., and Roshal, A.D. (2021). *Front. Chem.* 9: 63799.

11 Ferreres, F., Esteban, E., Carpena-Ruiz, R. et al. (1995). *Phytochemistry* 39: 1443–1446.
12 Du, Y., Wei, G., and Linhardt, R.J. (2004). *J. Org. Chem.* 69: 2206–2209.
13 Xu, Y.-M., Smith, J.A., Lannigan, D.A., and Hecht, S.M. (2006). *Bioorg. Med. Chem.* 14: 3974–3977.
14 Matthes, H.W.D., Luu, B., and Ourisson, G. (1980). *Phytochemistry* 19: 2643–2650.
15 Maloney, D.J. and Hecht, S.M. (2005). *Org. Lett.* 7: 1097–1099.
16 Shan, M. and O'Doherty, G.A. (2006). *Org. Lett.* 8: 5149–5152.
17 Shan, M. and O'Doherty, G.A. (2010). *Org. Lett.* 12: 2986–2989.
18 Mrozowski, R.M., Sandusky, Z.M., Vemula, R. et al. (2014). *Org. Lett.* 16: 5996–5999.
19 Li, M., Li, Y., Ludwik, K.A. et al. (2017). *Org. Lett.* 19: 2410–2413.
20 Shiozaki, M. (1999). *Tetrahedron: Asymmetry* 10: 1477–1482.
21 Watanabe, Y., Shiozaki, M., and Kamegai, R. (2001). *Carbohydr. Res.* 335: 283–289.
22 Li, M., Han, X., and Yu, B. (2003). *J. Org. Chem.* 68: 6842–6845.
23 Wei, G. and Yu, B. (2008). *Eur. J. Org. Chem.* 2008: 3156–3163.
24 de Sousa, E., Zanatta, L., Seifriz, I. et al. (2004). *J. Nat. Prod.* 67: 829–832.
25 Abdel-Ghani, A.E., Hafez, S.S., Aziz, E.M.A., and El-Shazly, A.M. (2001). *Alexandria J. Pharm. Sci.* 15: 103–108.
26 Urgaonkar, S. and Shaw, J.T. (2007). *J. Org. Chem.* 72: 4582–4585.
27 Hu, Y., Tu, Y.-H., Liu, D.-Y. et al. (2016). *Org. Biomol. Chem.* 14: 4842–4847.
28 Furuta, T., Nakayama, M., Suzuki, H. et al. (2009). *Org. Lett.* 11: 2233–2236.
29 Kondo, T., Oyama, K.-i., Nakamura, S. et al. (2006). *Org. Lett.* 8: 3609–3612.
30 Price, K.R., Colquhoun, I.J., Barnes, K.A., and Rhodes, M.J.C. (1998). *J. Agric. Food. Chem.* 46: 4898–4903.
31 Bouktaib, M., Atmani, A., and Rolando, C. (2002). *Tetrahedron Lett.* 43: 6263–6266.
32 Kajjout, M., Zemmouri, R., and Rolando, C. (2011). *Tetrahedron Lett.* 52: 4738–4740.
33 Kim, H.J., Woo, E.-R., Shin, C.-G., and Park, H. (1998). *J. Nat. Prod.* 61: 145–148.
34 Li, M., Han, X., and Yu, B. (2002). *Tetrahedron Lett.* 43: 9467–9470.
35 Candy, H.A., Laing, M., Weeks, C.M., and Kruger, G.J. (1975). *Tetrahedron Lett.* 16: 1211–1214.
36 Ren, X., Shen, L.-l., Muraoka, O., and Cheng, M. (2011). *J. Carbohydr. Chem.* 30: 119–131.
37 Liu, H., Orjala, J., Sticher, O., and Rali, T. (1999). *J. Nat. Prod.* 62: 70–75.
38 Yang, W., Sun, J., Lu, W. et al. (2010). *J. Org. Chem.* 75: 6879–6888.
39 Kaouadji, M., Morand, J.-M., and Garcia, J. (1993). *J. Nat. Prod.* 56: 1618–1621.
40 Li, Y., Yang, W., Ma, Y. et al. (2011). *Synlett* 7: 915–918.
41 Zhang, Q., Yang, M., Zhao, Y. et al. (2001). *Acta Pharmacol. Sin.* 36: 827.
42 Zhang, Y., Wang, K., Zhan, Z. et al. (2011). *Tetrahedron Lett.* 52: 3154–3157.
43 Li, G.Z., Chai, O.H., Lee, M.S. et al. (2005). *Biol. Pharm. Bull.* 28: 1864–1868.
44 Jian, J., Fan, J., Yang, H. et al. (2018). *J. Nat. Prod.* 81: 371–377.
45 Wang, H.-J. and Murphy, P.A. (1996). *J. Agric. Food. Chem.* 44: 2377–2383.

**46** Lewis, P., Kaltia, S., and Wahala, K. (1998). *J. Chem. Soc., Perkin Trans. 1* 16: 2481–2484.

**47** Veitch, N.C., Grayer, R.J., Irwin, J.L., and Takeda, K. (1998). *Phytochemistry* 48: 389–393.

**48** Sonmez, F., Nebioglu, M., Besoluk, S. et al. (2013). *Nat. Prod. Res.* 27: 631–638.

**49** Zhang, W.-D., Chen, W.-S., Wang, Y.-H. et al. (2000). *Chin. Tradit. Herb. Drugs* 31: 565–568.

**50** Liu, X., Wen, G.-E., Liu, J.-C. et al. (2019). *Carbohydr. Res.* 475: 69–73.

**51** Çalis, I., Heilmann, J., Tasdemir, D. et al. (2001). *J. Nat. Prod.* 64: 961–964.

**52** Du, Y., Wei, G., and Linhardt, R.J. (2003). *Tetrahedron Lett.* 44: 6887–6890.

**53** Ibrahim, L.F., Kawashty, S.A., El-Hagrassy, A.M. et al. (2008). *Carbohydr. Res.* 343: 155–158.

**54** Liu, Q., Li, W., Guo, T. et al. (2011). *Chem. Lett.* 40: 324–325.

**55** Chaboud, A., Raynaud, J., and Dellamonica, G. (1989). *Pharm. Acta Helv.* 64: 16–18.

**56** Marrassini, C., Davicino, R., Acevedo, C. et al. (2011). *J. Nat. Prod.* 74: 1503–1507.

**57** Sato, S. and Koide, T. (2010). *Carbohydr. Res.* 345: 1825–1830.

**58** Ho, T.C., Kamimura, H., Ohmori, K., and Suzuki, K. (2016). *Org. Lett.* 18: 4488–4490.

**59** Araho, D., Miyakoshi, M., Chou, W.-h. et al. (2005). *Nat. Med.* 59: 113–116.

**60** Wang, Y., Liu, M., Liu, L. et al. (2018). *J. Org. Chem.* 83: 4111–4118.

**61** Hondo, T., Yoshida, K., Nakagawa, A. et al. (1992). *Nature* 358: 515–518.

**62** Oyama, K.-i. and Kondo, T. (2004). *J. Org. Chem.* 69: 5240–5246.

**63** Misawa, K., Gunji, Y., and Sato, S. (2013). *Carbohydr. Res.* 374: 8–13.

**64** Šlapetová, T., Šmejkal, K., Innocenti, G. et al. (2009). *Carbohydr. Res.* 344: 1770–1774.

**65** Shang, C., Zhao, C., Nie, P. et al. (2021). *Tetrahedron* 91: 132216.

**66** Toki, K., Yamamoto, T., Terahara, N. et al. (1991). *Phytochemistry* 30: 3828–3829.

**67** Oyama, K.-i., Kawaguchi, S., Yoshida, K., and Kondo, T. (2007). *Tetrahedron Lett.* 48: 6005–6009.

**68** Ichiyanagi, T., Shida, Y., Rahman, M.M. et al. (2005). *J. Agric. Food. Chem.* 53: 145–150.

**69** Cruz, L., Mateus, N., and de Freitas, V. (2013). *Tetrahedron Lett.* 54: 2865–2869.

# 5

# Macrolide Glycosides

Macrolides belong to the polyketides/polyenes class of natural products, which are characterized by a macrolactone ring, typically consisting of 12–20 atoms and attached sugars [1–4]. The first macrolide antibiotic, pikromycin (**1**), was isolated from *Streptomyces erythreus* by Brockman and Henkel in 1950 (Figure 5.1) [5]. Subsequently, numerous structurally related macrolide antibiotics have been discovered from *Streptomyces.* Notably, erythromycins (**2**), known for their ability to inhibit ribosomal-dependent protein biosynthesis, have become the most well-known members of the macrolide antibiotic family. In 1997, apoptolidin A (**3**), a 20-membered macrolide glycoside antibiotic, was isolated from *Nocardiopsis* sp. by Hayakawa and coworkers, which has shown potent and selective induction of programmed cell death in rat glia cells transformed with the adenovirus E1A oncogene [6, 7]. Due to the vast structural diversity of macrolides and sugar residues, many macrolide glycosides have been discovered to date. These compounds commonly exhibit strong antibacterial activities and possess potent anti-inflammatory, antimalarial, and anticancer properties. The antibacterial macrolide glycosides bind near the PTC-loop with the 50S subunit in the bacterial ribosomal tunnel filled with water. Extensive studies have revealed the essential role of sugar moieties in the biological activities of these compounds.

## Erythromycin

Erythromycin A (**2**) was initially isolated from *Saccharopolyspora erythraea* by McGuire and coworkers in 1952 and was fully characterized in 1957 (Figure 5.2) [8]. Structurally, erythromtcin A consists of a 14-membered lactone with 10 asymmetric centers and two distinct sugar residues, namely L-cladinose and D-desosamine. Erythromycin A and its derivatives are included in the World Health Organization's List of Essential Medicines due to their antibiotic activities, which are attributed to their ability to inhibit ribosomal-dependent protein biosynthesis. Due to their significant antibiotic activities and complex structures, these natural products have been the focus of synthetic investigations for over two decades. The total synthesis of erythromycin A (**2**) was accomplished by the Woodward group in 1981 and the

*Carbohydrate Chemistry in the Total Synthesis of Naturally Occurring Glycosides*,
First Edition. Biao Yu and Xiaoyu Yang.
© 2024 WILEY-VCH GmbH. Published 2024 by WILEY-VCH GmbH.

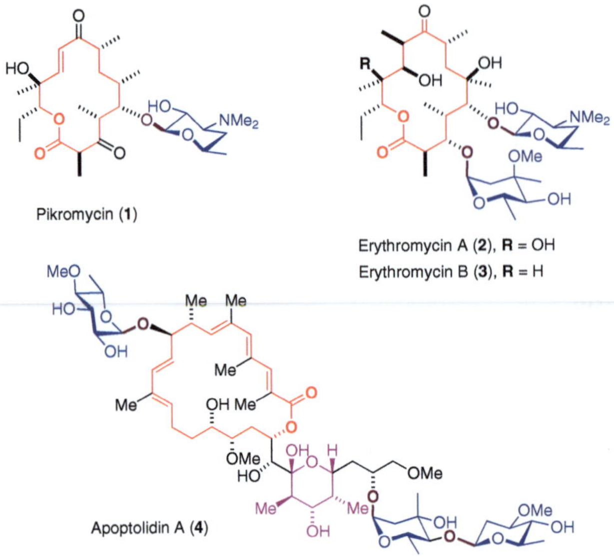


**Figure 5.1** Representative macrolide glycosides natural products.

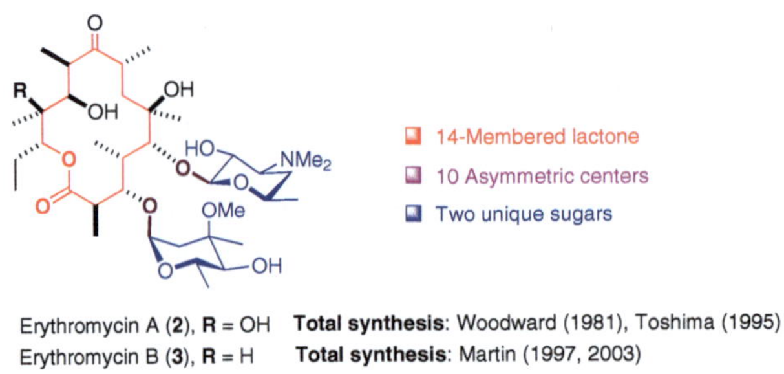


**Figure 5.2** Erythromycin A and B and their structural features.

Toshima group in 1995. Additionally, the Martin group achieved the total synthesis of erythromycin B (**3**) in 1997 and 2003.

**Total synthesis of erythromycin A by the Woodward group**. In 1981, Woodward and coworkers reported the first total synthesis of erythromycin A (**2**) (Scheme 5.1) [9–11]. The synthesis involved the aldol condensation of aldehyde fragment **5** and ketone **6**, followed by oxidation to form product **7**. Several steps were undertaken to convert compound **7** to thioester **8**, which was cyclized using Corey's method (toluene, 110 °C) to give the desired lactone **9** in 70% yield. The regioselective glycosylation of pentaol **9** with D-desosaminide sulfide **10**, in the presence of AgOTf, provided the desired β-glycoside **11** in 36% yield. Subsequently, selective coupling of **11** with L-cladinose-derived glycosyl sulfide **12**, under the promotion of $Pb(ClO_4)_2$, furnished the glycosylated product **13** in 20% yield (55% yield brsm) as the desired α-anomer. This stereochemical outcome was attributed to the involvement of $CH_3CN$ solvent, resulting in a double inversion during the

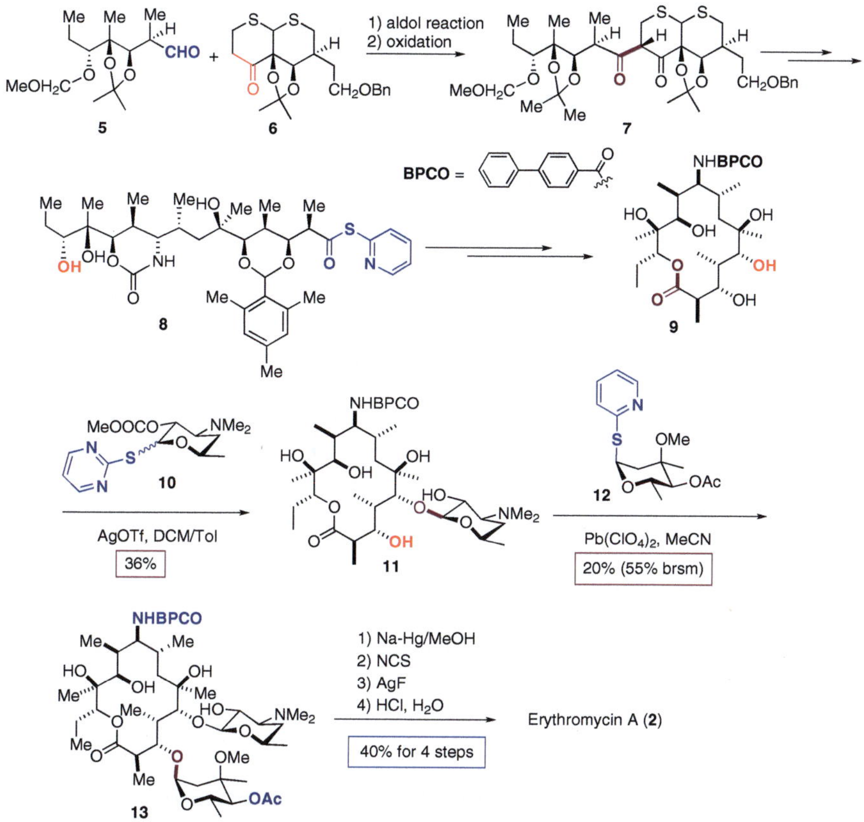

**Scheme 5.1** Total synthesis of erythromycin A (**2**) by Woodward and coworkers.

reaction. Finally, removal of the Ac group on the C-4″ hydroxyl of cladinose and the BPCO group on the C-9 amino group was achieved in a single step using Na–Hg/MeOH, producing (9*S*)-erythromyculamine, which was then converted into erythomycin A (**2**) through known sequential transformations from amine to ketone in 40% overall yield.

**Synthesis of erythromycin A by the Toshima group**. In 1995, Toshima and coworkers reported an efficient synthesis of erythromycin A (**2**) using a highly stereocontrolled glycosylation strategy with a 2,6-anhydro-2-thio sugar donor (Scheme 5.2) [12]. The synthetic approach began with the preparation of (9*S*)-9-dihydroerythronolide (**14**) from natural erythromycin A. After three steps of protecting group manipulation, the coupling of compound **15** with thioglycoside **10** using Woodward's procedure, facilitated by AgOTf and 4 Å molecular sieves, proceeded excellent regio- and stereoselectively to furnish the desired C5-glycosylated β-glycoside **16** as the sole anomer in 63% yield. The use of 4 Å molecular sieves was found to be crucial for achieving the high yield, and the carbomethoxy group at the C2′ position of the thioglycoside played a significant role in achieving excellent β-stereoselective glycosidation. Attempts to introduce the L-cladinose residue to

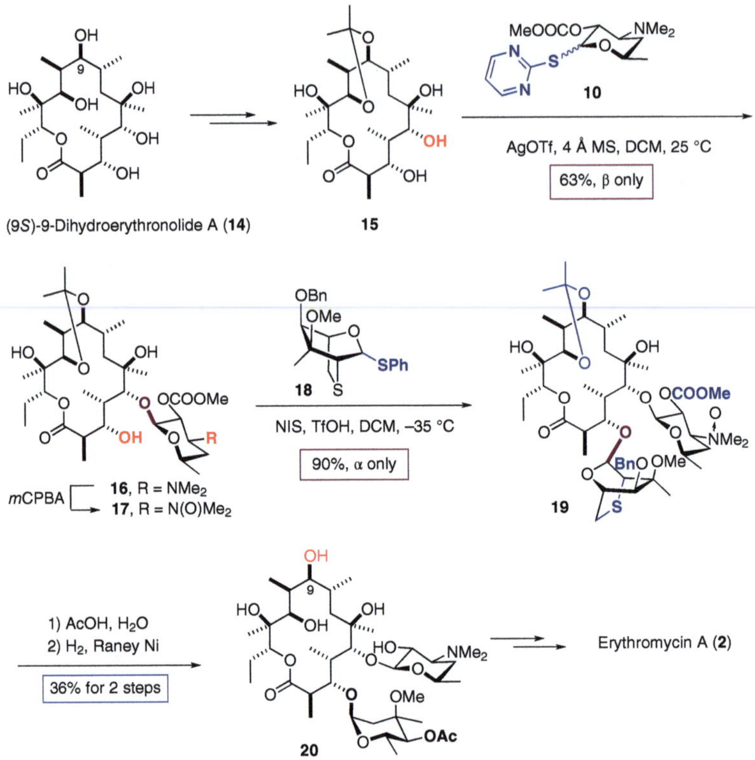


**Scheme 5.2** Total synthesis of erythromycin A (**2**) by Toshima and coworkers.

**16** using various glycosyl donors under different conditions were unsuccessful, predominantly due to the formation of *N*-formylated byproducts. Interestingly, after converting **16** into its *N*-oxide form, the glycosylation of **17** with 2,6-anhydro-2-thio sugar donor [13] **18** proceeded efficiently in the presence of NIS and TfOH in DCM at −35 °C, resulting in the desired α-glycoside **19** as the exclusive α-anomer in 90% yield. Subsequent steps involved the cleavage of the isopropylidene group under acidic conditions, followed by desulfurization, reduction of the *N*-oxide, and removal of the Bn and carbomethoxy groups via catalytic hydrogenation, leading to the formation of (9*S*)-9-dihydroerythromycin A (**20**). Lastly, selective oxidation of the C9-OH group accomplished the synthesis of erythromycin A (**2**).

**Total synthesis of erythromycin B by the Martin group**. In 1997, Martin and coworkers reported an asymmetric synthesis of erythromycin B (**3**) (Scheme 5.3) [14]. The synthesis began with the aldol condensation of ketone fragment **21** with aldehyde **22** under basic conditions, yielding the desired *syn*-addition product **23** with high diastereoselectivity. Following protecting group manipulations and functional group transformations, the formation of *seco*-acid **24** was achieved, which was subsequently converted into the desired macrolactone with a high yield of 93% through macrolactonization using the Yamaguchi protocol ($Cl_3C_6H_2COCl$, DMAP). Regioselective glycosylation of macrolactone **25** with pyrimidyl thioglycoside **10**, in

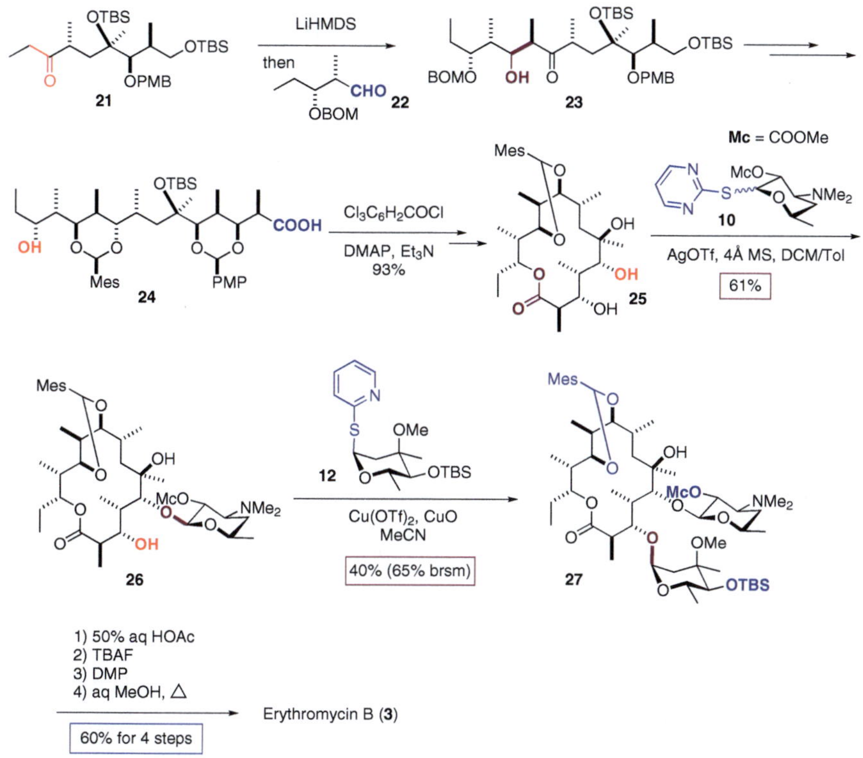

**Scheme 5.3** The first generation total synthesis of erythromycin B (**3**) by Martin and coworkers.

the presence of AgOTf, led to the formation of glycoside **26** in 61% yield as a single β-anomer. The introduction of the L-cladinose residue was achieved using thioglycoside **12** as the donor, in the presence of $Cu(OTf)_2$ and CuO in MeCN, providing the desired β-glycosylated product **27** in 40% yield (65% yield brsm). Finally, sequential hydrolysis of the mesitylene acetal, removal of the TBS group, selective oxidation of the C9 hydroxy group, and removal of the methyl carbonate accomplished the total synthesis of erythromycin B (**3**) in 60% yield (for four steps).

In 2003, Martin and coworkers reported an alternative synthetic route to erythromycin B (**3**), where the sugar residue was attached prior to the macrolactonization step (Scheme 5.4) [15]. The synthesis commenced with the condensation of ketone **28** with aldehyde **22** under basic conditions, resulting in the formation of the *syn*-aldol product **29**. Subsequent transformations were carried out, and the glycosylation of diol **30** with pyrimidyl thioglycoside **10** using AgOTf resulted in a mixture of the desired C5-glycosylated product **32** and the C6-glycosylated **31** (**32**/**31** = 1.2 : 1). Despite unsuccessful attempts to improve the regioselectivity or protect the C6 tertiary hydroxy group, the C5-glycoside **32** was isolated and utilized for further progression in the synthesis. After introducing the remaining three-carbon unit, efficient macrolactonization was achieved using the Yamaguchi protocol, yielding the mono-glycosylated macrolactone **26**. Finally, the same steps

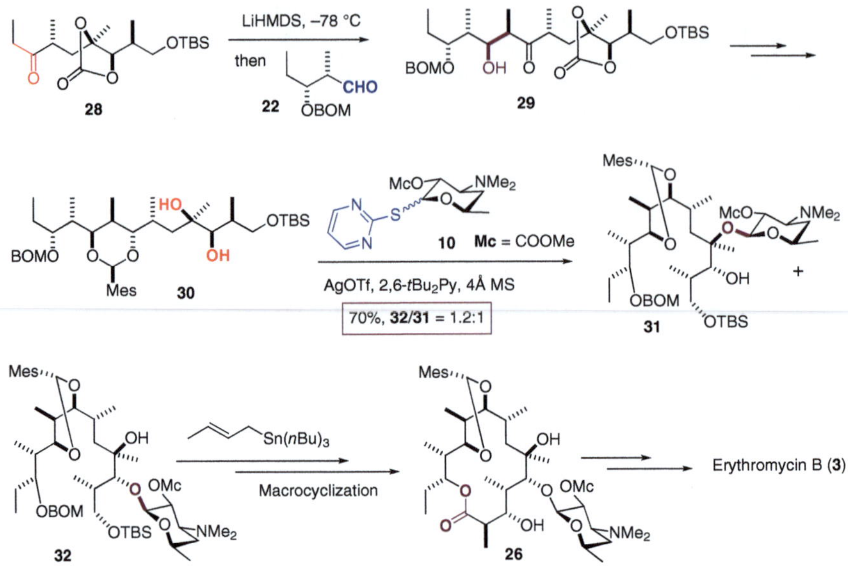

**Scheme 5.4** The second generation total synthesis of erythromycin B (**3**) by Martin and coworkers.

employed in the previous synthesis were adopted to successfully complete the total synthesis of erythromycin B (**3**).

## Apoptolidin A

Apoptolidin A (**4**) was first isolated from the soil bacteria *Nocardiopsis* sp. by Hayakawa et al. in 1997 (Figure 5.3) [6]. It functions as a potent and specific inhibitor of the mitochondrial F0F1-ATPase and exhibits the ability to induce apoptosis in cells transformed with the adenovirus E1A oncogene at concentrations as low as 10 nM, while having no impact on normal cells even at concentrations as high as 88 μM. Remarkably, this macrolide glycoside ranks among the top 0.1% of the most selective cell line cytotoxic agents. The presence of sugar residues has been demonstrated to be crucial for the potent bioactivity of apoptolidin A. Structurally,

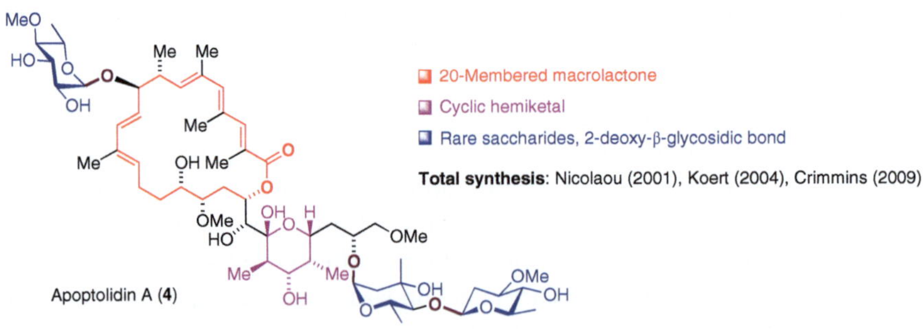


**Figure 5.3** Apoptolidin A (**4**) and its structural features.

apoptolidin A (**4**) consists of a 20-membered macrolactone linked to a side chain featuring a cyclic hemiketal. It carries a 6-deoxy-4-*O*-methyl-L-glucose attached at *O*-9 and a disaccharide composed of L-olivomycose and D-oleandrose linked at *O*-27. The remarkable biological activity and structural complexity of apoptolidin A make it an appealing target for total synthesis. Notably, the Nicolaou group (2001), the Koert group (2004), and the Crimmins group (2009) have successfully accomplished the total synthesis of apoptolidin A (**4**).

**Total synthesis of apoptolidin A by the Nicolaou group.** In 2001, Nicolaou et al. reported the first total synthesis of apoptolidin (**4**, Scheme 5.5) [16–18]. The

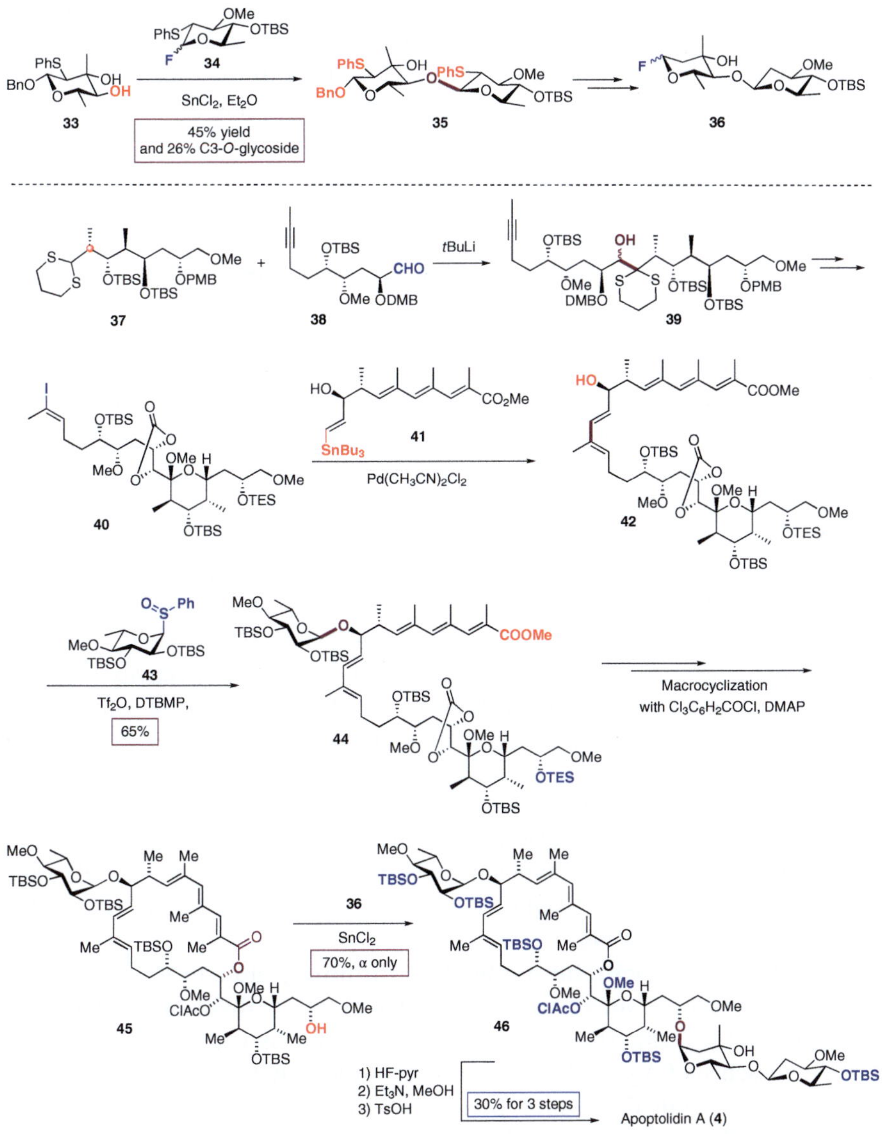

**Scheme 5.5** Total synthesis of apoptolidin A (**4**) by Nicolaou and coworkers.

disaccharide fragment was synthesized by glycosylation of acceptor **33** with glycosyl fluoride **34** in the presence of $SnCl_2$, resulting in the desired β-glycoside **35** in 45% yield. This low yield was attributed to the participation of the hindered tertiary C3 hydroxy group on **33** in the glycosidation reaction, leading to the formation of a significant amount of the C3-*O*-glycoside (26%). Removal of the phenylsulfanyl groups and treatment with DAST provided disaccharide fluoride donor **36**. The condensation of dithiane **37** with aldehyde **38** under basic conditions provided the coupled product **39**. After subsequent transformations and assembly of the six-membered ketal moiety, vinyl iodide **40** was obtained, which was then coupled with vinyl stannane **41** via Stille coupling to yield the desired C1–C28 segment **42**. The glycosylation of **42** with glycosyl sulfoxide **43** in the presence of $Tf_2O$ produced the desired α-glycoside **44** in 65% yield. Following further transformations, macrolactonization utilizing the Yamaguchi protocol resulted in macrolactone **45** with 63% yield. Subsequently, the coupling of **45** with disaccharide fluoride **36**, under the promotion of $SnCl_2$, furnished the protected apoptolidin A **46** in 70% yield with exclusive α-selectivity. Finally, sequential removal of the silyl groups (HF·pyr), ClAc group ($Et_3N$, MeOH), and methyl acetal (TsOH) completed the total synthesis of apoptolidin A (**4**) (30% overall yield for three steps).

**Total synthesis of apoptolidin A by the Koert group**. In 2004, Koert et al. reported the total synthesis of apoptolidin A (**4**), which involved an early introduction of sugar residue and late-stage macrolactonization of the fully glycosylated precursor (Scheme 5.6) [19, 20]. The north portion of the molecule was synthesized by glycosylation of **47** with glycosyl sulfoxide **48** containing a disilyl ether, utilizing $Tf_2O$ and DTBMP as catalysts, to provide the desired glycoside **49** in a yield of 65% and a moderate stereoselectivity of α/β = 85 : 15. Notably, the use of bis-TES or bis-TBS-protected glycosyl sulfoxide donors led to unsatisfactory results, either with TES group cleavage or lower stereoselectivity (α/β = 2 : 1). Several steps were performed to interchange the silyl groups and generate the ester moiety, ultimately yielding the north fragment **50**. The coupling of glycal acceptor **51** and glycosyl trichloroacetimidate **52**, under the mediation of TMSOTf at −60 °C, provided the corresponding disaccharide in high yield and excellent β-selectivity (>95 : 5). Subsequently, the coupling of disaccharide glycal **53** and acceptor **54** containing a six-membered ketal, in the presence of NIS, afforded compound **55** in 53% yield with high anomeric ratio (>95 : 5). After introducing the five-carbon segment **56**, the coupling of the north and south fragments under Cu(I)-catalysis provided compound **57**, which upon a ring-size-selective macrolactonization using Yamaguchi protocol furnished the desired 20-membered lactone in 75% yield. Finally, cleavage of all silyl ethers and methyl ketal by treatment with $H_2SiF_6$ in MeCN afforded apoptolidin A (**4**) in 71% yield.

**Total synthesis of apoptolidin A by the Crimmins group**. In 2009, Crimmins and coworkers reported an alternative approach to the synthesis of apoptolidin A (**4**, Scheme 5.7) [21–23]. The coupling of diol acceptor **58** with glycosyl bromide **59** in the presence of $Ag_2O$-silica gel proceeded smoothly to provide disaccharide **60** in 68% yield, with the desired β-anomer as the major isomer (β/α = 6 : 1). Subsequently, three steps of transformations were required to convert **60** into

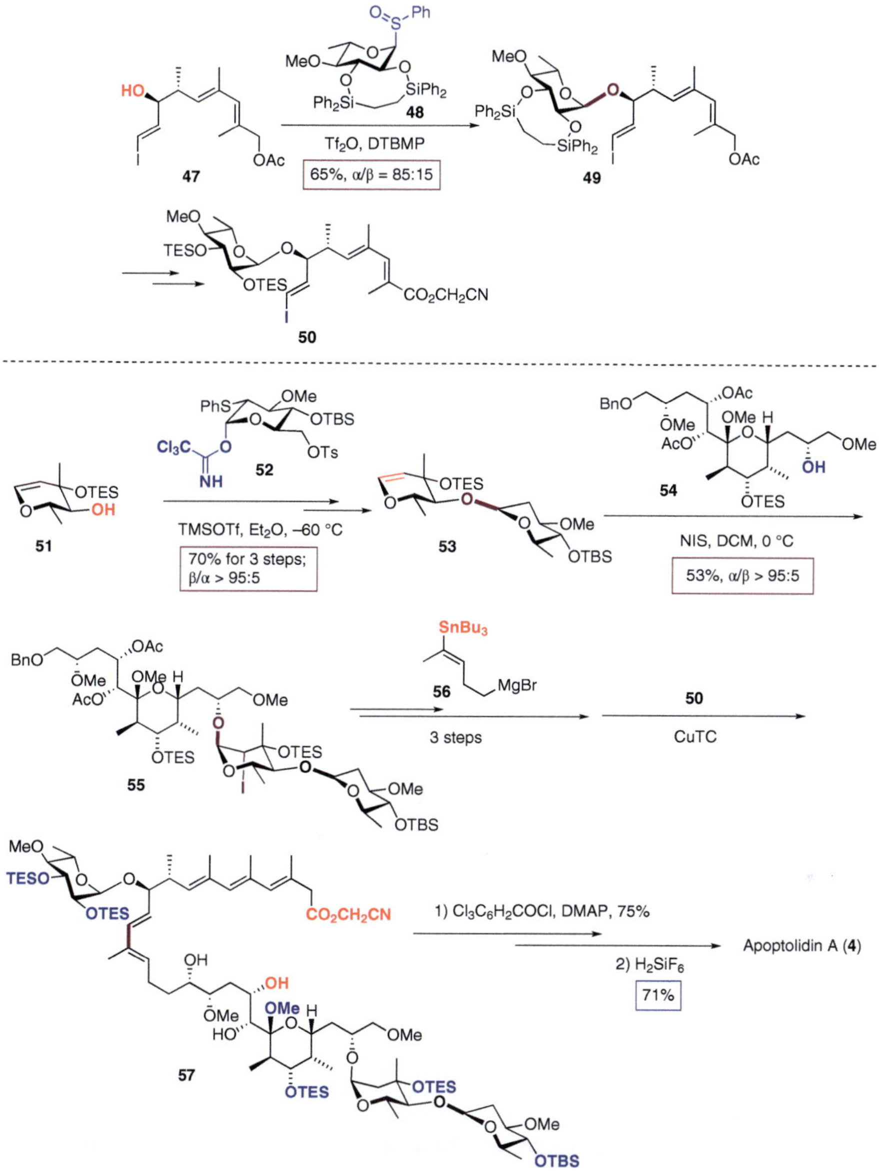


**Scheme 5.6** Total synthesis of apoptolidin A (**4**) by Koert and coworkers.

disaccharide fluoride **61**. Glycosylation of the south fragment **62** with glycosyl fluoride **61** under the action of $SnCl_2$ in $Et_2O$ provided the glycosylated product **63** in 81% yield (brsm) with high stereoselectivity ($\alpha/\beta = 7:1$). The coupling of diene **63** with alkene **64**, facilitated by the Grubbs heterocyclic carbene catalyst, provided the cross-metathesis product **65** in good yield with high *E*/*Z* selectivity. Subsequent glycosylation of **65** with glycosyl sulfoxide **66**, in the presence of $Tf_2O$, resulted in compound **67** in 40% yield with high stereoselectivity ($\alpha/\beta = 10:1$). Following ester

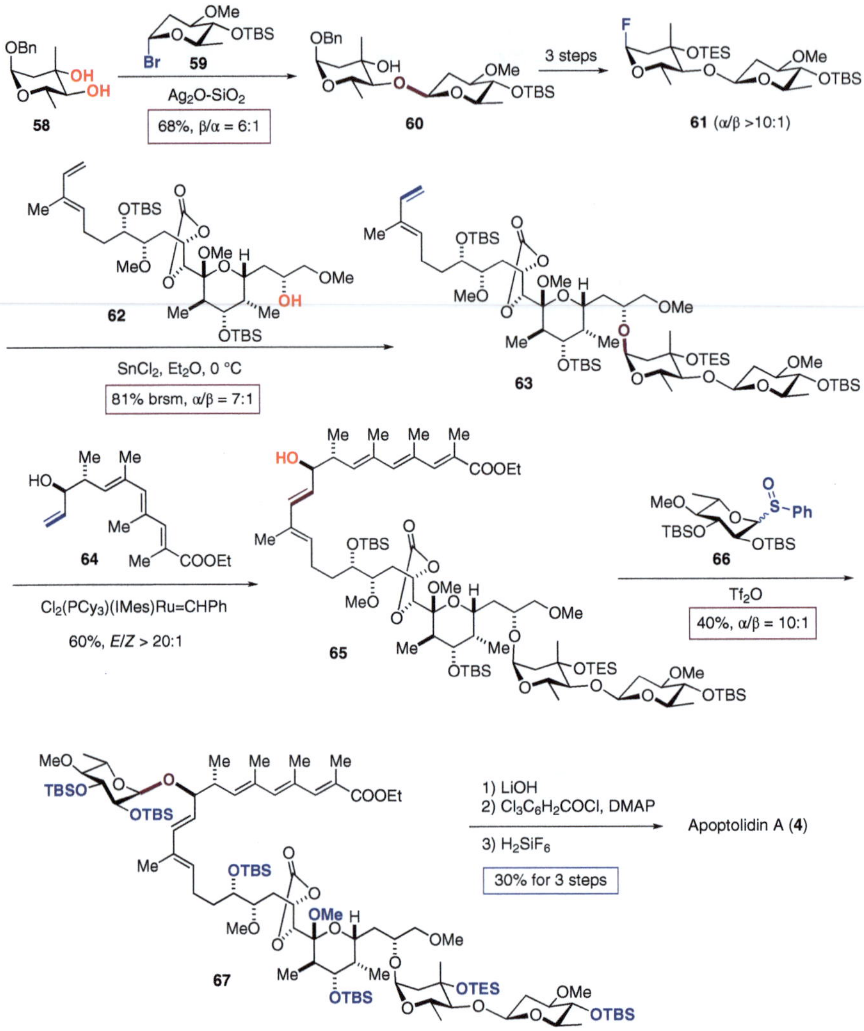

**Scheme 5.7** Total synthesis of apoptolidin A (**4**) by Crimmins and coworkers.

and carbonate group cleavage, macrolactonization via the Yamaguchi protocol led to the corresponding macrolactone, which underwent global deprotection upon treatment with $H_2SiF_6$, completing the total synthesis of apoptolidin A (**4**) (30% overall yield for the last three steps).

## Spinosyn A (Lepicidin A)

Spinosyn A (**68**), also known as lepicidin A and (+)-A83543A, represents the major component of a biosynthetic mixture produced by the soil microbe *Saccharopolyspora spinosa* (Figure 5.4) [24]. The mixture is currently marketed by Dow AgroSciences as an agricultural insecticide against a variety of insects.

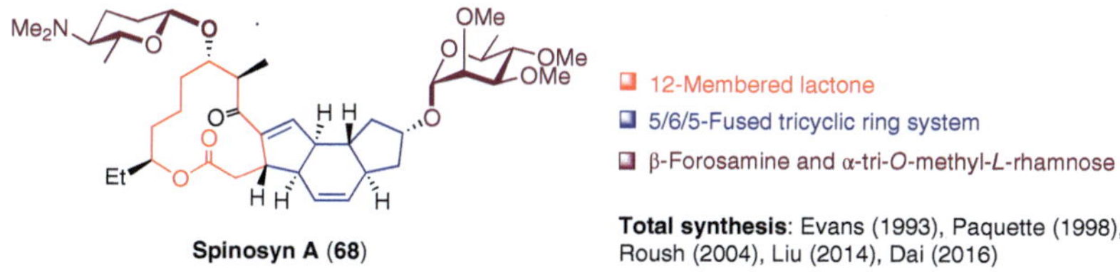


**Figure 5.4** Spinosyn A (**68**) and its structural features.

These compounds primarily act on insect nicotinic acetylcholine receptors in the nervous system while also serving as γ-aminobutyric acid neurotransmitter agonists. Notably, these insecticides possess several advantageous qualities, such as low toxicity to beneficial insects and rapid environmental degradation. Moreover, Spinosad, a significant organic and natural insecticide containing spinosyn A as its primary component, has US Food and Drug Administration (FDA) approval for treating human head lice. Structurally, spinosyn A (**68**) features a unique 5,6,5,12-fused tetracyclic ring system, with 17 stereogenic centers and multiple functional groups, along with two rare carbohydrate moieties, i.e. D-forosamine and tri-*O*-methyl-L-rhamnose. To date, spinosyn A (**68**) has been synthesized four times by the Evans group [25, 26], the Paquette group [27, 28], the Roush group [29], and the Dai group [30], respectively. In addition, a chemoenzymatic synthesis was reported by the Liu group in 2014 [31].

**Total synthesis of *ent*-spinosyn A by the Evans group**. In 1993, Evans and coworkers reported the total synthesis of the enantiomer of spinosyn A (*ent*-**68**, Scheme 5.8) [25, 26], as the absolute configuration of spinosyn A was not determined when the project commenced. The palladium-catalyzed Stille reaction between macrocycle stannane **69** and vinyl iodide **70** proceeded smoothly to provide triene **71** in 65% yield, which then underwent an intramolecular Diels–Alder reaction under the mediation of Lewis acid, yielding the desired cycloadduct **72** in 71% yield with high stereoselectivity. Through several transformations, aldehyde **73** underwent intramolecular aldol reaction under basic conditions, generating tetracyclic product **74**, which was further converted into aglycone acceptor **75**. The glycosylation of **75** with glycosyl acetate **76** in the presence of a catalytic amount of trityl perchlorate provided the desired α-anomer in 87% yield, with the β-anomer being isolated in 5% yield. Following the removal of the triisopropylsilyl (TIPS) group, the subsequent glycosylation of the resulting compound with glycosyl bromide **78** (4 equiv.) in the presence of silver zeolite led to the desired β-glycoside **79** in only 10% yield, whereas the α-anomer was isolated in 59% yield. Despite the low yield, a sufficient quantity of the fully glycosylated product **79** was obtained. The final steps involved the removal of the Fmoc protecting group through treatment with $Et_2NH$, followed by *N*-dimethylation, ultimately completing the total synthesis of *ent*-spinosyn A (*ent*-**68**).

**Total synthesis of spinosyn A by the Paquette group**. In 1998, Paquette and coworkers reported the synthesis of spinosyn A (**68**, Scheme 5.9) [27, 28]. The synthesis commenced with the 1,2-addition of the lithiated **80** with ketone **81**, which approached from the *endo* direction to deliver alcohol **82**. Oxy-Cope rearrangement

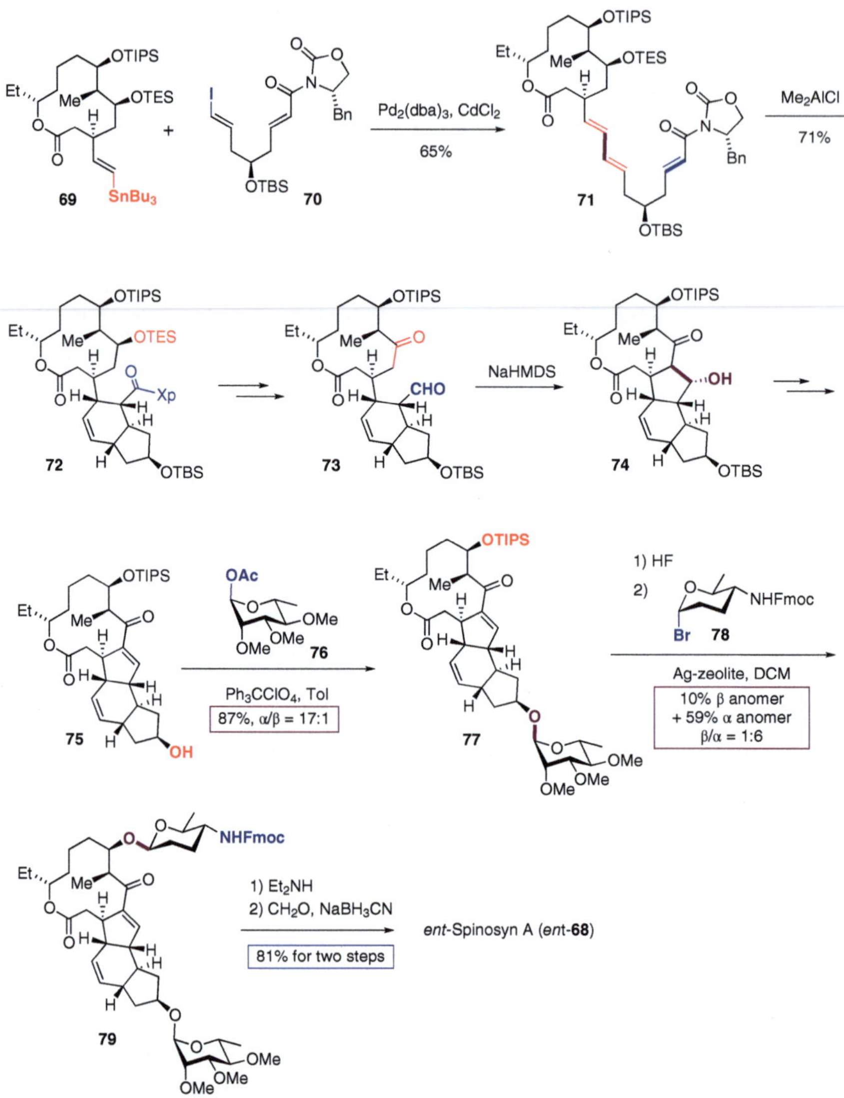

**Scheme 5.8** Total synthesis of *ent*-spinosyn A (*ent*-**68**) by Evans and coworkers.

of **82** in the presence of KH proceeded smoothly, yielding the tricyclic product **83** in 77% yield. Several steps were involved in modifying the tricyclic compounds, including epimerizations at C-7, C-11, and C-3, as well as the formation of the vinylstannane, leading to compound **85**. Stille coupling of **85** with acyl chloride **86** under the catalysis of $Pd_2(dba)_3{\cdot}CHCl_3$ provided ketone **87**. After generating the carboxyl group and releasing the hydroxyl group, macrolactonization was achieved using the Yamaguchi protocol, resulting in the formation of aglycone acceptor **88**. The glycosylation of **88** with glycosyl acetate **89**, enabled by trityl perchlorate, exclusively provided α-glycoside **90** in 85% yield. Following the removal of the *tert*-butyldiphenylsilyl (TBDPS) group, coupling of the resulting alcohol with

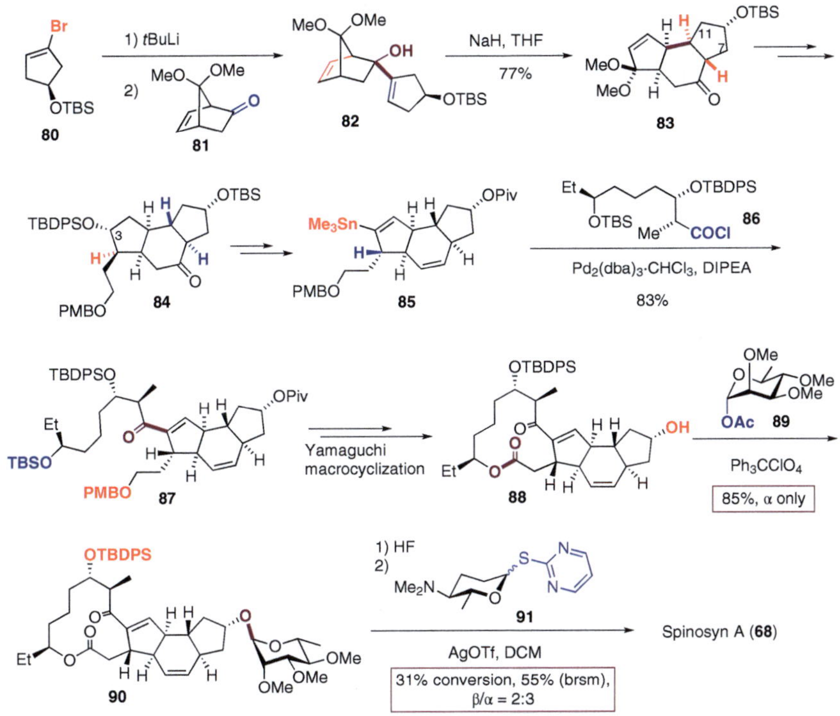


**Scheme 5.9** Total synthesis of spinosyn A (**68**) by Paquette and coworkers.

thioglycoside **91** in the presence of AgOTf provided a mixture of spinosyn A (**68**) and its anomer as a 2 : 3 mixture in 55% yield.

**Total synthesis of spinosyn A by the Roush group**. In 2004, the Roush group reported their total synthesis of spinosyn A (**68**, Scheme 5.10) [29, 32]. The synthesis commenced with the glycosylation of alcohol **92** with L-rhamnopyranosyl acetate **89** in the presence of TBSOTf, yielding α-glycoside **93** in 84% yield. After several steps of functional groups transformations, the Horner–Wadsworth–Emmons (HWE) reaction of aldehyde **94** with β-ketophosphonate **95** provided the corresponding triene product, which underwent selective Suzuki coupling with vinyl boronic acid **96** and subsequent oxidation, affording aldehyde **97**. The intramolecular HWE reaction efficiently enabled the macrocyclization of compound **97**, followed by a concomitant transanular Diels–Alder (TDA) reaction, resulting in a 75% yield of tricyclic product **98**. Subsequently, a vinylogous Morita–Baylis–Hillman (MBH) cyclization was used to give the desired tetracyclic product **99**. To enhance the efficiency and stereoselectivity while introducing the forosamine residue, the authors adopted an indirect strategy. Glycosylation of **99** with 2-*O*-acetyl-glycosyl trichloroacetimidate **100**, in the presence of TMSOTf, exclusively provided β-glycoside **101** in 90–97% yield, attributed to the participation of the C2″-*O*-Ac group. Finally, a series of five steps were employed to deoxygenate at the C2″ position and convert the C4″-$N_3$ group to C4″-$NMe_2$ group, ultimately completing the total synthesis of spinosyn A (**68**) in a 27% yield over the last five steps.

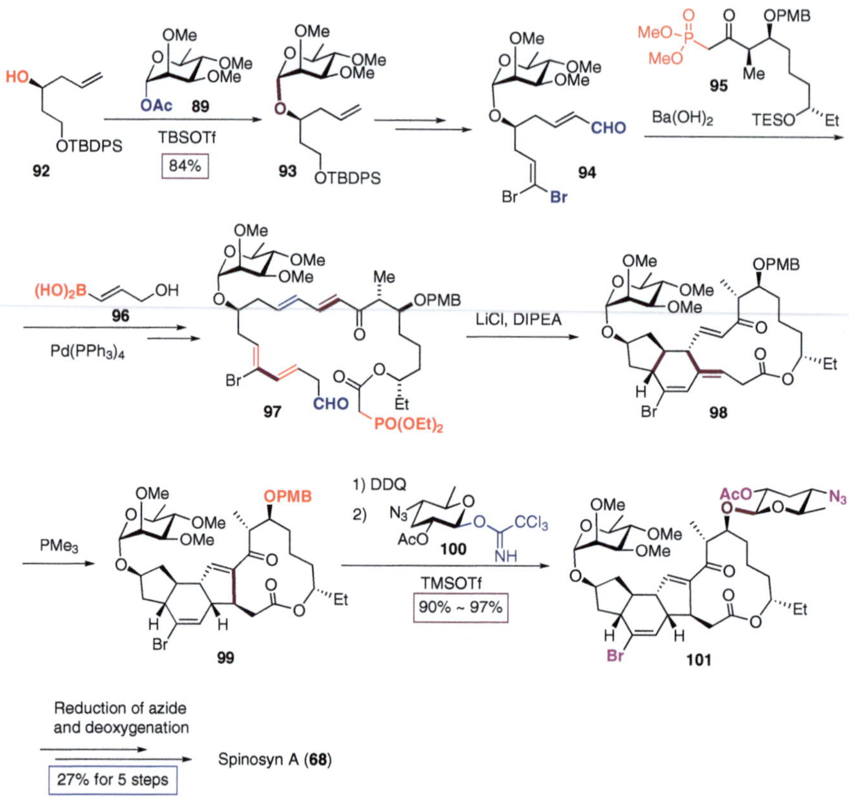

Scheme 5.10 Total synthesis of spinosyn A (**68**) by Roush and coworkers.

**Total synthesis of spinosyn A by the Dai group**. In 2016, Dai and coworkers reported a concise total synthesis of spinosyn A (**68**) accomplished in the longest linear sequence of 15 steps (Scheme 5.11) [30]. Stille coupling between vinyl iodide **102** with vinyl stannane **103**, catalyzed by $Pd_2(dba)_3$, afforded triene **104**. Following three steps of functional group transformations, the resulting triene **105** underwent an asymmetric intramolecular Diels–Alder reaction (IMDA) under the action of chiral secondary amine catalyst **106**, yielding bicyclic **107** as a single diastereomer. The coupling of aldehyde **107** with segment **108**, in the presence of *n*BuLi, led to the formation of acetate **109**. Subsequently, a one-step sequence involving gold-catalyzed rearrangement, oxidative selenide elimination, and TBS removal provided product **110**. Carbonylative Heck macrolactonization of **110** was performed using Pd-catalysis, which afforded tetracyclic **111** in 43% yield as a single diastereomer. After removal of the TBDPS group, glycosylation with glycosyl trichloroacetimidate **112** afforded compound **113** in 65% yield (2 steps). Finally, the β-D-forosamine motif was introduced through the utilization of *ortho*-alkynylbenzoate **114** as the glycosyl donor, under the promotion of glod(I) catalyst, resulting in the formation of spinosyn A (**68**). Notably, this glycosylation reaction, in a 71% yield with a β/α-selectivity ratio of 1 : 1, demonstrated the most favorable outcome for β-D-forosamine installation in the spinosyn A synthesis.

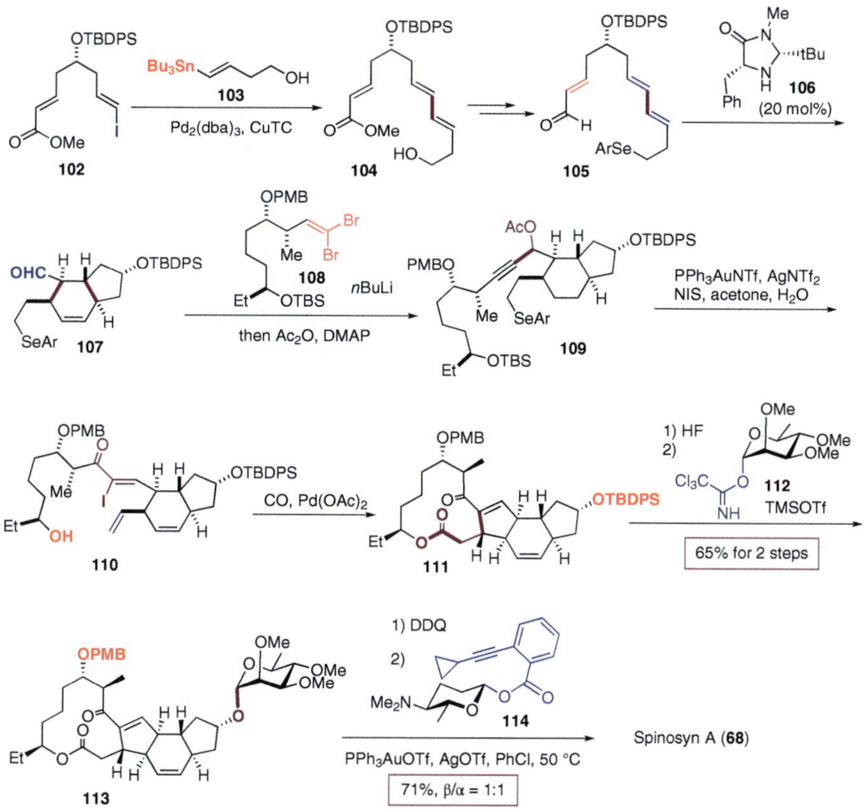

**Scheme 5.11** Total synthesis of spinosyn A (**68**) by Dai and coworkers.

## Tiacumicin B

Tiacumicin B (**115**), also known as fidaxomicin or lipiarmycin A3, was first isolated in 1975 from *Dactylosporangium aurantiacum*, an actinobacterium (Figure 5.5) [33]. It exerts its antibacterial activity by inhibiting RNA polymerase and displays remarkable efficacy against drug-resistant strains of *Mycobacterium tuberculosis*. In 2011, tiacumicin B obtained FDA approval for the treatment of nosocomial

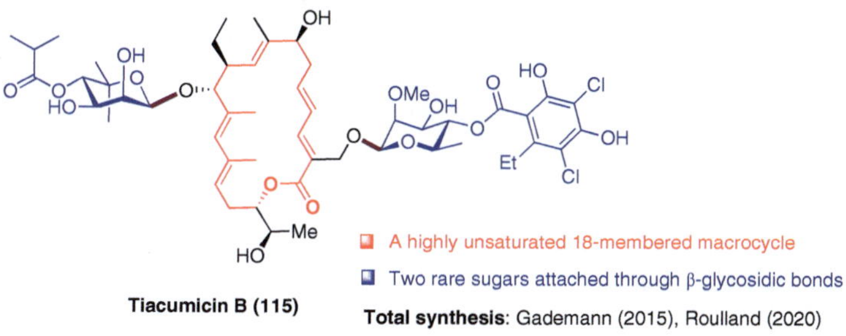


**Figure 5.5** Tiacumicin B (**115**) and its structural features.

(hospital-acquired) diarrhea associated with *Clostridium difficile* infections. Structurally, tiacumicin B (**115**) comprises an 18-membered macrolactone core adorned with two rare sugars, namely D-noviose and D-rhamnose, connected through β-glycosidic bonds. This molecule exhibits 14 stereogenic centers and a number of polysubstituted alkenes. Despite its isolation in the 1970s and its ongoing clinical use, the total synthesis of tiacumicin B has only been accomplished recently by the Gademann group in 2015 and the Roulland group in 2020.

**Total synthesis of tiacumicin B by the Gademann group.** In 2015, Gademann and coworkers reported the first total synthesis of tiacumicin B (**115**, Scheme 5.12) [34, 35]. The glycosylation of alcohol **116** with glycosyl bromide **117**, employing Helferich's conditions (HgO, $HgBr_2$), afforded the noviosylated fragment **118** in 63% yield with a good β/α selectivity (β/α = 3 : 1). Subsequently, the Suzuki coupling of vinyl iodide **118** with vinyl boronate **119**, in the presence of $Pd(PPh_3)_4$,

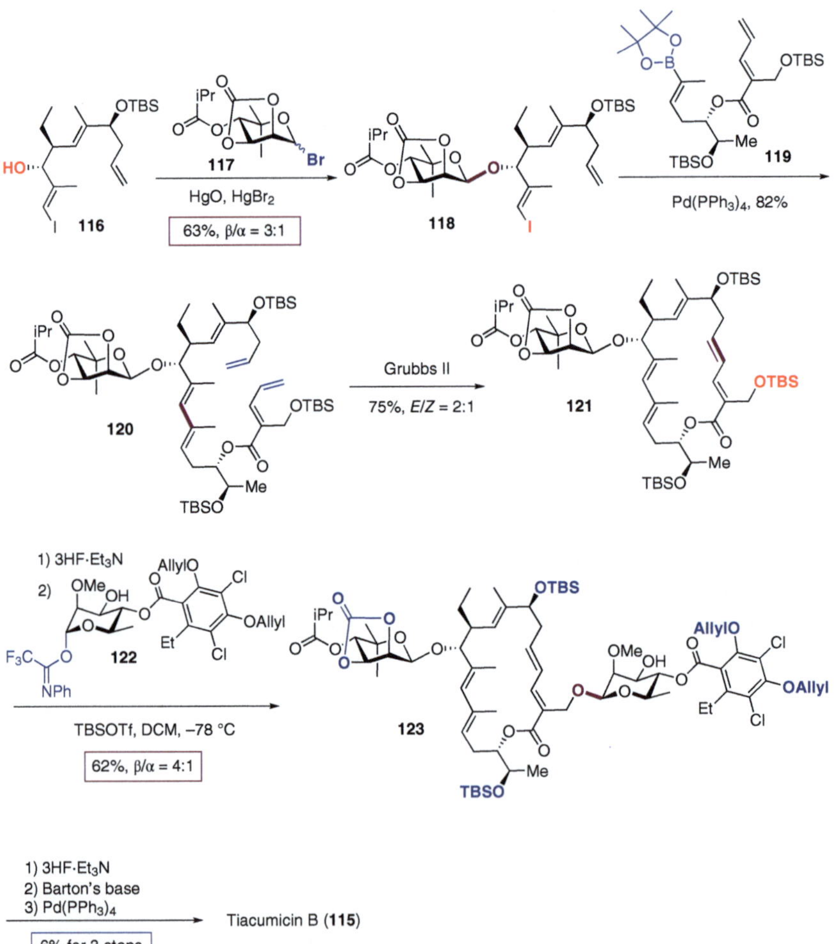


**Scheme 5.12** Total synthesis of tiacumicin B (**115**) by Gademann and coworkers.

provided **120** in high yield. The ring-closing metathesis reaction of **120**, utilizing the second-generation Grubbs catalyst, gave macrolactone **121** in 75% yield with an *E/Z* ratio of 2 : 1. Upon selective deprotection of the primary TBS ether, glycosylation with rhamnosyl trifluoroacetimidate donor **122**, under the catalysis TBSOTf, afforded the desired β-glycoside **123** in 62% yield with a satisfactory β-selectivity of 4 : 1. Finally, the sequential deprotection of the TBS groups (with HF·$Et_3N$),

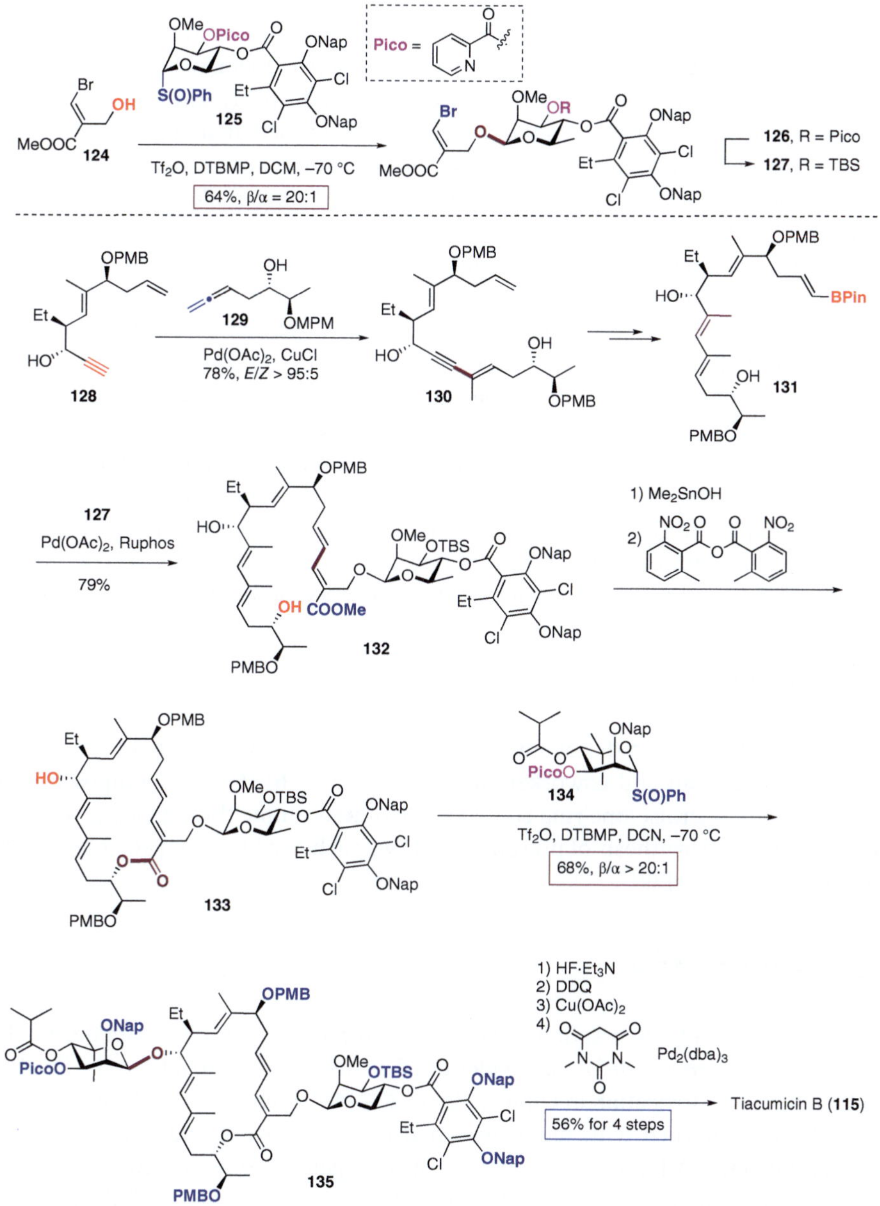

**Scheme 5.13** Total synthesis of tiacumicin B (**115**) by Roulland and coworkers.

the carbonate group (with Barton's base), and the allyl groups (with Pd-catalysis) completed the total synthesis of tiacumicin B (**115**) in 6% yield over three steps. The relatively low overall yield for these transformations was partly associated with the purification of the final product.

**Total synthesis of tiacumicin B by the Roulland group.** In 2020, Roulland and coworkers reported their total synthesis of tiacumicin B (**115**) employing a highly β-selective glycosylation protocol (Scheme 5.13) [36–38]. The glycosylation of alcohol **124** with rhamnosyl sulfoxide **125**, in the presence of $Tf_2O$, afforded the desired β-glycoside **126** in 64% yield with high stereoselectivity (β/α = 20 : 1). It is noteworthy that the presence of the C3′-*O*-picoloyl (Pico) group in donor **125** played a crucial role in achieving the high facial selectivity, which effectively directed the nucleophilic attack to the β-face through a hydrogen-bond-mediated aglycone delivery (HAD). Following two steps of protecting group manipulations, the key fragment **127** was obtained. On the other hand, Pd/Cu-catalyzed alkyne/allene cross-coupling between alkyne **128** and allene **129** provided enyne **130** in good yield. Further manipulation of functional groups led to vinyl boronate **131**, which underwent Suzuki coupling with vinyl bromide **127** under Pd-catalysis, resulting in ester **132**. Selective hydrolysis of the methyl ester and subsequent macrolactonization of the resulting acid, using the Shiina protocol, afforded macrocycle **133** in good yield (72%). Subsequently, glycosylation of macrolactone alcohol **133** with noviosyl sulfoxide **134**, in the presence of $Tf_2O$, provided the desired glycoside **135** in 68% yield with an excellent β-selectivity (β/α > 20/1). Notably, the HAD approach was employed once again with a picoloyl group installed at C3″-OH of the donor. Finally, sequential removal of the TBS group (with $HF{\cdot}Et_3N$), the *para*-methoxybenzyl (PMB) groups (with DDQ), the Pico group (with $Cu(OAc)_2$), and the Nap groups (with Pd-catalyzed condition) completed the synthesis of tiacumicin B (**115**) in 56% yield in four steps.

## Pikromycin

Pikromycin (**1**), the first discovered macrolide antibiotic, was isolated from *S. erythreus* by Brockman and coworkers in 1950 (Figure 5.6) [5]. It is composed of a 14-membered macrolactone ring connected to D-desosamine through a β-glycosidic linkage. The total synthesis of pikromycin (**1**) was achieved by Kang and coworkers in 2012 (Figure 5.6) [39].

## Polycavernoside A

Polycavernoside A (**136**) was first isolated by Yasumoto et al. in 1993 from the edible red alga *Polycavernosa tsudai*, being identified as the causative toxin responsible for sudden human intoxication in Guam (Figure 5.7) [40]. It consists of a 16-membered macrolactone adorned with tetrasubstituted tetrahydropyran and five-membered hemiacetal rings. Additionally, it features a conjugated triene side chain attached at C15 and a highly *O*-methylated L-fucosyl-D-xylose disaccharide linked at C5. The total synthesis of polycavernoside A (**136**) has been achieved by five research groups, including the Murai group (1998) [41, 42], the Paquette group (1999)

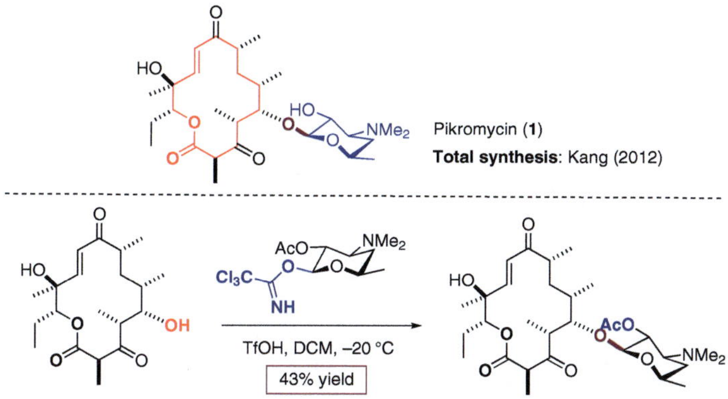


**Figure 5.6** Pikromycin (**1**) and the glycosylation step in the synthesis.

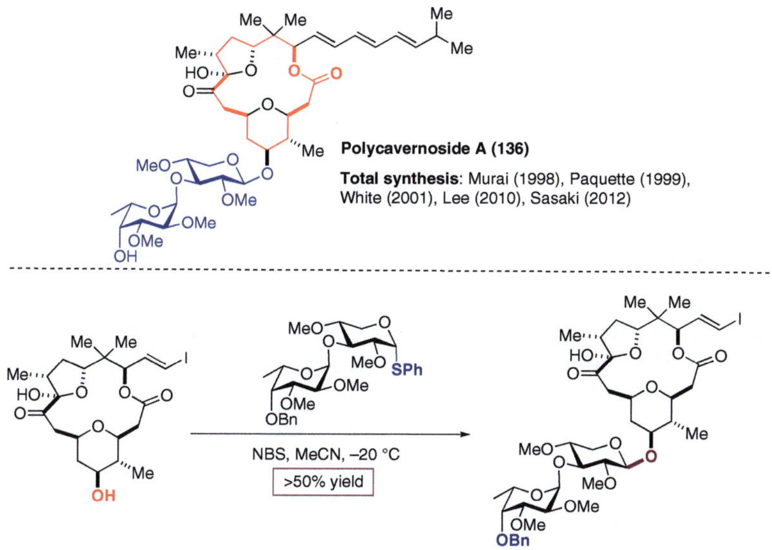


**Figure 5.7** Polycavernoside A (**136**) and the glycosylation step in the total synthesis.

[43–45], the White group (2001) [46, 47], the Lee group (2010) [48], and the Sasaki group (2012) [49].

## Auriside A

Auriside A (**137**) was initially isolated from the Japanese sea hare *Dolabella auricularia* by Yamada and coworkers in 1996 (Figure 5.8) [50]. Initial biological assays highlighted the significant cytotoxicity of auriside A, with $IC_{50}$ values against HeLa S3 cell lines measured at 0.17 μg/ml. Structurally, auriside A (**137**) is a 14-membered macrolide glycoside featuring a six-membered hemiacetal ring, an *E*-trisubstituted enone with an *E,E*-bromodiene side chain, and two L-rhamnose-derived saccharides connected through α-glycosidic linkages. The total synthesis of auriside A was

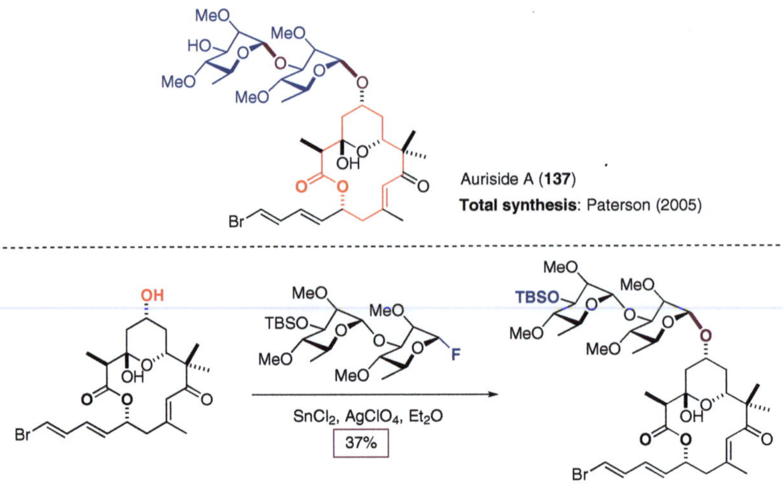


**Figure 5.8** Aurisides A (**137**) and the key glycosylation step in the total synthesis.

realized by Paterson and coworkers using a late-stage glycosylation with disaccharide fluoride donor in 2005 [51].

## Lyngbyaloside B

Lyngbyaloside B (**138**) was isolated from marine cyanobacterium *Lyngbya* sp. in 2002 (Figure 5.9) [52]. It demonstrated cytotoxicity against the human epidermoid

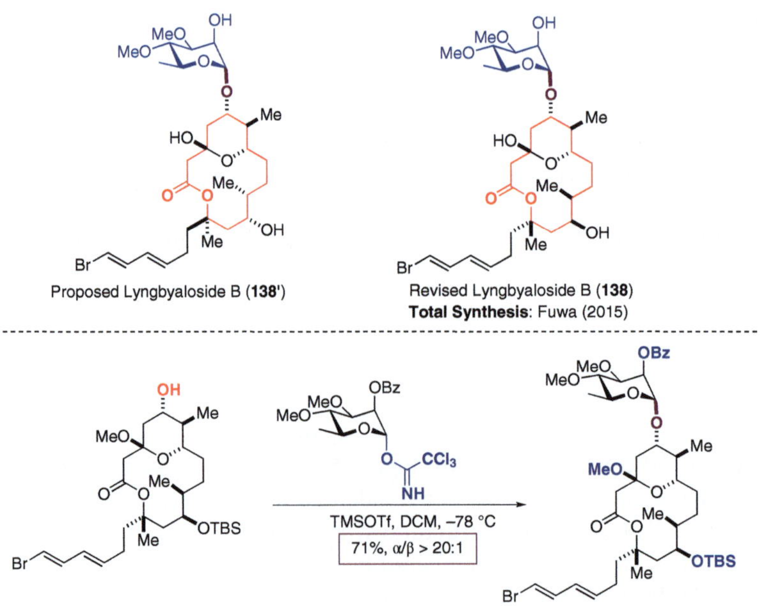


**Figure 5.9** Lyngbyaloside B (**138**) and the glycosylation step in the total synthesis.

carcinoma KB cell line and colon adenocarcinoma LoVo cell line at low micromolar concentrations. Structurally, Lyngbyaloside B features a 14-membered macrolactone with an unusual backbone that includes an acylated tertiary alcohol and a 3,4-di-*O*-methyl-L-α-rhamnopyranoside. The total synthesis and structural revision of lyngbyaloside B (**138**) have been achieved by the Fuwa group by using glycosyl imidate donor in 2015 [53, 54].

## Avermectin

Avermectins were isolated from an actinomycete called *Streptomyces avermitilis* MA 4680 (NRRL 8165) in 1979 (Figure 5.10) [55]. These compounds exhibit high activity against two major types of parasites: nematodes and arthropods, and their insecticidal activity is achieved by disrupting invertebrate neurotransmission. Structurally, avermectins consist of a 16-membered macrolactone ring, a distinct spiroketal motif, and a disaccharide composed of two α-L-oleandrosides. The total synthesis of avermectin $B_{1a}$ (**139**) has been achieved by the Hanessian group (1986) [56], the Ley group (1991) [57], and the White group (1995) [58]. Additionally, a total synthesis of avermectin $A_{1a}$ (**140**) has been realized by the Danishefsky group in 1989 [59].

## Formamicin

Formamicin (**141**) was isolated by Igarashi and coworkers in 1997 from an actinomycete strain (MK27-91F2) derived from a soil sample (Figure 5.11) [60]. It exhibited potent activity against phytopathogenic fungi and moderate activity toward gram-positive bacteria. Moreover, it demonstrated strong cytotoxicity against murine tumor cell lines. Structurally, formamicin consists of a seven-membered formyl acetal unit embedded within a 16-membered macrocycle, along with a hemiketal side chain that contains a 2,6-dideoxy-β-glucopyranoside unit. The total synthesis of formamicin was achieved by the Roush group in 2004 through late-stage glycosylation with a C-2 iodo-substituted glycosyl fluoride donor [61–63].

## Tylosin

Tylosin (**142**), which was first isolated in 1961, is considered as one of the most important and complex 16-membered-ring macrolide antibiotics (Figure 5.12). It is widely used both as a nutrient and a therapeutic agent [64]. Structurally, tylosin consists of a 16-membered macrolactone ring with a dienone motif and two saccharides. The total synthesis of tylosin (**142**) was realized by the Tatsuta group in 1982 by using glycosyl bromide and glycal as donors [65, 66]. In the same year, the Nicolaou group reported the synthesis of a major degradation product of tylosin, namely *O*-mycinosyltylonolide (**143**) using an early-stage glycosylation strategy [67, 68].

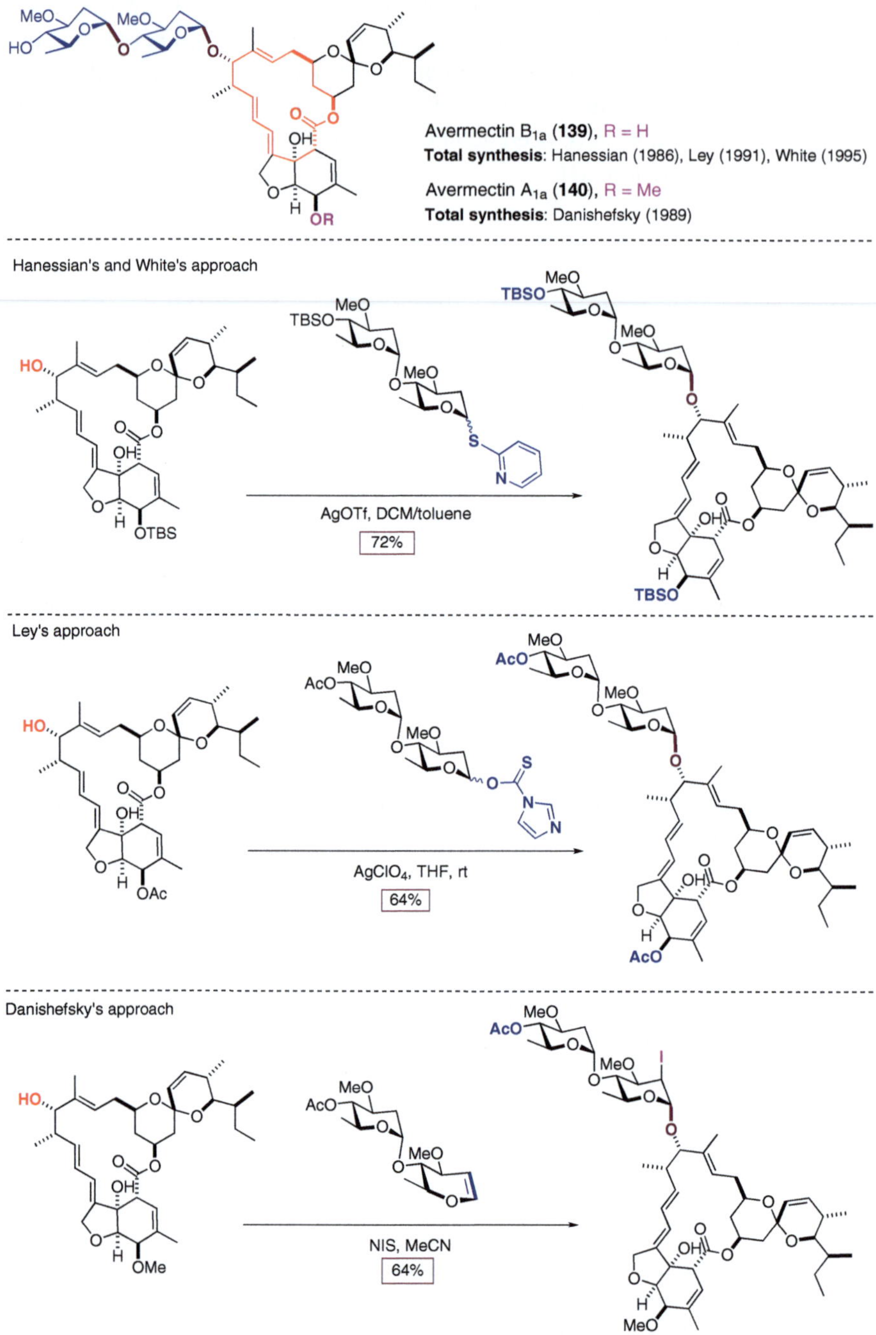


**Figure 5.10** Avermectin $B_{1a}$ (**139**) and $A_{1a}$ (**140**) and the key glycosylation steps used in the synthesis.

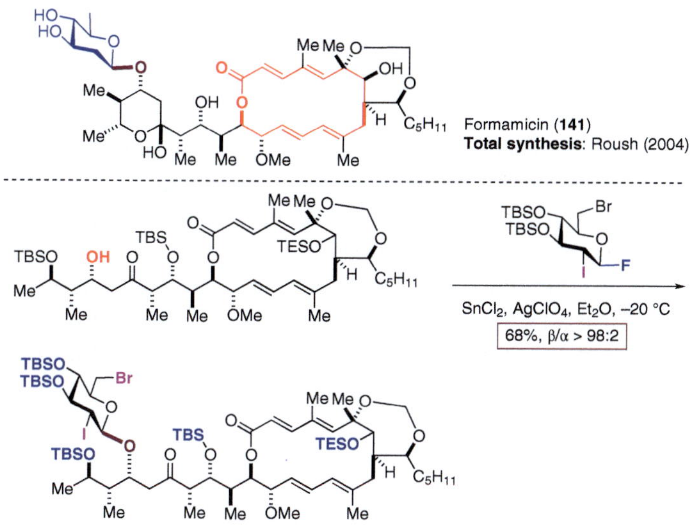


**Figure 5.11** Formamicin (**141**) and the glycosylation step in the total synthesis.

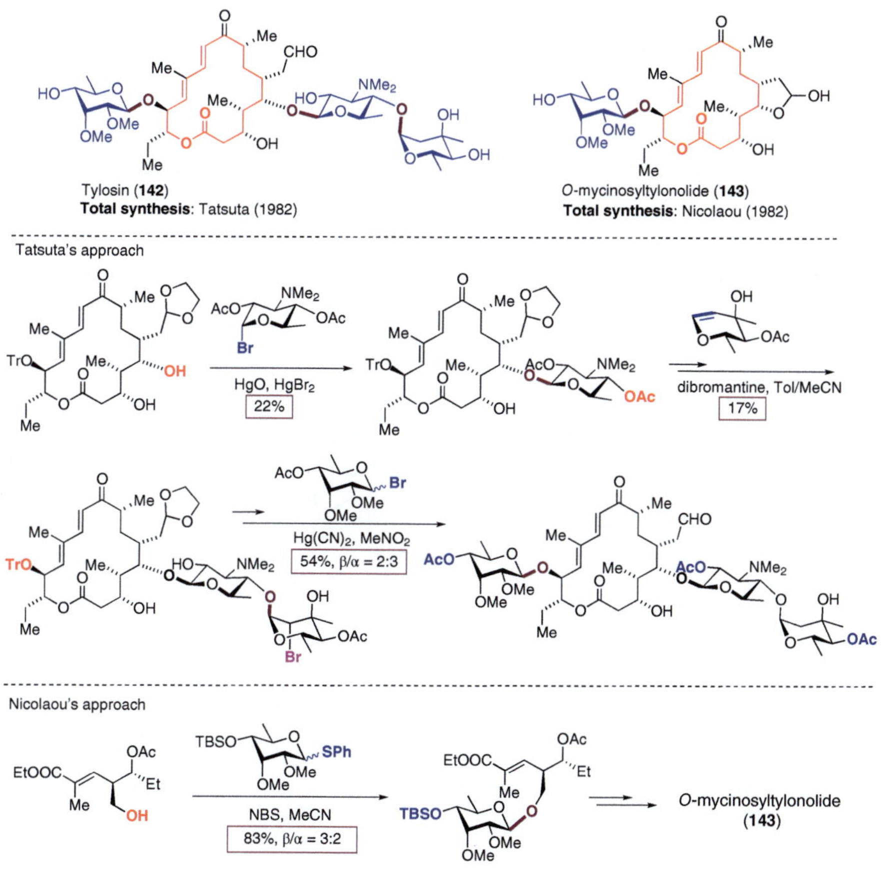


**Figure 5.12** Tylosin (**142**) and *O*-mycinosyltylonolide (**143**) and the key glycosylation steps in the total synthesis.

## Mangrolide

Mangrolide A (**144**) and D (**145**) were isolated from the fermentation broth of a rare marine actinomycete strain *Actinoalloteichus spitensis* in 2013 (Figure 5.13) [69]. These compounds exhibited potent activity against gram-negative bacteria, including *Acinetobacter baumannii* and *Pseudomonas aeruginosa*. Structurally, mangrolides feature a highly unsaturated 18-membered macrocycle decorated with uncommon saccharides. In mangrolide A (**144**), D-2,4-dimethyl quinovose and D-mycaminose are connected through β-glycosidic bonds, whereas in mangrolide D (**145**), a 4′-epi-vancosamine is attached via α-glycosidic bond. The total synthesis of mangrolide A (**144**) was accomplished by the Gademann group (2018) [70], while

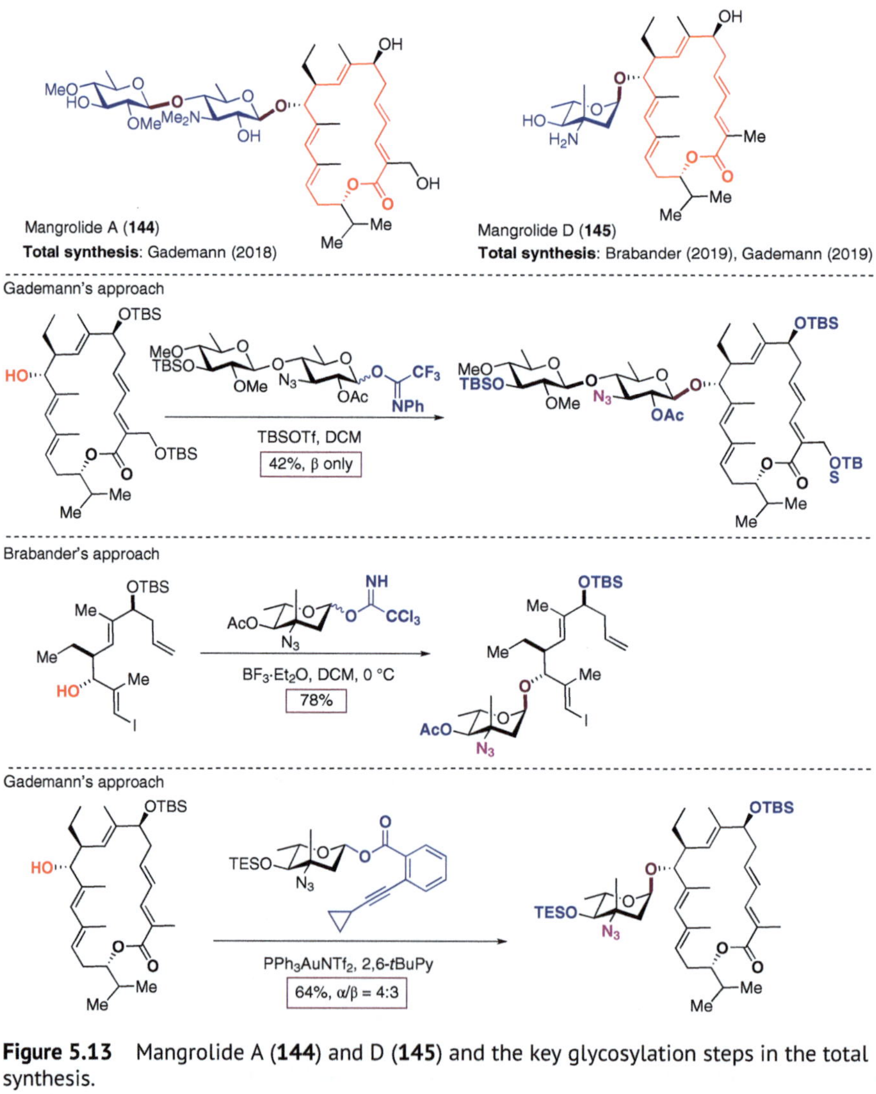


**Figure 5.13** Mangrolide A (**144**) and D (**145**) and the key glycosylation steps in the total synthesis.

the total synthesis of mangrolide D (**145**) was achieved by the Brabander group (2019) [71] and the Gademann group (2019) [72].

## Amphotericin B

Amphotericin B (**146**) is the most prominent mycosamine polyene macrolide and holds significant importance in medicine (Figure 5.14) [73]. Its various formulations have proven successful in treating fungal infections, such as aspergillosis, candidemia, mucormycosis, and multidrug-resistant forms of leishmania. Structurally, amphotericin B (**146**) features a 38-membered polyene macrolactone ring, which is linked to a mycosamine unit through a β-glycosidic bond. The total synthesis of amphotericin B (**146**) has been achieved by Nicolaou and coworkers in 1988 [74–76]. In addition, a total synthesis of 35-deoxy amphotericin B methyl ester (**147**) was reported by Carreira and coworkers in 2008 [77, 78]. It is worth noting that the glycosylation of the macrolactone aglycone required the use of a mycosamine C2′-epimer to establish a β-mannoside-type glycosidic bond, which required a late-stage inversion of the C2′-OH group to accomplish the total synthesis of amphotericin B.

## Aldgamycins

Aldgamycins were isolated from *Streptomyces* sp. HK-2006-1 by Gao and coworkers in 2016 (Figure 5.15) [79]. These compounds showed antimicrobial activities

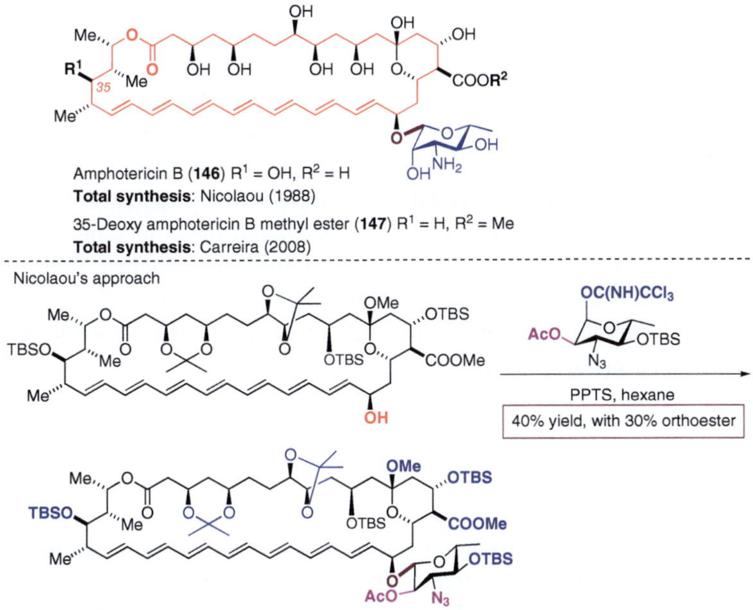


**Figure 5.14** Amphotericin B (**146**) and 35-deoxy amphotericin B methyl ester (**147**) and the glycosylation step in the total synthesis.

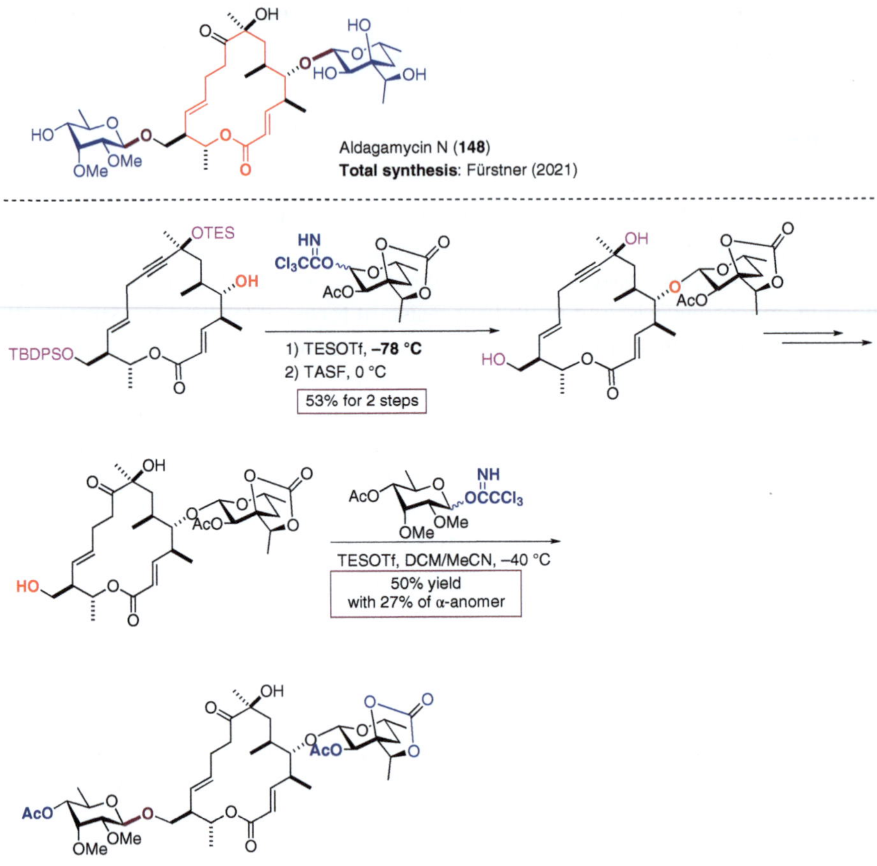


**Figure 5.15** Aldgamycin N (**148**) and the glycosylation steps in the total synthesis.

against gram-positive *Staphylococcus aureus* 209P, gram-negative *Escherichia coli* ATCC0111, as well as *Candida albicans* FIM709 and *Aspergillus niger* R330 fungi. Most of these compounds also demonstrated antibacterial activity against *S. aureus*. Structurally, aldgamycin N (**148**) consists of a 16-membered macrolactone moiety with a unique aldgatose-type branched oxtose and a β-D-mycinose attached. The total synthesis of aldgamycin N (**148**) was realized by Fürstner and coworkers in 2021, in which the *O*-glycosylation with glycosyl imidate donors has been employed for the attachment of the two saccharide segments [80].

## Mycinamicin IV

Mycinamicins are a novel class of 16-membered macrolide antibiotics produced by *Micromonospora qriseorubida* sp. nov., which exhibited potent antibacterial activity against gram-positive bacteria. Among this family, mycinamicin IV (**149**) consists of a 16-membered macrolactone with attached D-desosamine and D-mycinose at the

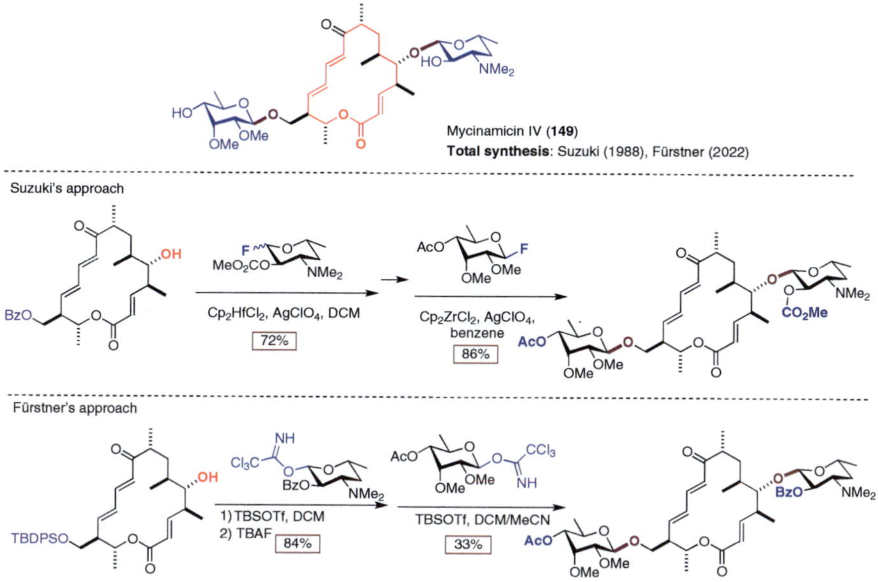


**Figure 5.16** Mycinamicin IV (**149**) and the glycosylation steps in the total synthesis.

C-5 OH and C-14 $CH_2OH$ positions, respectively (Figure 5.16) [81, 82]. In 1988, Suzuki and colleagues reported the first total synthesis of mycinamicin IV (**149**), which involved late-stage glycosylation using $Cp_2MCl_2$–$AgClO_4$ (M = Zr, Hf) as promoters with glycosyl fluoride donors [83, 84]. In 2022, Fürstner and colleagues presented their approaches to the total synthesis of mycinamicin IV (**149**), employing late-stage glycosylation with glycosyl imidate donors [85].

## References


1 Masamune, S., Bates, G.S., and Corcoran, J.W. (1977). *Angew. Chem. Int. Ed. Engl.* 16: 585–607.
2 Weymouth-Wilson, A.C. (1997). *Nat. Prod. Rep.* 14: 99–110.
3 Lee, K., Lanier, M.L., Kwak, J.-H. et al. (2016). *Nat. Prod. Rep.* 33: 1393–1424.
4 Dash, A.K., Hussain, N., and Mukherjee, D. (2020). *Carbohydrates in Drug Discovery and Development* (ed. V.K. Tiwari), 481–522. Elsevier.
5 Brockmann, H. and Henkel, W. (1950). *Naturwissenschaften* 37: 138–139.
6 Woo Kim, J., Adachi, H., Shin-Ya, K. et al. (1997). *J. Antibiot.* 50: 628–630.
7 Hayakawa, Y., Kim, J.W., Adachi, H. et al. (1998). *J. Am. Chem. Soc.* 120: 3524–3525.
8 Wiley, P.F., Gerzon, K., Flynn, E.H. et al. (1957). *J. Am. Chem. Soc.* 79: 6062–6070.
9 Woodward, R.B., Logusch, E., Nambiar, K.P. et al. (1981). *J. Am. Chem. Soc.* 103: 3210–3213.

10 Woodward, R.B., Au-Yeung, B.W., Balaram, P. et al. (1981). *J. Am. Chem. Soc.* 103: 3213–3215.

11 Woodward, R.B., Logusch, E., Nambiar, K.P. et al. (1981). *J. Am. Chem. Soc.* 103: 3215–3217.

12 Toshima, K., Nozaki, Y., Mukaiyama, S. et al. (1995). *J. Am. Chem. Soc.* 117: 3717–3727.

13 Toshima, K., Mukaiyama, S., Nozaki, Y. et al. (1994). *J. Am. Chem. Soc.* 116: 9042–9051.

14 Martin, S.F., Hida, T., Kym, P.R. et al. (1997). *J. Am. Chem. Soc.* 119: 3193–3194.

15 Hergenrother, P.J., Hodgson, A., Judd, A.S. et al. (2003). *Angew. Chem. Int. Ed.* 42: 3278–3281.

16 Nicolaou, K.C., Li, Y., Fylaktakidou, K.C. et al. (2001). *Angew. Chem. Int. Ed.* 40: 3849–3854.

17 Nicolaou, K.C., Li, Y., Fylaktakidou, K.C. et al. (2001). *Angew. Chem. Int. Ed.* 40: 3854–3857.

18 Nicolaou, K.C., Fylaktakidou, K.C., Monenschein, H. et al. (2003). *J. Am. Chem. Soc.* 125: 15433–15442.

19 Wehlan, H., Dauber, M., Mujica Fernaud, M.-T. et al. (2004). *Angew. Chem. Int. Ed.* 43: 4597–4601.

20 Wehlan, H., Dauber, M., Fernaud, M.T.M. et al. (2006). *Chem. Eur. J.* 12: 7378–7397.

21 Crimmins, M.T., Christie, H.S., Long, A., and Chaudhary, K. (2009). *Org. Lett.* 11: 831–834.

22 Crimmins, M.T., Christie, H.S., Chaudhary, K., and Long, A. (2005). *J. Am. Chem. Soc.* 127: 13810–13812.

23 Crimmins, M.T. and Long, A. (2005). *Org. Lett.* 7: 4157–4160.

24 Kirst, H.A., Michel, K.H., Martin, J.W. et al. (1991). *Tetrahedron Lett.* 32: 4839–4842.

25 Evans, D.A. and Black, W.C. (1992). *J. Am. Chem. Soc.* 114: 2260–2262.

26 Evans, D.A. and Black, W.C. (1993). *J. Am. Chem. Soc.* 115: 4497–4513.

27 Paquette, L.A., Collado, I., and Purdie, M. (1998). *J. Am. Chem. Soc.* 120: 2553–2562.

28 Paquette, L.A., Gao, Z., Ni, Z., and Smith, G.F. (1998). *J. Am. Chem. Soc.* 120: 2543–2552.

29 Mergott, D.J., Frank, S.A., and Roush, W.R. (2004). *Proc. Natl. Acad. Sci. U.S.A.* 101: 11955.

30 Bai, Y., Shen, X., Li, Y., and Dai, M. (2016). *J. Am. Chem. Soc.* 138: 10838–10841.

31 Kim, H.J., Choi, S.-h., Jeon, B.-s. et al. (2014). *Angew. Chem. Int. Ed.* 53: 13553–13557.

32 Winbush, S.M., Mergott, D.J., and Roush, W.R. (2008). *J. Org. Chem.* 73: 1818–1829.

33 Parenti, F., Pagani, H., and Beretta, G. (1975). *J. Antibiot.* 28: 247–252.

34 Kaufmann, E., Hattori, H., Miyatake-Ondozabal, H., and Gademann, K. (2015). *Org. Lett.* 17: 3514–3517.

**35** Miyatake-Ondozabal, H., Kaufmann, E., and Gademann, K. (2015). *Angew. Chem. Int. Ed.* 54: 1933–1936.

**36** Jeanne-Julien, L., Masson, G., Astier, E. et al. (2017). *Org. Lett.* 19: 4006–4009.

**37** Norsikian, S., Tresse, C., François-Eude, M. et al. (2020). *Angew. Chem. Int. Ed.* 59: 6612–6616.

**38** Tresse, C., François-Heude, M., Servajean, V. et al. (2021). *Chem. Eur. J.* 27: 5230–5239.

**39** Oh, H.-S. and Kang, H.-Y. (2012). *J. Org. Chem.* 77: 1125–1130.

**40** Yotsu-Yamashita, M., Haddock, R.L., and Yasumoto, T. (1993). *J. Am. Chem. Soc.* 115: 1147–1148.

**41** Fujiwara, K., Amano, S., and Murai, A. (1995). *Chem. Lett.* 24: 855–856.

**42** Fujiwara, K., Murai, A., Yotsu-Yamashita, M., and Yasumoto, T. (1998). *J. Am. Chem. Soc.* 120: 10770–10771.

**43** Johnston, J.N. and Paquette, L.A. (1995). *Tetrahedron Lett.* 36: 4341–4344.

**44** Paquette, L.A., Barriault, L., and Pissarnitski, D. (1999). *J. Am. Chem. Soc.* 121: 4542–4543.

**45** Paquette, L.A., Barriault, L., Pissarnitski, D., and Johnston, J.N. (2000). *J. Am. Chem. Soc.* 122: 619–631.

**46** White, J.D., Blakemore, P.R., Browder, C.C. et al. (2001). *J. Am. Chem. Soc.* 123: 8593–8595.

**47** Blakemore, P.R., Browder, C.C., Hong, J. et al. (2005). *J. Org. Chem.* 70: 5449–5460.

**48** Woo, S.K. and Lee, E. (2010). *J. Am. Chem. Soc.* 132: 4564–4565.

**49** Kasai, Y., Ito, T., and Sasaki, M. (2012). *Org. Lett.* 14: 3186–3189.

**50** Sone, H., Kigoshi, H., and Yamada, K. (1996). *J. Org. Chem.* 61: 8956–8960.

**51** Paterson, I., Florence, G.J., Heimann, A.C., and Mackay, A.C. (2005). *Angew. Chem. Int. Ed.* 44: 1130–1133.

**52** Luesch, H., Yoshida, W.Y., Harrigan, G.G. et al. (2002). *J. Nat. Prod.* 65: 1945–1948.

**53** Fuwa, H., Yamagata, N., Okuaki, Y. et al. (2016). *Chem. Eur. J.* 22: 6815–6829.

**54** Fuwa, H., Okuaki, Y., Yamagata, N., and Sasaki, M. (2015). *Angew. Chem. Int. Ed.* 54: 868–873.

**55** Burg, R.W., Miller, B.M., Baker, E.E. et al. (1979). *Anrimicrob. Agents Chemother.* 15: 361.

**56** Hanessian, S., Ugolini, A., Dube, D. et al. (1986). *J. Am. Chem. Soc.* 108: 2776–2778.

**57** Ley, S.V., Armstrong, A., Díez-Martín, D. et al. (1991). *J. Chem. Soc., Perkin Trans. 1* 667–692.

**58** White, J.D., Bolton, G.L., Dantanarayana, A.P. et al. (1995). *J. Am. Chem. Soc.* 117: 1908–1939.

**59** Danishefsky, S.J., Armistead, D.M., Wincott, F.E. et al. (1989). *J. Am. Chem. Soc.* 111: 2967–2980.

**60** Garashi, M., Kinoshita, N., Ikeda, T. et al. (1997). *J. Antibiot.* 50: 926.

**61** Powell, N.A. and Roush, W.R. (2001). *Org. Lett.* 3: 453–456.

**62** Durham, T.B., Blanchard, N., Savall, B.M. et al. (2004). *J. Am. Chem. Soc.* 126: 9307–9317.

**63** Savall, B.M., Blanchard, N., and Roush, W.R. (2003). *Org. Lett.* 5: 377–379.

**64** McGune, J.M., Bonlece, W.S., Higgens, C.E. et al. (1961). *Antibiot. Chemother.* 11: 320.

**65** Tatsuta, K., Amemiya, Y., Kanemura, Y., and Kinoshita, M. (1981). *Tetrahedron Lett.* 22: 3997–4000.

**66** Tatsuta, K., Amemiya, Y., Kanemura, Y. et al. (1982). *Tetrahedron Lett.* 23: 3375–3378.

**67** Nicolaou, K.C., Pavia, M.R., and Seitz, S.P. (1982). *J. Am. Chem. Soc.* 104: 2027–2029.

**68** Nicolaou, K.C., Seitz, S.P., and Pavia, M.R. (1982). *J. Am. Chem. Soc.* 104: 2030–2031.

**69** Jamison, M.T. (2013). PhD thesis, The University of Texas Southwestern Medical Center at Dallas (US).

**70** Hattori, H., Roesslein, J., Caspers, P. et al. (2018). *Angew. Chem. Int. Ed.* 57: 11020–11024.

**71** Gong, J., Li, W., Fu, P. et al. (2019). *Org. Lett.* 21: 2957–2961.

**72** Hattori, H., Hoff, L.V., and Gademann, K. (2019). *Org. Lett.* 21: 3456–3459.

**73** Sternberg, T.H., Wright, E.T., and Oura, M. (1955–1956). *Antibiot. Annu.* 566–573.

**74** Nicolaou, K.C., Daines, R.A., Uenishi, J. et al. (1988). *J. Am. Chem. Soc.* 110: 4672–4685.

**75** Nicolaou, K.C., Daines, R.A., Chakraborty, T.K., and Ogawa, Y. (1988). *J. Am. Chem. Soc.* 110: 4685–4696.

**76** Nicolaou, K.C., Daines, R.A., Ogawa, Y., and Chakraborty, T.K. (1988). *J. Am. Chem. Soc.* 110: 4696–4705.

**77** Szpilman, A.M., Cereghetti, D.M., Wurtz, N.R. et al. (2008). *Angew. Chem. Int. Ed.* 47: 4335–4338.

**78** Szpilman, A.M., Manthorpe, J.M., and Carreira, E.M. (2008). *Angew. Chem. Int. Ed.* 47: 4339–4342.

**79** Wang, C.-X., Ding, R., Jiang, S.-T. et al. (2016). *J. Nat. Prod.* 79: 2446–2454.

**80** Fürstner, A. and Späth, G. (2021). *Angew. Chem. Int. Ed.* 60: 7900–7905.

**81** Hayashi, M., Ohno, M., Kinoshita, K. et al. (1981). *J. Antibiot.* 34: 345.

**82** Kinoshita, K., Satoi, S., Hayashi, M., and Nakatsu, K. (1989). *J. Antibiot.* 43: 1003–1005.

**83** Keisuke, S., Takashi, M., Katsuhiko, T. et al. (1987). *Chem. Lett.* 16: 113–116.

**84** Matsumoto, T., Maeta, H., Suzuki, K., and Tsuchihashi, G. (1988). *Tetrahedron Lett.* 29: 3575–3578.

**85** Späth, G. and Fürstner, A. (2022). *Chem. Eur. J.* 28: e202104400.

# 6

# Nucleosides

Primary nucleosides are crucial cellular components, fulfilling integral roles in DNAs and RNAs, while also exhibiting diverse functions across fundamental pathways, including acting as metabolite carriers, energy donors, second messengers, enzyme cofactors, and carriers for posttranslational modifications. Correspondingly, nucleoside and nucleotide natural products exhibit a wide range of biological activities, including antibacterial, antifungal, antitumor, antiviral, herbicidal, and insecticidal properties, and hold significant promise as drug candidates [1–7].

Nucleoside natural products are a diverse group of secondary metabolites derived mostly from microbial organism, which encompass a diverse array of structural modifications derived from nucleosides and nucleotides. These modifications often involve integrating structural elements from higher monosaccharides, disaccharides, peptides, and lipids, resulting in the formation of complex molecular architectures (Figure 6.1). For instance, polyoxins (e.g. polyoxin J, **1**) were initially isolated from the culture broth of *Streptomyces cacaoi* var. *asoensis* in the late 1960s by Isono and coworkers [8]. These compounds show prominent activity against phytopathogenic fungi while being nontoxic to bacteria, plants, or animals. This remarkable property stems from the ability of polyoxins to inhibit the chitin synthase, thereby preventing the biosynthesis of chitin, which is a fundamental component of the fungal cell wall structure. The first total synthesis of polyoxin J (**1**) was achieved by Emoto and coworkers in 1973, marking a significant milestone in the total synthesis of complex nucleoside natural products [9].

Nucleoside natural products exhibit complex structures and contain sensitive functional groups, making them challenging targets for organic synthesis [10–14]. Generally, two strategies are employed for the synthesis of complex nucleosides: (i) direct elaboration from simple nucleosides and (ii) preparation of easily definable components, including nucleobases, higher monosaccharides, and peptidyl portions, followed by strategic coupling to assemble the targets (Figure 6.2).

## Tunicamycin

Tunicamycins (e.g. tumicamycin V, **2**) were initially isolated from the fermentation broths of *Streptomyces lysosuperficus* and later from *Streptomyces chartreusis*

*Carbohydrate Chemistry in the Total Synthesis of Naturally Occurring Glycosides*, First Edition. Biao Yu and Xiaoyu Yang.
© 2024 WILEY-VCH GmbH. Published 2024 by WILEY-VCH GmbH.

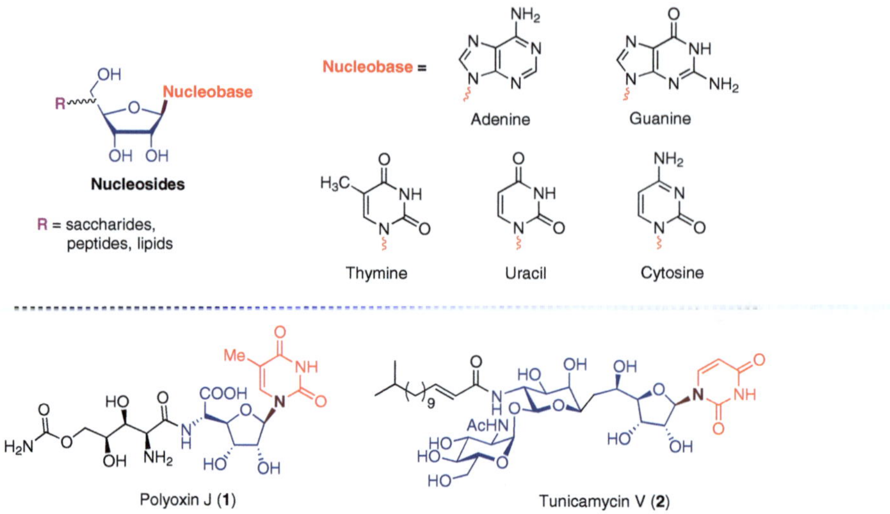


**Figure 6.1** Typical structures and representative examples of nucleoside natural products.

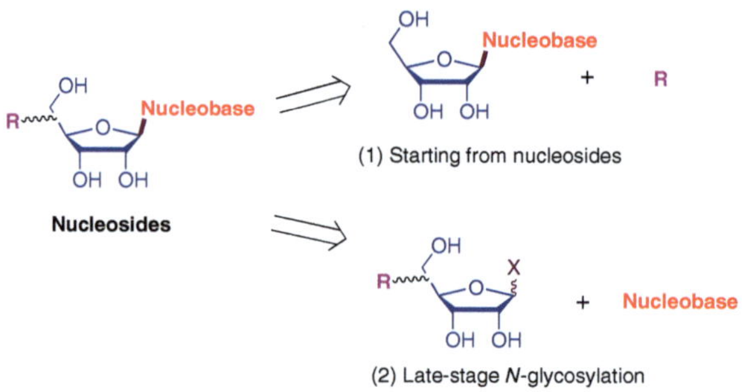


**Figure 6.2** Two general strategies for the synthesis of nucleoside natural products.

(Figure 6.3) [15]. These compounds can inhibit enzymatic reactions catalyzed by bacterial phospho-*N*-acetylmuramyl-pentapeptide transferase (MraY), which is responsible for the biosynthesis of peptidoglycan, and undecaprenyl-phosphate α-*N*-acetylglucosaminyl 1-phosphate transferase (WecA). Structurally, tunicamycins consist of five units: uracil, ribose, galactosamine (GalN), *N*-acetylglucosamine (GlcNAc), and fatty acid. Notably, they feature a unique tail-to-tail C–C linkage between the ribose and GalN units (C5′–C6′), resulting in the formation of an undecose known as tunicamine, as well as a head-to-head glycosidic linkage connecting the GalN and GlcNAc units (C11′-*O*–C1″).

**Total synthesis of tunicamycin V (2) by the Suami group.** In 1983, Suami and coworkers reported the first total synthesis of tunicamycin (Scheme 6.1) [16–18]. The synthesis of the tail-to-tail disaccharide fragment began with the Henry reaction between aldehyde **3** and nitro-ribose derivative **4** in the presence of KF. This reaction afforded the desired coupled product **5** in 51% yield, which existed as a mixture of four stereoisomers. Subsequent functional group transformations and protecting group manipulations provided undecosyl acetate **6**, which was then coupled with 2,4-bis[(trimethylsilyl)oxy]pyrimidine **7** in the presence of $SnCl_4$, resulting in the

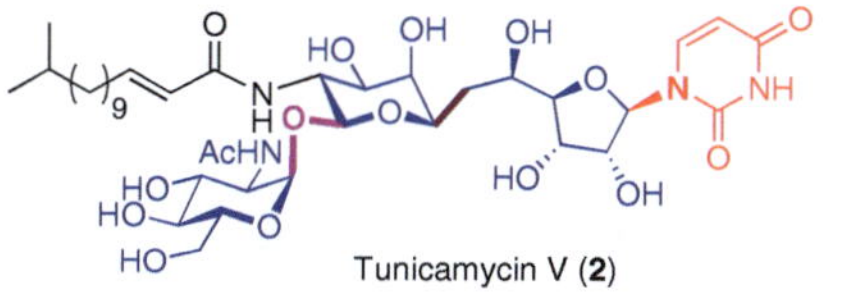


- **Five structural units**: uracil, ribose, galactosamine (GalN), *N*-acetylglucosamine (GlcNAc) and a fatty acid
- Tail-to-tail C-C linkage between the ribose and GalN units
- Head-to-head glycosidic linkage between the GalN and GlcNAc units

**Total synthesis:** Suami (1983), Myers (1993), Yu (2015), Ichikawa (2018), Kurosu (2022)
**Formal synthesis**: Danishefsky (1985, 1989)

**Figure 6.3** Tunicamycin (**2**) and its structure features.

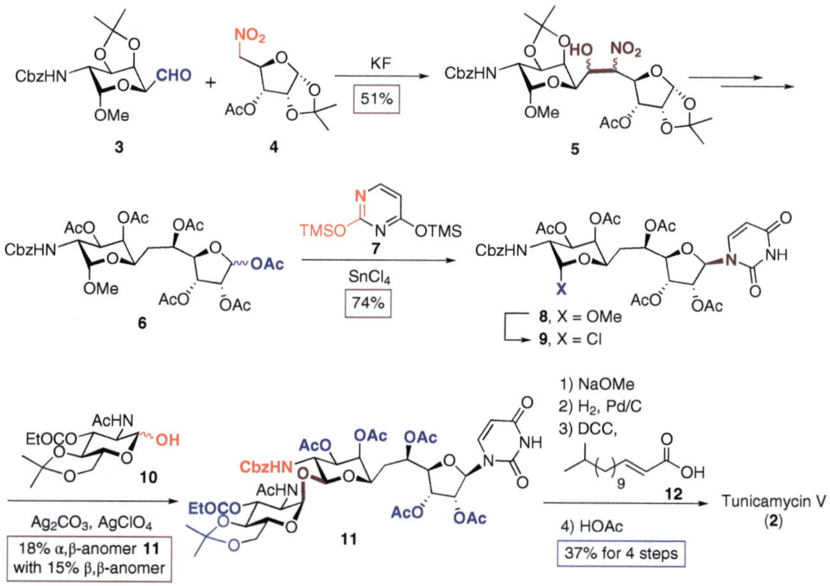


**Scheme 6.1** Synthesis of tunicamycin V (**2**) by Suami and coworkers.

desired product **8** in 74% yield. The conversion of methyl galactoside into galactosyl chloride **9** was followed by glycosylation with 2-aminoglucose **10** in the presence of $Ag_2CO_3$ and $AgClO_4$. However, the desired α,β-anomer **11** was obtained in only an 18% yield, while the undesired β,β-anomer was isolated in a yield of 15%. Finally, successive removal of acetyl groups (with NaOMe), *N*-Cbz group (with $H_2$, Pd/C), selective condensation with fatty acid **12**, and removal of the acetonide group (with HOAc) accomplished the total synthesis of tunicamycin V (**2**) in 37% overall yield for four steps.

**Synthetic studies toward tunicamycin V (2) by the Danishefsky group.** In 1985, Danishefsky and coworkers reported a formal synthesis of tunicamycins (Scheme 6.2) [19]. The oxa-Diels–Alder reaction between ribose-derived aldehyde **13** and 1,3-butadiene **14**, under the catalysis of $Eu(fod)_3$, afforded dihydropyranone

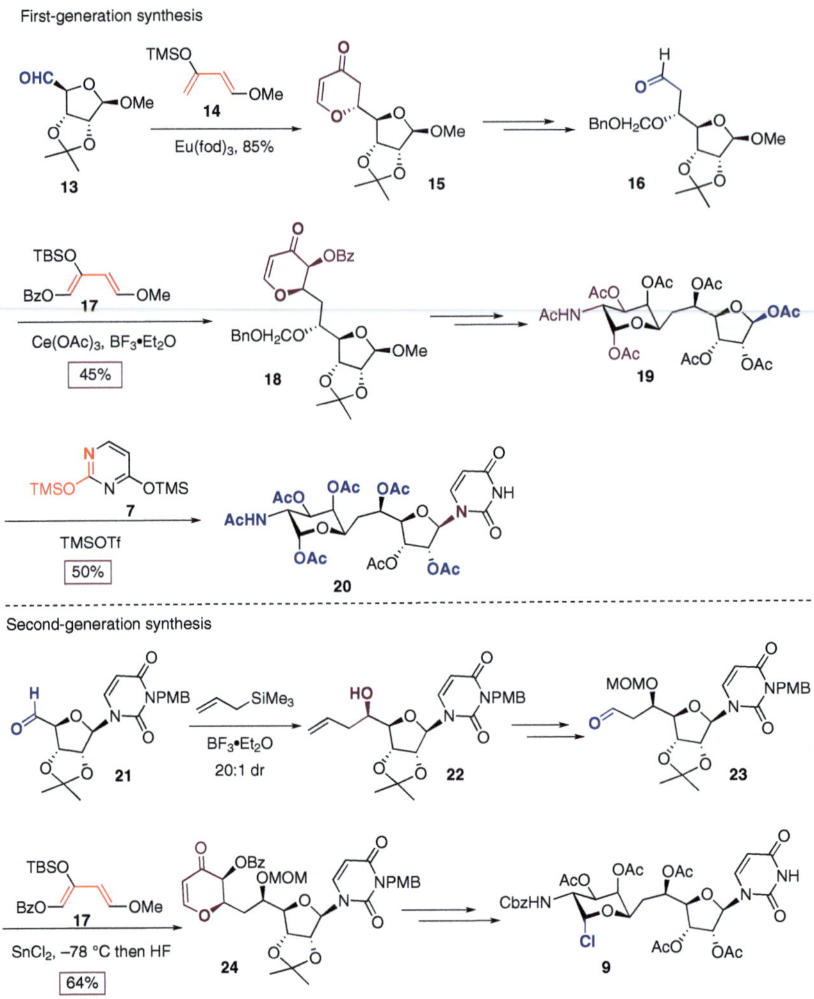


**Scheme 6.2** Synthetic studies toward tunicamycin V (**2**) by Danishefsky and coworkers.

**15** in 85% yield, with the C-5′ hydroxyl group being introduced with the desired configuration. Oxidative cleavage of the dihydropyranone moiety, followed by oxidative state adjustment, afforded aldehyde **16**. Cyclization of aldehyde **16** with 1,3-diene **17**, catalyzed by $Ce(OAc)_3$-$BF_3{\cdot}OEt_2$, furnished dihydropyranone **18** as a single diastereomer in a 45% yield, which established the desired configuration at C-4″ and C-5″ of the GalN moiety. Stereoselective reduction of the ketone moiety and installation of the amino group on the GalN residue led to undecosyl acetate **19**, which was then glycosylated with 2,4-bis[(trimethylsilyl)oxy]pyrimidine **7** under the promotion of TMSOTf. This reaction yielded the desired (peracetyl-tunicaminyl)uracil **20** in 50% yield. This synthetic route encountered two main challenges: (i) regioselectivity issue in the final glycosylation step, resulting in the observation of products with both "ribosyl" and "galactocyl" sectors in **19**, and (ii) difficulty in distinguishing the protecting group of the nitrogen on the galactosamine residue from those on the oxygens in compound **20**. Consequently, the authors reported the second-generation synthesis in 1989 [20]. The coupling of uridine-derived aldehyde **21** with allyltrimethylsilane in the presence of $BF_3{\cdot}Et_2O$ yielded the homoallylic alcohol **22** with excellent diastereoselectivity (20 : 1). Following functional group transformations, the cycloaddition between aldehyde **23** with diene **17** catalyzed by $SnCl_2$ and treatment with HF afforded dihydropyranone **24** in 64% yield. Upon introduction of the amino group and adjustment of the protecting groups, compound **24** was successfully converted into galactosyl chloride **9**, a key intermediate in Suami's total synthesis of tunicamycin V (**2**). However, Danishefsky and coworkers noted that the glycosylation between **9** and 2-aminoglucoside **10** could not be achieved in their hands on a small scale.

**Total synthesis of tunicamycin V (2) by the Myers group.** In 1993, Myers and coworkers reported their total synthesis of tunicamycin V (**2**), employing silicon-mediated reductive coupling of an aldehyde and allylic alcohol for the construction of the undecose core (Scheme 6.3) [21–23]. The glycosylation reaction between anomeric mixtures of **25** ($\beta/\alpha > 10:1$) and imidate **26** (synthesized from glucose) under the action of HOTf (20 mol%) afforded disaccharide **27** in 70% yield, in which the undesired α,α-diastereomer was isolated in approximately 10% yield. The 2-phthalimido group in **25** was believed to play a crucial role in imparting steric destabilization to the undesired α-anomer. Steps were employed to convert **27** into allylic alcohol **28**, which was then coupled with an appropriately protected uridine 5′-aldehyde **29** (PhSeH, $Me_2SiCl_2$, pyridine) to generate *O*-silylhemiselenoacetal

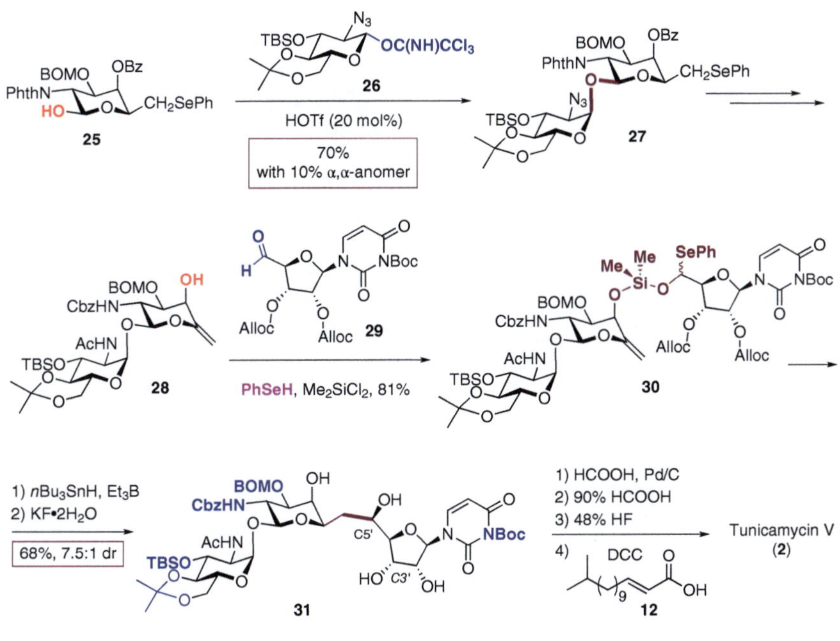


**Scheme 6.3** Total synthesis of tunicamycin V (**2**) by Myers and coworkers.

**30** in 81% yield. Upon removal of the Alloc groups, free-radical cyclization of hemiselenoacetal **30** ($n$Bu$_3$SnH, Et$_3$B) led to the cyclic product with the desired configuration at C5′ (68%, 7.5 : 1 dr), which afforded tetraol **31** after cleavage of the siloxane linker. Notably, the proposed hydrogen bonding between the C3′-hydroxyl group of the derived 5′-radical and the glucosyl amide oxygen was found to be crucial for producing the desired stereochemistry at C5′, as cyclization of diols in methanol showed only a 1.6 : 1 diastereoselectivity. Finally, sequential deprotection of the BOM and Cbz groups (under transfer hydrogenolysis), removal of Boc and acetonitride groups (under acidic conditions), and TBS group (with HF), followed by selective *N*-acylation with fatty acid **12**, accomplished the total synthesis.

**Total synthesis of tunicamycin V (2) by the Yu group.** In 2015, Yu and coworkers reported a modular approach to the total synthesis of tunicamycins V (**2**, Scheme 6.4) [24]. Silyl enol ether **32**, derived from D-GalN, was coupled with aldehyde **33** through a chelation-controlled Mukaiyama aldol reaction in the presence of SnCl$_4$, which provided the desired β-hydroxyketone **34** in 89% yield with 13 : 1 diastereoselectivity, establishing the C5′-OH with the desired stereochemistry. A number of steps were involved in converting **34** into hemiacetal

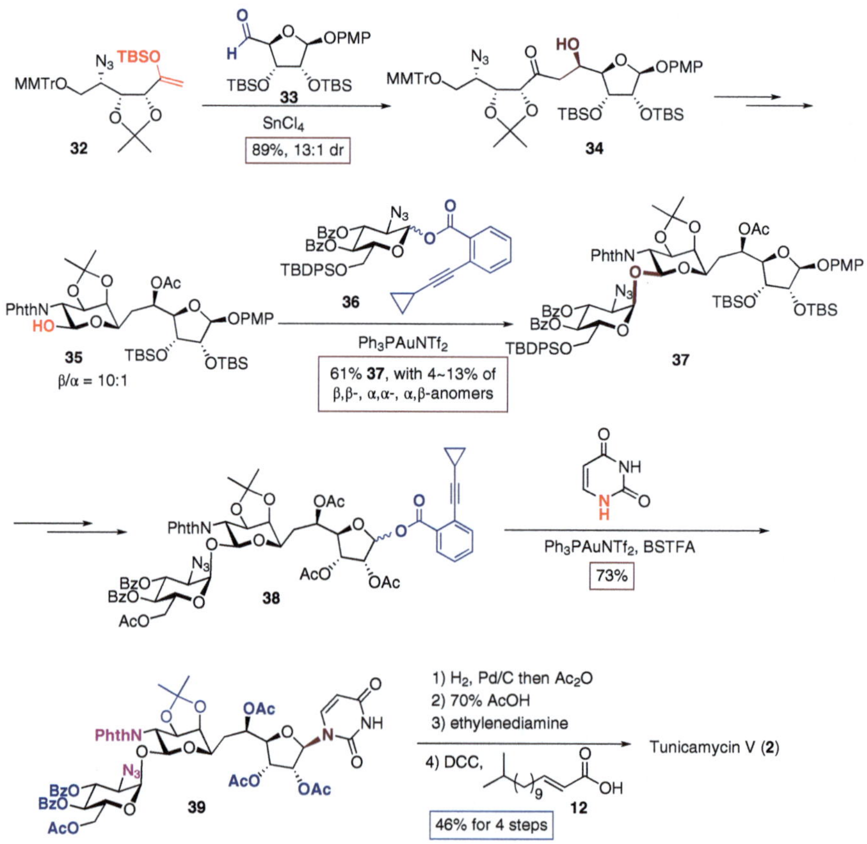


**Scheme 6.4** Total synthesis of tunicamycin V (**2**) by Yu and coworkers.

**35**, predominantly favoring the desired β anomer (β/α = 10 : 1). The coupling of **35** with GlcN *ortho*-alkynylbenzoate donor **36**, catalyzed by $PPh_3AuNTf_2$ (0.4 equiv.), led to formation of the desired β,α-trehalose **37** in a satisfactory 61% yield, with the undesired β,β-, α,α-, and α,β-trehaloses isolated in 4–13% yields. The installation of the non-participating azide at the C2 position and the bulky TBDPS group at the 6-hydroxy moiety were believed to facilitate the α-selective glycosylation. Further steps were employed to transfer **37** into trisaccharide *ortho*-alkynylbenzoate donor **38**, which was condensed with silylated uracil (generated *in situ* by treatment of uracil with BSTFA) in the presence of $PPh_3AuNTf_2$ (0.2 equiv.), yielding the desired nucleoside **39** in a satisfactory 73% yield. The final stage of the synthesis involved four straightforward steps: (i) conversion of the azido group into the acetamide; (ii) cleavage of the isopropylidene acetal; (iii) removed of the *N*-Phth group, benzoyl and acetyl groups; and (iv) selective *N*-acylation with fatty acid **12** under the promotion of DCC, thus completing the total synthesis of tunicamycin V (**2**) in 46% yield over four steps.

**Total synthesis of tunicamycin V (2) by the Ichikawa group.** In 2018, Ichikawa and coworkers reported the total synthesis of tunicamycin V (**2**, Scheme 6.5) [25]. Their synthesis began with the Mukaiyama aldol reaction between uridine-5′-aldehyde derivative **40** and silyl enol ether **41** in the presence of $BF_3{\cdot}Et_2O$, which afforded the desired *5′R*-product **42** in 64% yield with excellent diastereoselectivity (98 : 2). Asymmetric reduction of the ketone moiety followed by furan oxidation and introduction of the carbamate motif provided product **43**. The [3,3]-sigmatropic rearrangement of **43**, followed by dihydroxylation and further transformations, gave rise to tunicaminyluracil derivative **44**, which

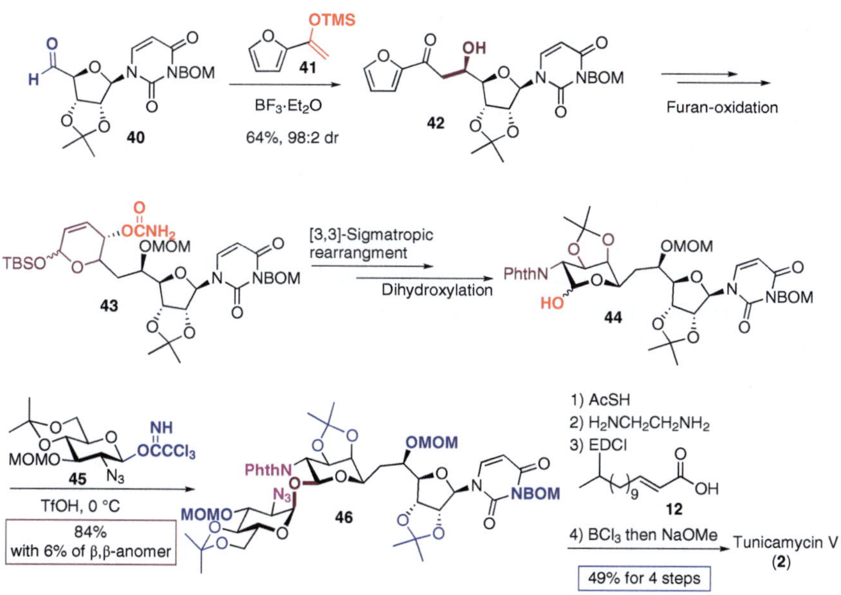


**Scheme 6.5** Total synthesis of tunicamycin V (**2**) by Ichikawa and coworkers.

existed predominantly as β-anomer (β/α = 10 : 1). The glycosylation of **44** with 2-azidoglucosyl trichloroacetimidate donor **45**, in the presence of TfOH, resulted in the desired glycoside **46** in 84% yield, along with 6% of the unwanted β,β-anomer. Alternatively, the head-to-head glycosidic bond could be constructed by glycosylation of a 2-azidoglucosyl acceptor and tunicaminyl trichloroacetimidate donor. However, this alternative glycosylation only provided the undesired β,β-anomer in 54% yield, without detection of the desired product **46**. Finally, the transformation of the azido group to the acetamide group, removal of the phthaloyl group, followed by condensation with fatty acid **12**, and global deprotection of six acid-labile protecting groups accomplished the total synthesis in 49% overall yield over four steps.

Recently, Kurosu and coworkers reported a concise total synthesis of tunicamycin V (**2**) using only 15 chemical steps from D-galactal, resulting in an overall yield of 21% [26]. The Büchner–Curtius–Schlotterbeck (BCS) reaction between diazoalkyl-disaccharide and uridine 5′-aldehyde was employed as the key step for the coupling of the complex disaccharide and uridine segments. This synthetic approach demonstrated remarkable flexibility, enabling the synthesis of a series of analogs of tunicamycin V.

## Hikizimycin

Hikizimycin (**47**), also known as anthelmycin, was isolated from the fermentation broth of *Streptomyces A-5*, an organism obtained from a soil sample collected at the Hikizi riverside in Kanagawa, Japan in 1971 (Figure 6.4) [27]. This compound exhibited inhibitory activity against protein synthesis by preventing peptide-forming reactions, making it a potent anthelmintic agent against various common parasites, as well as an antibiotic agent. Structurally, hikizimycin (**47**) consists of a cytosine base, a 3-amino-3-deoxyglucose sugar (kanosamine), and a 4-aminoundecose sugar (hikosamine) with one amino group and 10 hydroxy groups.

**Total synthesis of Hikizimycin (47) by the Schreiber group.** In 1990, Schreiber and coworkers reported the first total synthesis of hikizimycin (**47**, Scheme 6.6) [28, 29]. The synthesis began with a Horner–Wadsworth–Emmons (HWE) reaction between diisopropyl tartrate **48** and phosphate **49** in the presence of DIBAl-H, resulting in the formation of the coupled product **50**. Iterative chain extension via HWE reaction and dihydroxylation led to tetraol **51**, providing the necessary carbon framework of the hikosamine motif. Cyclization of **51** via a

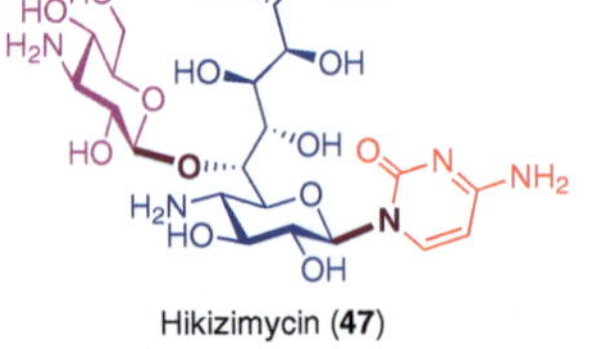


Hikizimycin (**47**)

A cytosine base

A 4-aminoundecose sugar (hikosamine) with 1 amino and 10 hydroxy groups

A 3-amino-3-deoxyglucose sugar (kanosamine)

**Total synthesis**: Schreiber (1990), Inoue (2020)

**Figure 6.4** Hikizimycin (**47**) and its structural features.

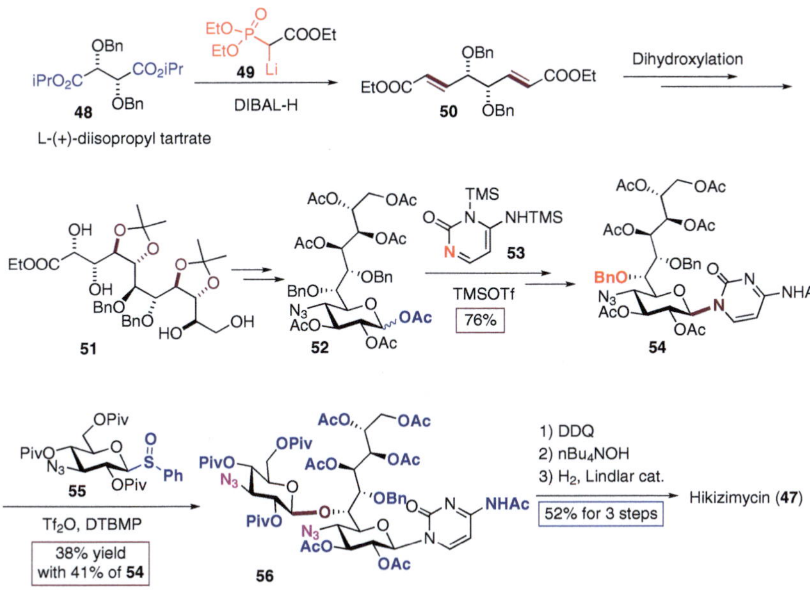

**Scheme 6.6** Total synthesis of hikizimycin (**47**) by Schreiber and coworkers.

six-membered lactol formation, followed by the introduction of the azido group, furnished hikosamine derivative **52**. The glycosylation of bis(tri-methylsilyl)cytosine **53** with hikosamine derivative **52** in the presence of TMSOTf yielded the desired nucleoside in 76% yield at 127 °C, which was subsequently acetylated to give compound **54**. Site-selective unmasking of the C6-hydroxyl group was achieved by oxidative debenzylation with DDQ, which was then coupled with glycosyl sulfoxide **55** in the presence of $Tf_2O$, yielding the desired β-disaccharide **56** in 38% yield, along with 41% of **54** recovered. Finally, sequential debenzylation with DDQ, complete deacylation with refluxing $nBu_4NOH$, and reduction of the azido groups via hydrogenation with Lindlar catalyst accomplished the total synthesis of hikizimycin (**47**) in 52% overall yield over three steps.

**Total synthesis of hikizimycin (47) by the Inoue group.** In 2020, Inoue and coworkers reported a convergent total synthesis of hikizimycin (**47**, Scheme 6.7) [30]. Their synthesis began with D-galactose derivative **57**, which was converted into α-alkoxyacyltelluride **58** via several steps of transformations. Subsequently, the intermolecular radical addition of **58** to aldehyde **59** was performed in the presence of $Et_3B$ and air, which successfully delivered the coupled alcohol **60** in 65% with moderate stereoselectivity, favoring the desired **60α** isomer (**60α**/**60β** = 2.2 : 1). Protecting group manipulations on **60** led to hikosamine derivative **61**, which was then glycosylated with bis-TMS-cytosine **62** in the presence of TMSOTf at 130 °C to provide nucleoside **63** in 19% yield (four steps from the mixture of **60α** and **60β**). Selective removal of the C-6 benzyl group with DDQ furnished the C6-alcohol, which was then coupled with 3-azido-glycosyl trichloroacetimidate **64** in the presence of TMSOTf, leading to the desired β-disaccharide **65** in 39% yield, with the unwanted α-anomer isolated in 22% yield. Finally, the removal of the seven benzoyl groups,

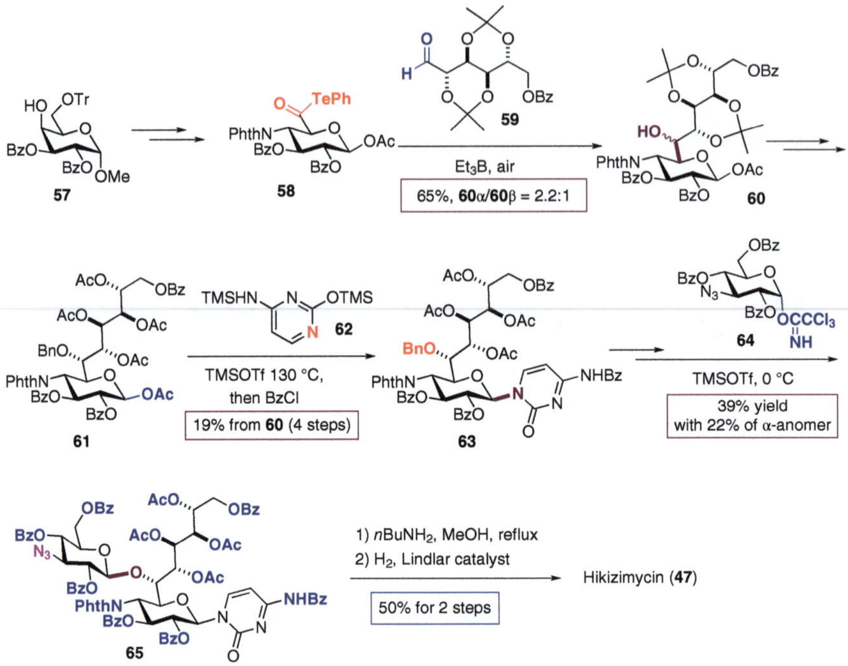


**Scheme 6.7** Total synthesis of hikizimycin (**47**) by Inoue and coworkers.

four acetyl groups, and one phthaloyl group was achieved by treating the compound with $nBuNH_2$ in refluxing MeOH. Subsequent reduction of the azido group through catalytic hydrogenation completed the total synthesis of hikizimycin (**47**).

## Herbicidins

Herbicidins, specifically herbicidin B (**66**) and C (**67**), were initially isolated from the strains of *Streptomyces* in 1976 (Figure 6.5) [31]. These compounds displayed potent inhibitory activity against *Xanthomas oryzae*, a bacterium responsible for leaf blight infection in rice crops. In addition, herbicidins also exhibited reduced seed germination, hindered algal growth, and selective toxicity toward dicotyledonous plants, while displaying no toxicity against animals. Structurally, herbicidins possess a number of intriguing characteristics, including a common undecose sugar moiety composed of a linear carbon chain that folds into a tricyclic furano-pyrano-pyran skeleton. Adenine serves as the nucleobase, residing in a sterically congested concave position. Additionally, the hemiketal at C-7 connects the pyrano-pyran system in a manner that orients all of its substituents axially. The different members of herbicidins vary from one another due to distinct methylation and esterification patterns.

**Synthesis of herbicidin glycoside by the Gallagher group.** In 1993, Gallagher and coworkers reported the synthesis of herbicidin glycoside (**73**) using

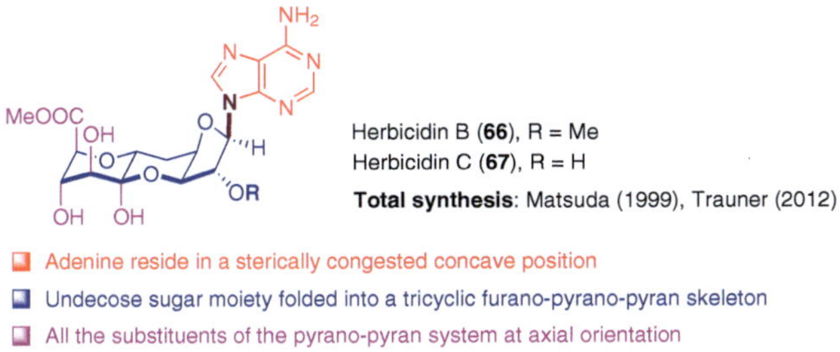


**Figure 6.5** Herbicidin B (**66**) and C (**67**) and their structural features.

a carbohydrate-based ketone enolate as a preformed heterocyclic building block (Scheme 6.8) [32]. Their synthesis began with 1,5 : 3,6-dianhydrohexulose derivative **68**, which could be obtained in two steps from 1,5 : 3,6-dianhydro-D-mannitol. The *t*BuOK-mediated condensation of **68** with D-glucose-derived aldehyde **69** proceeded efficiently to give the coupled enone **70** in 60% yield. Face-selective hydrogenation of the enone set the required stereochemistry at C-6, accompanied by concurrent *O*-debenzylation, resulting in the formation of the key furano-pyrano-pyran core of herbicidins (**71**) in 60% yield. The subsequent pivotal step involved the oxidation and cleavage of the ether bridge, which required discrimination between $-CH_2-O$ and >CH–O functionalities. Encouragingly, free-radical bromination proceeded efficiently and regioselectively, providing α-bromoether **72** in 65% yield. Finally, further oxidation to methyl ester and global deprotection yielded herbicidin glycoside **73**, which existed as a 1 : 1.5 mixture of the α- and β-anomer.

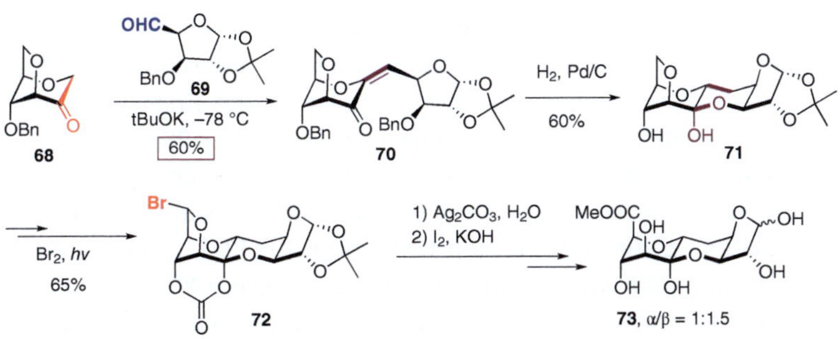


**Scheme 6.8** Synthesis of herbicidin glycoside **73** by Gallagher and coworkers.

**Total synthesis of Herbicidin B (66) by the Matsuda group.** In 1999, Matsuda and coworkers reported the first total synthesis of herbicidin B (**66**), utilizing a novel aldol-type *C*-glycosidation reaction promoted by samarium diiodide ($SmI_2$) as a key step (Scheme 6.9) [33]. To accomplish this, they prepared a key component, the 1-β-D-xylosyladenine 5′-aldehyde unit **75**, in seven steps from 2′-*O*-methyladenosine **74**. The other coupling partner, 1-phenylthio-2-ulose derivative **77**, was synthesized from glycal **76** through several steps of transformations. With these two coupling partners in hand, the $SmI_2$-promoted aldol reaction between phenylthioulose **77**

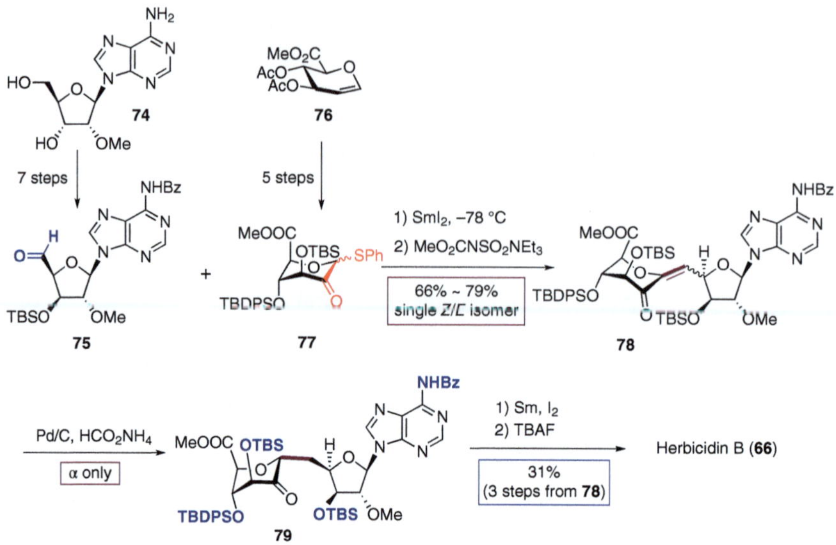

**Scheme 6.9** Total synthesis of herbicidin B (**66**) by Matsuda and coworkers.

and aldehyde **75** was performed at −40 °C to give the coupled β-hydroxyl ketone in 65% yield as a mixture of two diastereoisomers. Subsequent treatment with Burgess's inner salt ($MeO_2CNSO_2NEt_3$) further transformed the diastereoisomeric mixture into the corresponding enone **78** in 66–79% yields as a single isomer, while the alkene geometry remained unknown. Reduction of the enones using the $HCOONH_4$/Pd-C system yielded the desired α-anomer **79**, without observing the corresponding β-anomer. Finally, removal of the *N*-Bz group with $SmI_2$/MeOH and the silyl ether groups with TBAF led to formation of the internal ketal linkage between the 3′- and 7′-positions, ultimately completing the first synthesis of herbicidin B (**66**) in overall 31% yield (three steps from compound **78**).

**Total synthesis of herbicidin C (67) by the Trauner group.** In 2012, Trauner and coworkers reported their total synthesis of herbicidin C (**61**) with late-stage glycosylation as the pivotal step (Scheme 6.10) [34]. Their synthesis began with the Hosomi–Sakurai allylation of anhydro glucose **80**, affording the known *C*-glycoside **81** in 66% yield with complete α-stereoselectivity. Subsequent Grubbs' cross-metathesis of **81** with methyl acrylate furnished the unsaturated ester **82**, which was followed by Sharpless dihydroxylation to give ester **83** with high diastereoselectivity (14 : 1 dr). Saponification of the ester and formation of the corresponding Weinreb amide, followed by reaction with vinylmagnesium bromide, provided vinyl ketone **84**. Reduction of ketone **84** by *S*-configurated Corey–Bakshi–Shibata (CBS) reagent (*S*)-**85** gave alcohol **86** in 68% yield with the desired stereochemistry in high diastereoselectivity (14 : 1 dr). Several transformations were carried out, including removal of the acetonide group in **87**, followed by ozonolysis and acetylation, provided undecose **88**, which featured the core furano-pyrano-pyran framework of herbicidin C. The crucial late-stage glycosylation of **88** with adenine **89** in the presence of TMSOTf produced β-nucleoside **90** in

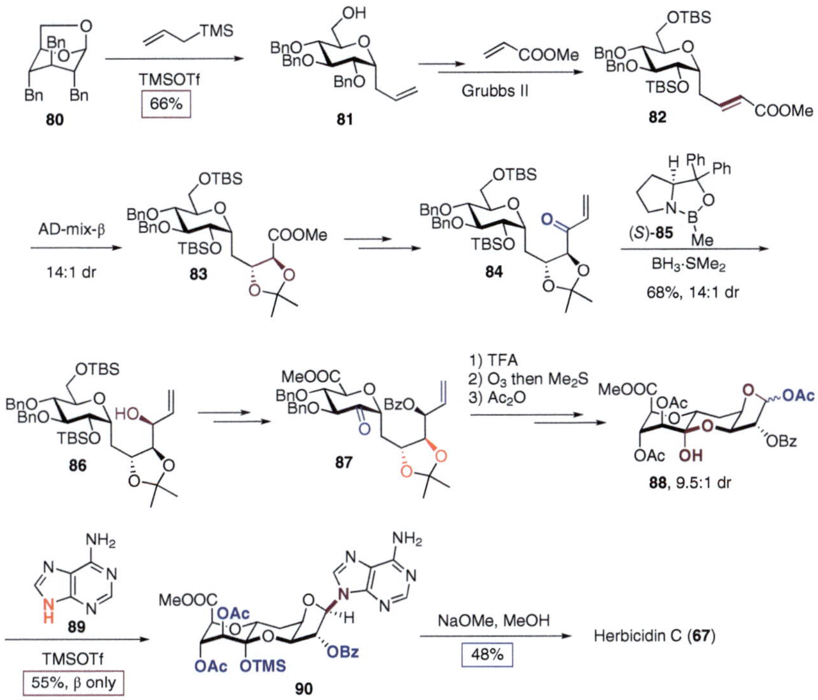

**Scheme 6.10** Total synthesis of herbicidin C (**67**) by Trauner and coworkers.

55% yield, without detecting the undesired α-anomer. Finally, global deprotection using NaOMe/MeOH accomplished the total synthesis of herbicidin C (**67**).

## A201A

Nucleoside antibiotic A201A (**91**) was initially isolated from a strain of *Streptomyces capreolus* in 1976 [35], and it was later re-discovered from the fermentation of a new strain of *Actinomycetes thermotolerans* collected from deep-sea sediment (Figure 6.6) [36]. A201A exhibited potent antibacterial activity against gram-positive bacteria

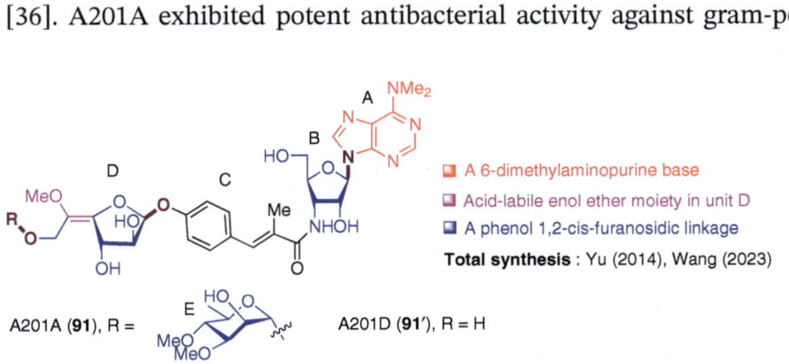

**Figure 6.6** Nucleoside antibiotic A201A (**91**) and its structural features.

and most gram-negative anaerobic bacteria. Notably, this structurally significant natural product is recognized as a mimic of the 3′-end of aminoacyl-*t*RNA. Structurally, A201A (**91**) consists of five distinct units: 6-dimethylaminopurine, 3-amino-3-deoxyribose, α-methyl-*p*-coumaric acid, an unnamed hexofuranose, and 3,4-di-*O*-methyl-D-rhamnose, which are connected linearly via one amide and three glycosidic linkages. The synthesis of A201A presents a challenge, primarily involving the construction of unit D, which contains an acid-labile enol ether moiety, and the highly stereoselective formation of the phenol 1,2-*cis*-furanosidic linkage.

**Total synthesis of A201A (91) by the Yu group.** In 2014, Yu and coworkers reported the first total synthesis of nucleoside antibiotic A201A (**91**) using a

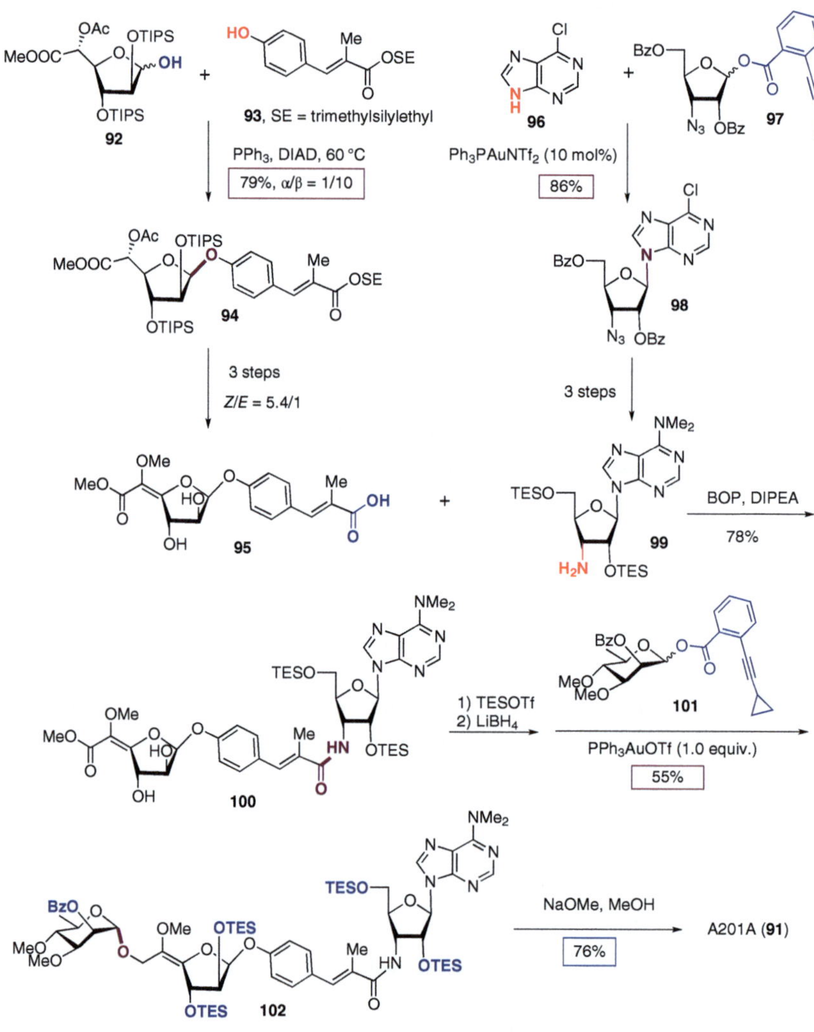


**Scheme 6.11** Total synthesis of antibiotic A201A (**91**) by Yu and coworkers.

highly modular and linear manner (Scheme 6.11) [37]. The challenging phenol 1,2-*cis*-furanosidic linkage was constructed via Mitsunobu glycosylation between furanose **92** and phenol **93** in 79% yield, with high diastereoselectivity (β/α = 10/1). Notably, the introduction of a triisopropylsilyl (TIPS) group at the 2-OH position favored a 1,2-*trans*-activation and subsequent $S_N2$ substitution, giving the desired 1,2-*cis* glycosylation product. With three-step transformation of **94**, the desired exocyclic enol ether **95** was obtained as the major isomer with a 5.4/1 *Z/E* ratio. On the other hand, the glycosylation between 6-chloro-purine **96** and ribofuranosyl *ortho*-alkynylbenzoate **97**, under the catalysis of $Ph_3PAuNTf_2$ (0.1 equiv.), proceeded smoothly to provide β-glycoside **98** cleanly in 86% yield. Subsequent steps of transformation resulted in the condensation of amine **99** with acid **95** in the presence of BOP, leading to the formation of the desired amide **100** in 78% yield. Reduction of the ester group of **100** yielded alcohol, followed by the challenging final glycosylation step with glycosyl *ortho*-alkynylbenzoate **101**. Conventional catalytic conditions using 0.1 equiv. of Au(I) catalyst resulted in sluggish reaction kinetics, potentially due to interception of the catalytic cycle by the basic nitrogen atoms in **100**. To overcome this, a stoichiometric amount of $Ph_3PAuOTf$ was utilized as promoter, enabling the desired α-rhamnoside **102** to be obtained in a 55% yield. Alternative Lewis acid-catalyzed glycosidation reactions with imidate donors were also attempted but led to the decomposition of the enol ether moiety. Finally, global deprotection of the TES and Bz groups with NaOMe (76%) accomplished the synthesis of A201A (**91**).

**Total synthesis of A201A (91) by the Wang group.** In 2023, Wang and coworkers reported the total synthesis of antibiotics A201A (**91**) along with A201D (**91′**) and A201E, employing a bioinspired synthetic route (Scheme 6.12) [38]. The pivotal L-galactofuranose was prepared from readily available D-galactose using a C1 and C6 head-to-tail oxidation-state-switch method, which was then converted into the corresponding glycosyl thioether **103**. Subsequent glycosylation of phenol **104** with **103**, in the presence of NIS and TfOH, afforded the desired 1,2-*cis*-furanoside **105** in 84% yield, in which the remote 2-quinolinecarbonyl group assisted the aglycone delivery process. Notably, it was mentioned that the direct glycosylation of **103** with adenosine-containing acceptor did not provide the desired product, probably due to the steric hindrance of **103** and the basic adenosine moiety of the acceptor. Followed by the hydrolysis of allyl ester, the amide formation with the protecting-group-free **106** yielded the coupled product **107** with high efficiency. After successful construction of the *Z*-type methyl enol moiety, global removal of the protecting groups finished the synthesis of A201D (**91′**). On the other hand, the late-stage glycosylation of alcohol **108** with **109** using Yu's glycosylation method delivered the expected α-glycosylation product **110** in 61% yield. Notably, direct glycosylation of A201D with the corresponding glycosyl fluoride donor was attempted, which only afforded A2021A in 17% yield. Finally, sequential removal of the *O*-silyl groups and the *O*-acetyl group on **110** accomplished the second total synthesis of A201A (55% yield for two steps).

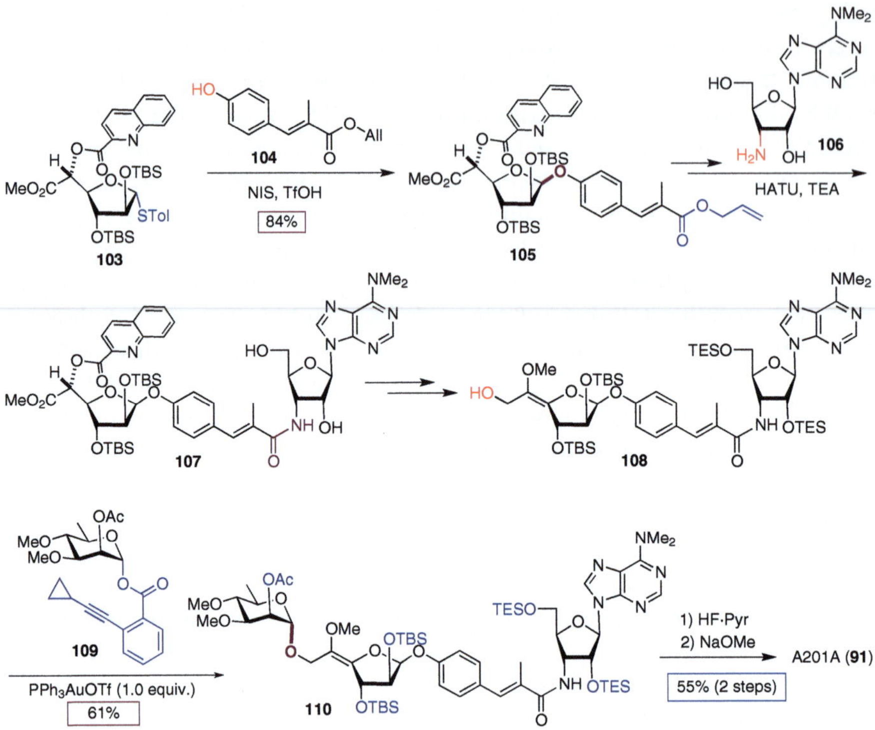


**Scheme 6.12** Total synthesis of antibiotic A201A (**91**) by Wang and coworkers.

## Amipurimycin

The complex peptidyl nucleoside antibiotic amipurimycin was originally isolated from *Streptomyces novoguineensis* sp. nov., which was discovered in soil samples collected from New Guinea during the 1970s (Figure 6.7) [39, 40]. This antibiotic exhibited potent *in vitro* antifungal activity and demonstrated curative effects against rice blast disease at concentrations ranging from 10 to 20 ppm. Despite its early discovery, the complete structure elucidation of amipurimycin proved challenging. In

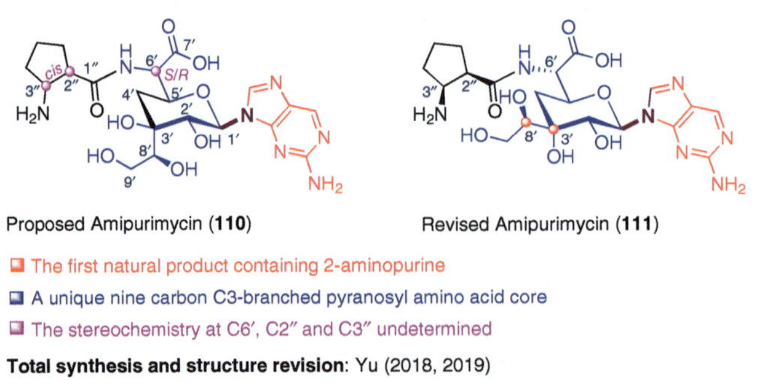


**Figure 6.7** Proposed and revised structure of amipurimycin and their structural features.

1982, the structure of amipurimycin was proposed to be a peptidyl nucleoside (**110**) featuring a distinct nine-carbon C3-branched pyranosyl amino acid core. However, the stereochemistry at C6′, C2″, and C3″ remained undetermined.

**Total synthesis of the proposed structure of amipurimycin by the Yu group.** In 2018, the Yu group reported the first total synthesis of the originally proposed structure of amipurimycin (**110**, Scheme 6.13) [41]. The synthesis commenced with the aldol reaction between ketone **113**-(*R*) (derived from D-serine) and 3-oxo-furanoside **112** (derived from L-Arabinose), facilitated by NaHMDS, yielding the desired adduct **114** in a stereocontrolled manner. The subsequent chelation-controlled reduction of the ketone group in **114** delivered the required 7*S*-ol with high diastereoselectivity (>12 : 1 dr), which was then converted into

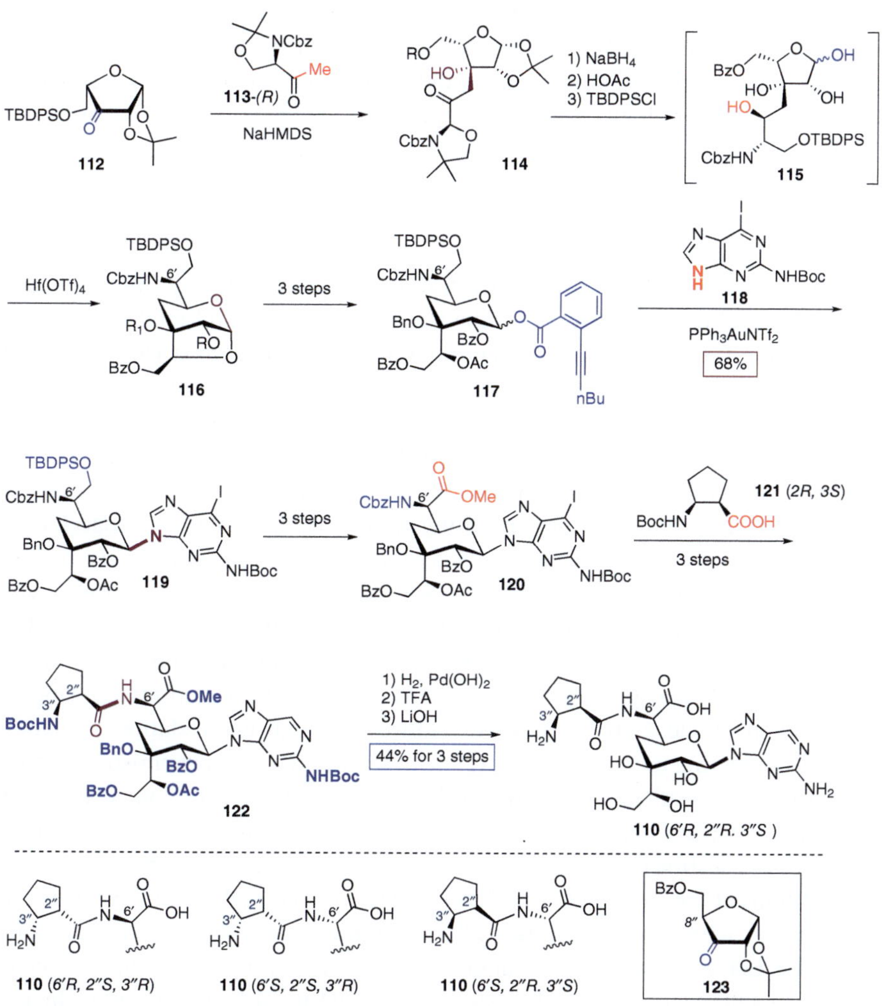


**Scheme 6.13** Total synthesis of the four diastereomers of the proposed structure of amipurimycin (**110**) by Yu and coworkers.

the [3,2,1]-bicyclic acetal **116** using $Hf(OTf)_4$, through the intermediate **115**. Further transformations were employed to convert **116** into the desired glycosyl *o*-alkynylbenzoate **117**. As anticipated, the glycosylation of the 6-iodopurine **118** with **117** proceeded smoothly under $PPh_3AuNTf_2$ catalysis (0.2 equiv.), yielding the desired β-$N^9$-nucleoside **119** as the sole anomer in a good 68% yield. After subsequent functional group manipulations, coupling of the amine group in **120** with amino acid **121**-(*2R,3S*) afforded the expected amide **122**. Finally, global deprotection of **121** was achieved (44% for three steps) through catalytic hydrogenation of the Bn group, acid cleavage of the two *N*-Boc groups, subsequent removal of the Bz groups and the methyl ester under basic conditions, ultimately accomplishing the total synthesis of amipurimycin **110** (*6′R,2″R,3″S*). In a similar manner, the other three diastereomers, namely **110** (*6′R,2″S,3″R*), **110** (*6′S,2″R,3″S*), and **110** (*6′S,2″S,3″R*) were synthesized without complication. However, none of these four diastereoisomers displayed $^1H$ and $^{13}C$ NMR data that matched with those reported for natural amipurimycin. Moreover, to validate the stereochemistry at C8′, the four 8′*R* epimers of the originally proposed amipurimycin were prepared. Unfortunately, the $^1H$ and $^{13}C$ NMR data of the 8′*R* diastereoisomers also deviated from those reported for natural amipurimycin. Based on these results, the authors proposed that the chiral center at the tertiary C3′ in the originally proposed structure of amipurimycin should possess the opposite configuration.

**Total synthesis and structural revision of amipurimycin by the Yu group.** In 2019, Yu and coworkers reported the structural revision and total synthesis of amipurimycin (Scheme 6.14) [42]. Through a meticulous comparison of the $^1H$ NMR spectra of the previously synthesized epimers with the authentic sample, the authors revised the structure of amipurimycin to **111**, which features 3′*S*, 6′*S*, 8′*R*, 2″*R*, and 3″*S* configurations. Analogously, the synthesis commenced with an aldol reaction between 3-oxo-furanoside **124** (derived from D-arabinose) and ketone **113**-(*R*) (derived from D-serine), resulting in the formation of nonasaccharide **125** (89%) with complete stereocontrol. Reduction of the ketone moiety on **125** using $Zn(BH_4)_2$ and $CeCl_3$ generated 7*S*-ol **126** with high diastereoselectivity (dr = 12.4 : 1). The synthesis then proceeded with the conversion of furanoside **126** to the fused pyranoside **127**, which adopted a twist-boat conformation, following the previously established procedure. After steps of protecting group manipulations, the inversion of the 2-OH in **128** was realized through an oxidation–reduction sequence, providing nonapyranoside **129** with the desired stereochemistry. Further transformations resulted in the synthesis of *o*-alkynylbenzoate **130**, which was then coupled with 6-iodopurine **118** in the presence of $Ph_3PAuNTf_2$ (0.2 equiv.), furnishing the desired β-$N^9$-nucleoside **131** as the single anomer in a high 81% yield. Modifications of the C6′ and C7′ position on **131** were carried out, and subsequent condensation with amino acid **121**-(*2R,3S*) afforded the fully protected peptidyl nucleoside **133**. Finally, the removal of the two *N*-Boc groups with trifluoroacetic acid (TFA) and, subsequently, the esters with LiOH furnished amipurimycin **111** (75% for two steps). Importantly, the NMR spectroscopic data of the synthetic **111** matched well with those recorded for the natural product.

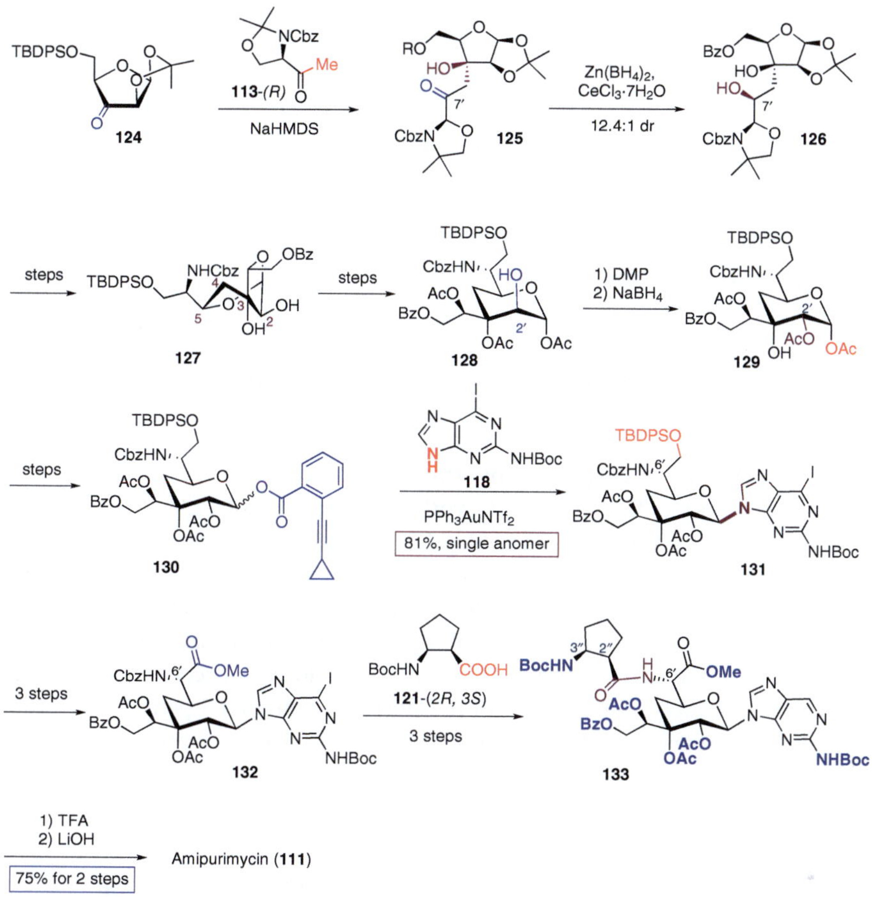

**Scheme 6.14** Total synthesis of the revised amipurimycin (**111**).

## Caprazamycin

Caprazamycins, such as caprazamycin A (**135**), were first isolated from *Streptomyces* sp. *MK730–62F2* by Igarashi and coworkers in 2003 (Figure 6.8) [43]. These compounds have shown excellent antimycobacterial activity *in vitro* against drug-susceptible and multidrug-resistant *Mycobacterium tuberculosis* strains, with no significant toxicity in mice. Structurally, carprazamycins consist of uridine, aminoribose, a distinctive diazepanone (a seven-membered ring containing two nitrogen atoms) and a fatty acid side chain featuring the vulnerable β-acyloxyester moiety. (+)-Caprazol (**134**) represents the core structure of caprazamycins and is a minor component of caprazamycin production through fermentation. Although caprazol itself only exhibited weak anti-tuberculosis (TB) activity, most of its molecular skeleton is shared by CPZEN-45, a compound exhibiting enhanced antibacterial activity against extensively multidrug-resistant TB (XDR-TB) (see Figure 6.8).

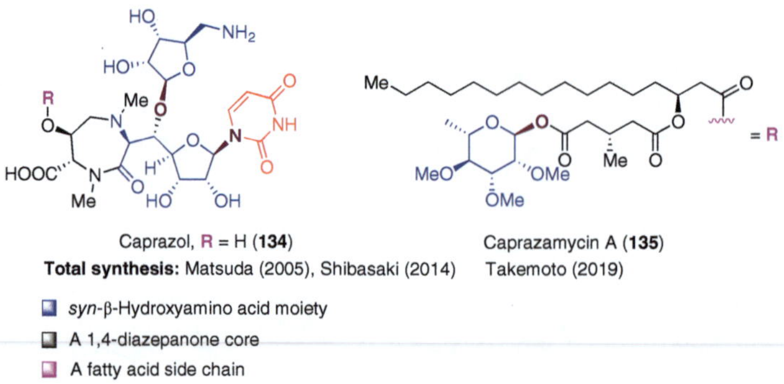


**Figure 6.8** Structure of caprazamycin A (**135**) and caprazol (**134**) and their structural features.

**Total synthesis of carprazol by the Matsuda group.** In 2005, the Matsuda group reported the first total synthesis of carprazol (**134**, Scheme 6.15) [44, 45]. Starting from uridine-derived aldehyde **136**, two-carbon elongation with $PPh_3$=CHCOOMe, followed by Sharpless asymmetric aminohydroxylation using $(DHQD)_2AQN$ as the chiral ligand, led to the formation of **137** with 86 : 14 dr. Subsequent glycosylation of **137** with ribosyl fluoride **138**, facilitated by $BF_3 \cdot Et_2O$ at −30 °C, afforded the desired β-anomer in 80% yield with excellent β selectivity (β/α = 96 : 4). Notably, the presence of the sterically hindered 3-pentylidene group played a crucial role in the diastereoselectivity of this reaction, as the corresponding

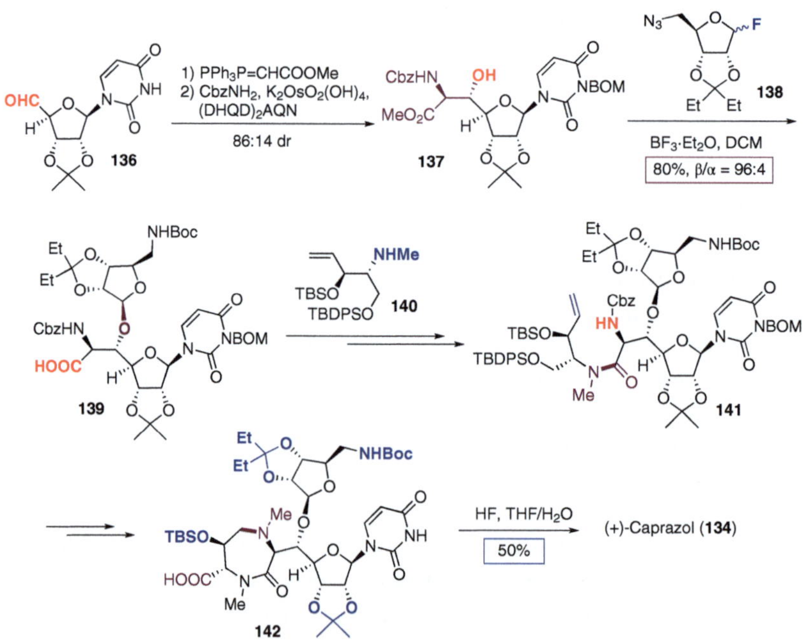


**Scheme 6.15** Total synthesis of carprazol (**134**) by Matsuda and coworkers.

ribosyl fluoride bearing an isopropylidene group resulted in the glycosylation with 73 : 27 (β/α) diastereoselectivity. Following several steps of functional group transformations, the coupling of acid **139** with amine **140** afforded amide **141**, which underwent intramolecular reductive amination to provide the diazepanone moiety. Finally, global deprotection of *N*-Boc, *O*-TBS, isopropylidene, and 3-pentylidene groups in **142** with aqueous HF provided the targeted caprazol (**134**) in 50% yield.

**Total synthesis of carprazol by the Watanabe and Shibasaki group.** In 2014, the Watanabe and Shibasaki group reported an alternative route for the total synthesis of carprazol (**134**, Scheme 6.16) [46]. Their syntheses began with the diastereoselective isocyanoacetate aldol reaction between uridine-derived aldehyde **143** and isocyanoacetate **144** enabled by Cu-catalysis, which provided the desired β-hydroxy-α-amino ester **145** in 67% yield with 88 : 12 dr. The ribosyl moiety was successfully introduced to give the desired β-anomer **146** using Matsuda's approach, with good yield and excellent diastereoselectivity. After converting the azido group to NHBoc, the coupling of the carboxylic acid with amine **147** (prepared through an enantioselective *anti*-nitroaldol reaction catalyzed by an Nd/Na-chiral amide ligand) gave access to amide **148**. Subsequent intramolecular reductive amination yielded the seven-membered diazepanone containing **149**, which underwent carboxylic acid formation and global deprotection with HF aqueous solution, ultimately achieving the total synthesis of carprazol (**134**).

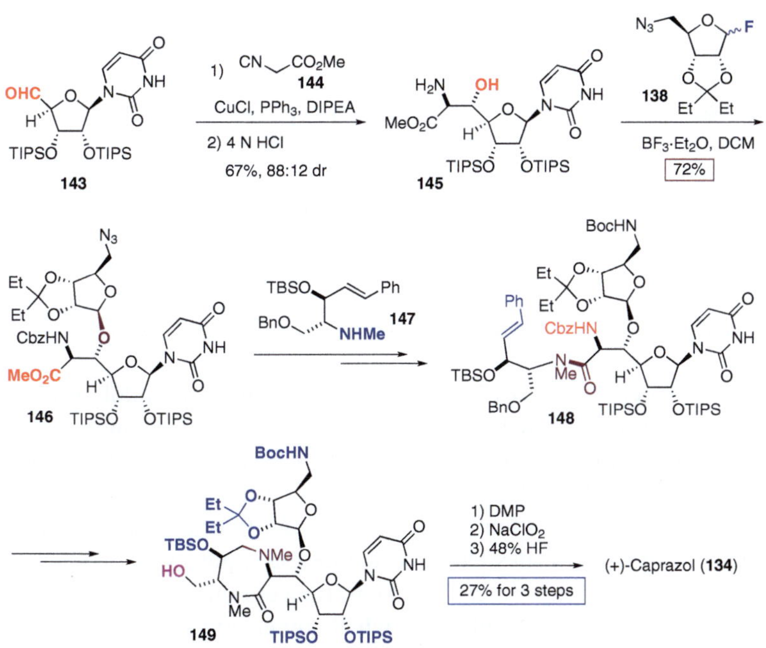


**Scheme 6.16** Total synthesis of Carprazol (**134**) by Watanabe, Shibasaki, and coworkers.

**Total synthesis of carprazamycin A by the Takemoto group.** In 2019, the Takemoto group reported the first total synthesis of caprazamucin A (**135**,

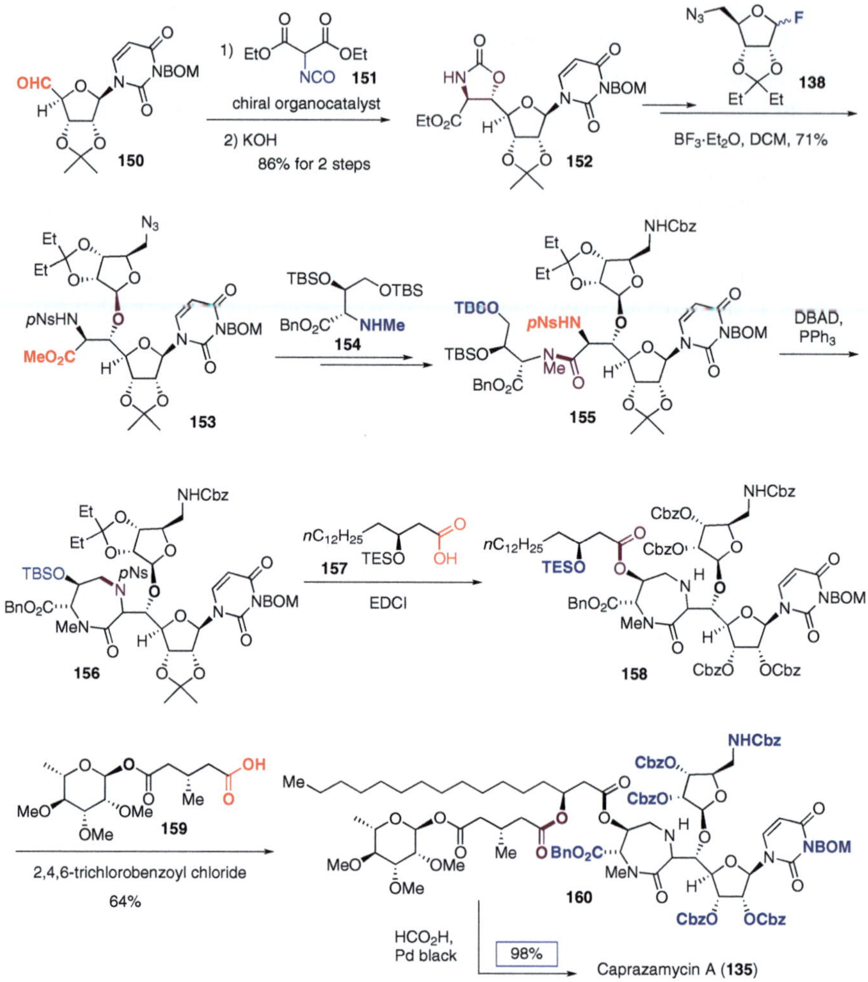

**Scheme 6.17** Total synthesis of carprazamycin A (**135**) by Takemoto and coworkers.

Scheme 6.17) [47]. Their syntheses began with the asymmetric aldol reaction between aldehyde **150** and diethyl isocyanomalonate **151** enabled by a chiral thiourea catalyst, which was followed by selective hydrolysis and decarboxylation to give the thermodynamically stable *trans*-oxazolodinone **152** with high yield and diastereoselectivity. The glycosylation of **152** with ribosyl fluoride **138**, following Matsuda's approach, afforded the desired disaccharide **153** in good yield. Following several steps of functional groups interconversion, the coupling of the carboxylic acid group with amine **154** afforded amide **155**, which underwent an intramolecular Mitsunobu reaction to generate the 1,4-diazepanone skeleton (**156**). The successful coupling of protected carprazol with fatty acid side chain **157** afforded the desired acylated product **158**, without the undesired side reactions such as β-elimination, epimerization, and retro-aldol reaction on the 1,4-diazepanone. The introduction of the rhamnose-containing fatty acid side chain **159** was realized by Yamaguchi conditions (enabled by 2,4,6-trichlorobenzoy chloride), resulting in **160** in 64% yield. Finally, the global deprotection of five Cbz groups, one BOM group, and one Bn group in the protected caprazamycin A **160** was achieved using Pd black and formic acid, which resulted a near-quantitative clean deprotection without olefin reduction, thus accomplishing the first total synthesis of caprazamycin A (**135**).

## Polyoxin J

Polyoxin J (**1**) was originally discovered in the culture broth of *S. cacaoi* var. *asoensis* in the late 1960s by Isono and coworkers (Figure 6.9) [8]. Alongside other polyoxin family members, these compounds exhibited prominent activity against phytopathogenic fungi, while nontoxic to bacteria, plants, and mammals. Consequently, polyoxins have been widely used as highly effective agricultural fungicides. The characteristic structural features of polyoxins include a peptide

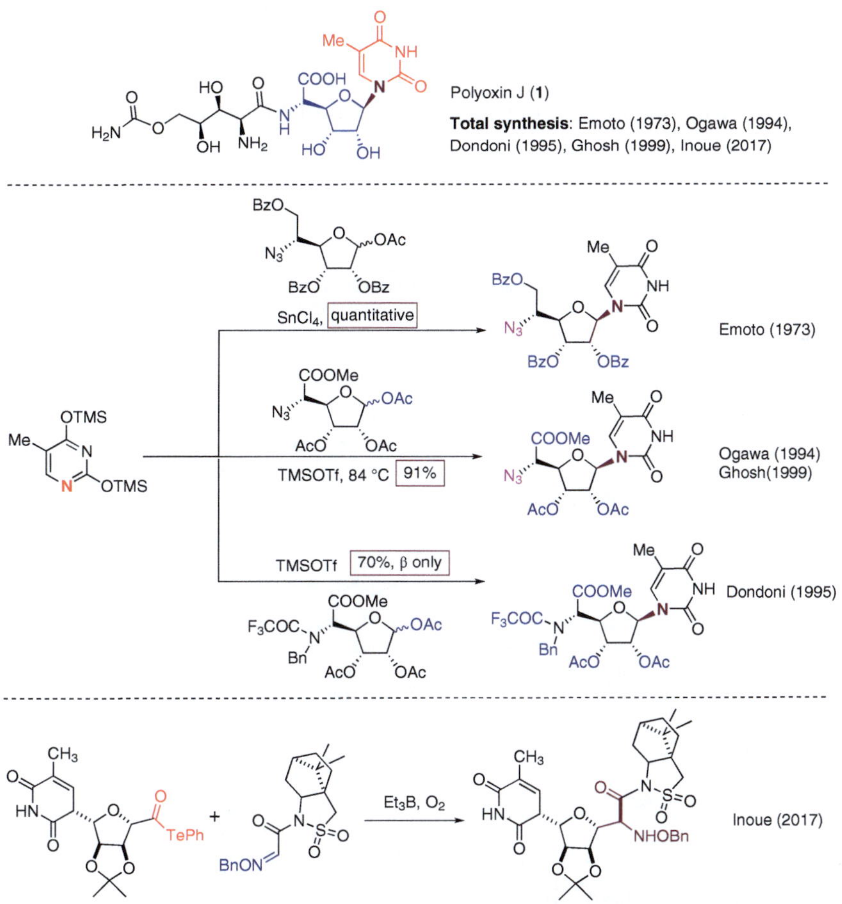


**Figure 6.9** Polyoxin J (**1**) and the glycosylation steps in the total synthesis.

linkage between the nucleoside α-amino acid and polyhydroxynorvaline. Since the first total synthesis of polyoxin J by the Emoto group in 1973 [9], this compound has been synthesized by numerous research groups. The Ogawa group [48], the Ghosh group [49], and the Dondoni group [50, 51] realized the total synthesis by late-stage glycosylation between 5-methyl-2,4-bis(trimethylsilyloxy)pyrimidine and the modified furanose unit, while the Inoue group accomplished the total synthesis through the modification of a commercially available ribonucleoside [52].

## Octosyl Acid A

Octosyl acid A (**161**) was isolated from *S. cacaoi* var. *asensis* by Isono and coworkers in 1975 (Figure 6.10) [53]. While octosyl acid A itself does not possess any biological activity, its adenine analog obtained through transglycosylation has shown to be an inhibitor of cyclic-AMP phosphodiesterase. Structurally, octosyl acid A (**161**) displays a unique 3,7-anhydrooctofuranose scaffold, wherein a furanose ring is *trans*-fused with a pyran ring. The Danishefsky group (1986) [54] and the Ogawa and Suami group (1988) [55] achieved the total synthesis of octosyl acid A (**161**) utilizing late-stage glycosylation, while the Hanessian group (1986) [56] accomplished the synthesis involving modifications on a readily available nucleoside.

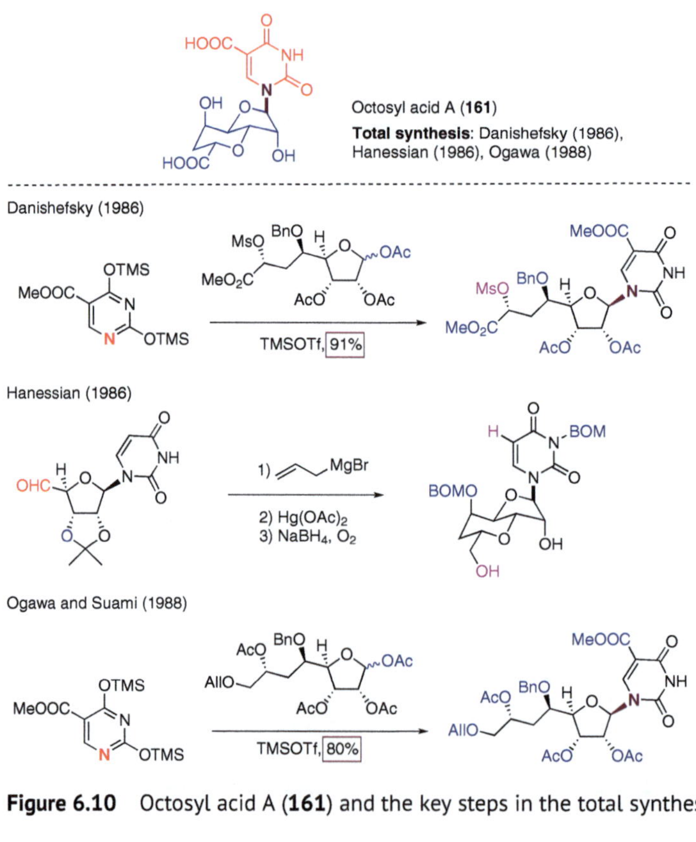


**Figure 6.10** Octosyl acid A (**161**) and the key steps in the total synthesis.

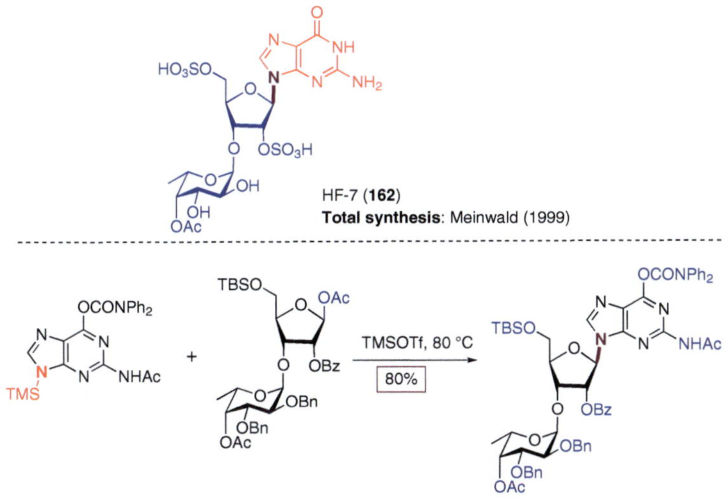


**Figure 6.11** HF-7 (**162**) and the key glycosylation step in the synthesis.

## HF-7

HF-7 (**162**), a novel neuroactive compound, was isolated in 1993 from the venom of a funnel-web spider, *Hololena curta* (Figure 6.11) [57]. This compound has been shown to selectively block kainate receptors and, although with modest affinity, L-type calcium channels. HF-7 holds potential therapeutic implications for the treatment of global cerebral ischemia, a form of excitotoxicity mediated by non-NMDA receptors, which can occur following incidents such as cardiac arrest, drowning, or carbon monoxide poisoning. Structurally, HF-7 consists of an unusual acetylated, disulfated fucopyranosyl guanosine, where the acetate ester is attached at the 4-position of an α-linked fucose ring and two sulfates attached to the ribose ring. In 1999, the Meinwald group reported the first total synthesis of HF-7 through a late-stage glycosylation strategy [58].

## Malayamycin

Malayamycin (**163**) was isolated from the soil organism of *Streptomyces malaysiensis* by a group at the Syngenta Crop Protection laboratories in Jealott's Hill, UK (Figure 6.12) [59]. Structurally, malayamycin possesses a 5-pyrimidinyl β-*C*-glycosidic bond, a *trans*-fused bicyclic perhydrofuropyran motif, and a *cis*-vicinal amino alcohol moiety. In 2003, the Hanessian group reported the first total synthesis, structural identity, and stereochemical confirmation of malayamycin A (**163**) [60]. The *C*-glycosidic bond was formed through a sequential process involving the addition of 2,4-dimethoxy-5-lithiopyrimidine with D-ribonolactone, L-Selectride reduction, and Mitsunobu reaction of the resulting diol precursor.

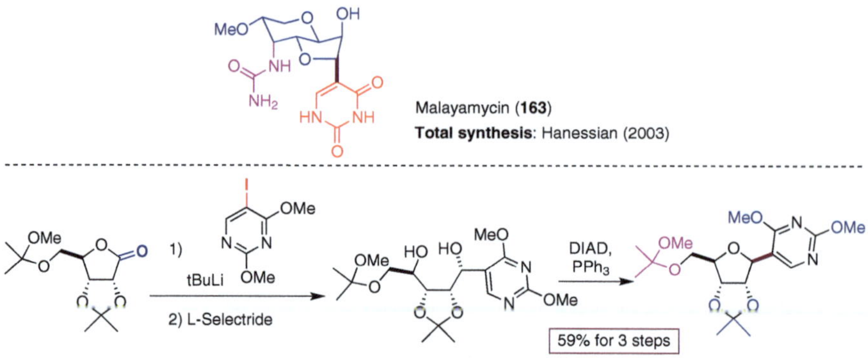


**Figure 6.12** Malayamycin (**163**) and the key glycosylation step in the total synthesis.

## Capuramycin

Capuramycin (**164**), a novel nucleoside antibiotic, was isolated from the culture filtrate of *Streptomyces griseus 446S3* in 1986 (Figure 6.13) [61]. It showed selective antimicrobial activities with minimum inhibitory concentration (MIC) values of 3.1 μg/ml against *Mycobacterium smegmatis* and 12.5 μg/ml and *Streptococcus pneumonia*. Structurally, capuramycin comprises a 4-deoxy-4-hexenuronic acid linked to an L-taluronamide nucleoside via a glycosidic bond, and a peptide bond connecting to 2(*S*)-aminocaprolactam. In 1994, the Knapp group reported the first total synthesis of capuramycin (**164**) through a stepwise glycosylation strategy, with the structure and absolute stereochemistry being confirmed [62]. In 2009, the Kurosu group reported another synthesis of capuramycin through a strategy involving modification of a readily available nucleoside [63, 64]. In 2021, Xiao and coworkers demonstrated an orthogonal one-pot synthesis of capuramycin using glycosyl *ortho*-(1-phenylvinyl)benzoates (PVB) donors, which efficiently furnished two glycosidic bonds in one pot [65].

## Muraymycin

Muraymycins are a group of aminoribosyl–uridyl peptides isolated from *Streptomyces* spp. by McDonald and coworkers in 2002 [66], which consist of 19 congeners identified to date (Figure 6.14). Among them, muraymycin A1 (**165**) stands out as one of the most potent members, exhibiting bactericidal activity against both gram-positive and gram-negative bacteria, with demonstrated efficacy in mice models infected with *Staphylococcus aureus* ($ED_{50}$ = 1.1 mg/kg). Structurally, these compounds feature a 5′-*O*-aminoribosyl-5′-*C*-glycyluridine moiety, along with an amino acid-urea-amino acid motif involving an L-*epi*-capreomycidine (L-*epi*-Cpm),

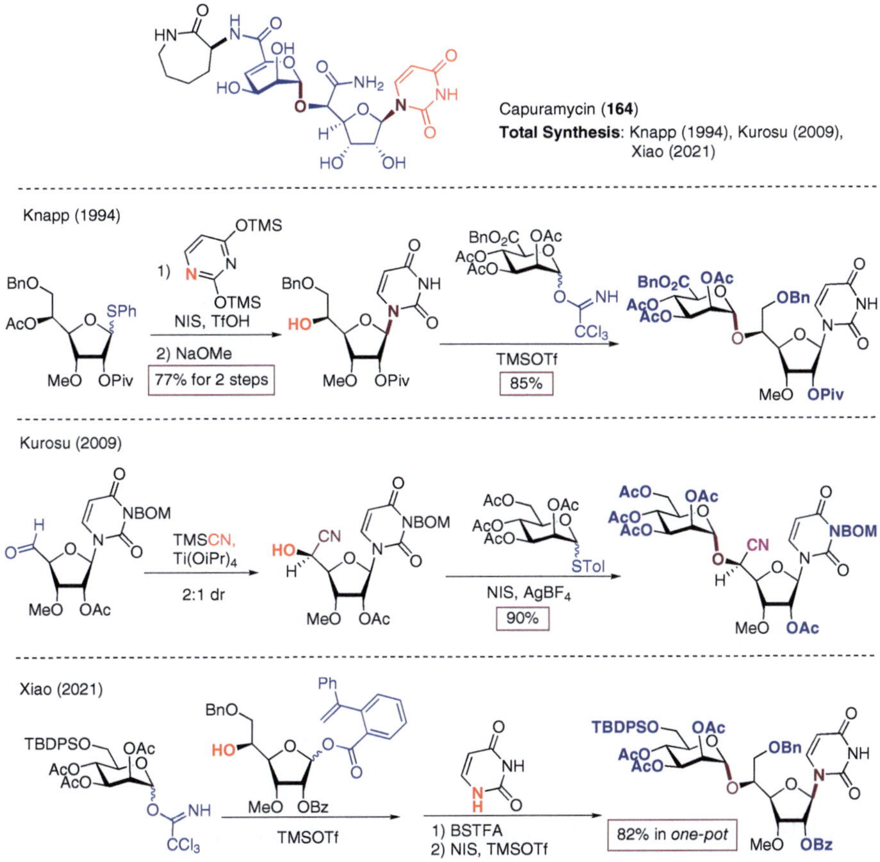


**Figure 6.13** Capuramycin (**164**) and the key glycosylation steps in the total synthesis.

which represents a cyclic guanidine amino acid. In 2010, the Matsuda group reported the total synthesis of muraymycin D2 (**167**) [67], employing a similar route as depicted in the synthesis of caprazol (Scheme 6.15) [44]. In 2016, the Kurosu group reported the first total synthesis of muraymycin D1 (**166**), featuring a selective β-ribosylation of the C2-methylated amino ribose and asymmetric Strecker and alkynation reaction [68]. Notably, acid-cleavable protecting groups for the secondary alcohol and uridine ureido nitrogen were employed, enabling their simultaneous deprotection alongside the Boc and *t*Bu groups.

## Plicacetin, Streptcytosine A, and Amicetin

Amicetin (**170**, Figure 6.15) was isolated in the early 1950s as the first member of the amicetin family antibiotics [69], which are produced in the fermentation broth

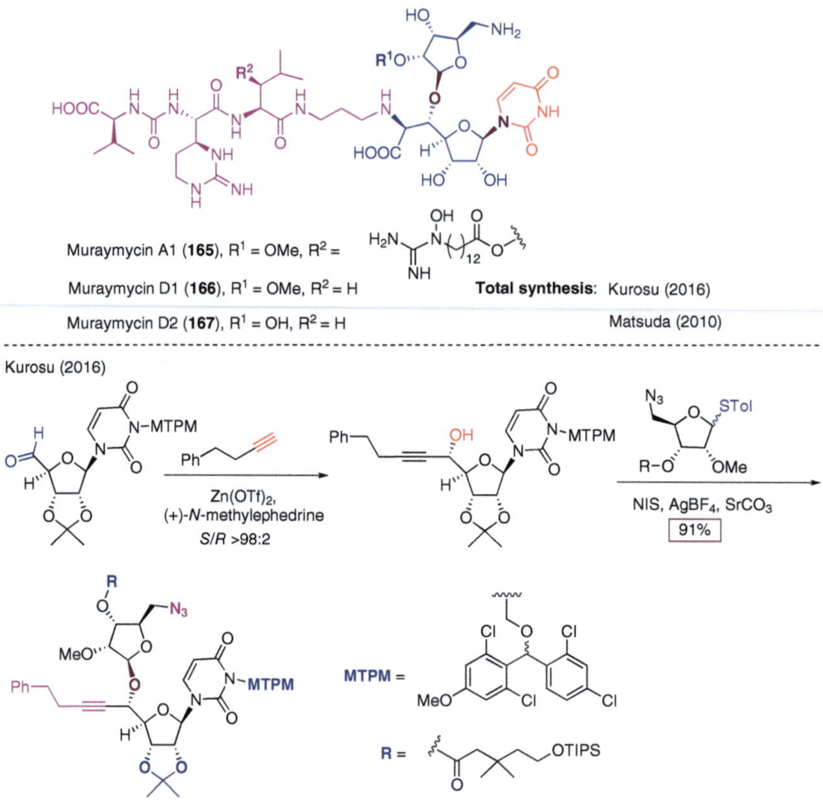


**Figure 6.14** Muraymycins and the glycosylation steps in the total synthesis.

of various species of *Streptomyces, Arthrobacter*, and *Nocardia*. Structurally, these antibiotics share a common disaccharide-pyrimidine nucleoside motif known as cytosamine. In this motif, amosamine is α-(1 → 4)-linked to amicetose, which in turn is β-(1 → N1)-linked to a cytosine nucleus. Notably, these compounds exhibit antibacterial and antiviral activities against tuberculosis, herpes, and polio, both *in vitro* and *in vivo*. In 1972, Stevens and coworkers reported the first total synthesis of plicacetion (**168**), wherein the key glycosidic bond was constructed between a nucleosidic alcohol and glycosyl chloride [70]. In 2001, the Sugimura group reported a formal total synthesis of plicacetion (**168**) using a similar strategy, but employed glycosyl fluoride as the glycosyl donor [71]. In 2018 and 2021, the Yu group reported the total synthesis of plicacetion (**168**), streptcytosin A (**169**), and amicetin (**170**) utilizing the gold(I)-catalyzed glycosylation between cytosine and disaccharide donor [72, 73].

## A-94964

A-94964 (**171**), a structurally unique nucleoside antibiotic, was isolated by Daiichi-Sankyo researchers from the fermentation broth of *Streptomyces*

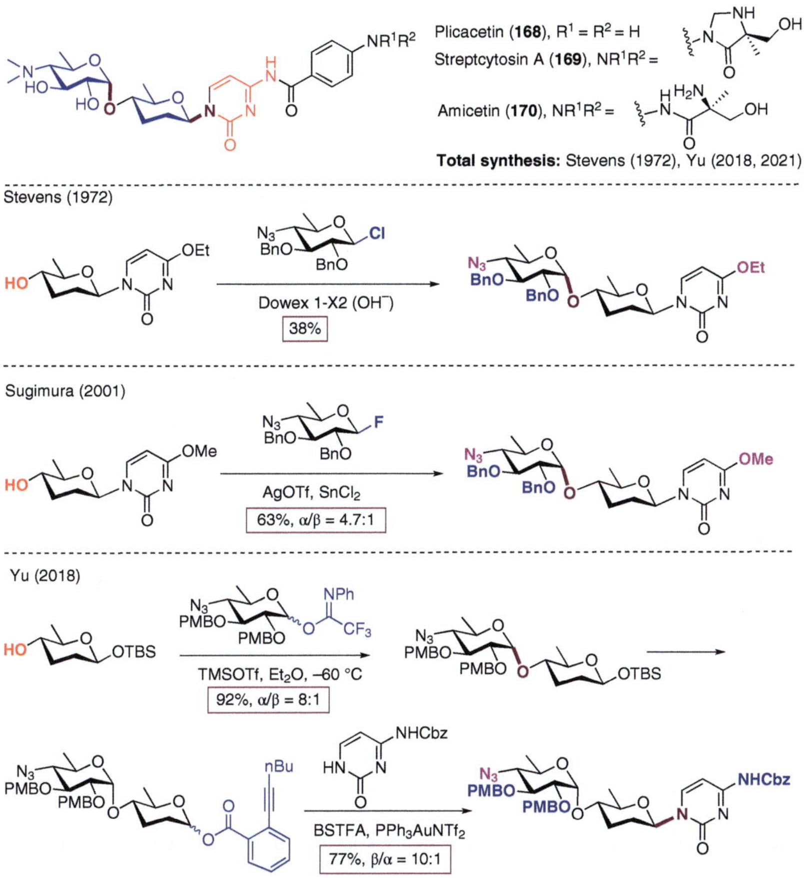


**Figure 6.15** Plicacetin (**168**), streptcytosine A (**169**), and amicetin (**170**) and the glycosylation steps in the total synthesis.

sp. SANK 60404 in 2008 (Figure 6.16) [74]. This compound exhibits potent phospho-*N*-acetylmuramyl-pentapeptide translocase (MraY) inhibitory activity ($IC_{50}$ = 1.1 μg/ml) and represents a novel scaffold mimicking the bisubstrate transition state of MraY and relevant glycosyltransferases. Structurally, A-94964 possesses several distinct features: (i) a nucleoside core with an unprecedented octuronic acid moiety, (ii) an additional hexopyranose decoration at the C6′ position of the octuronic acid moiety, and (iii) a phosphodiester connection at C7′ linked with an aminohexose *N*-acylated with a glycine, followed by an unsaturated and branched fatty acid. However, the stereochemistry of this intriguing antibiotic remained undetermined. In 2022, the Yu group reported a collective total synthesis of eight diastereoisomers of A-94964, along with the full stereochemistry assignment of A-94964 by detailed NMR analysis [75]. The study revealed that this nucleoside antibiotic contains a C6′(*S*), C7′(*S*)-octuronic acid uridine core, to which an α-D-mannopyranosyl residue is attached at C6′, and an

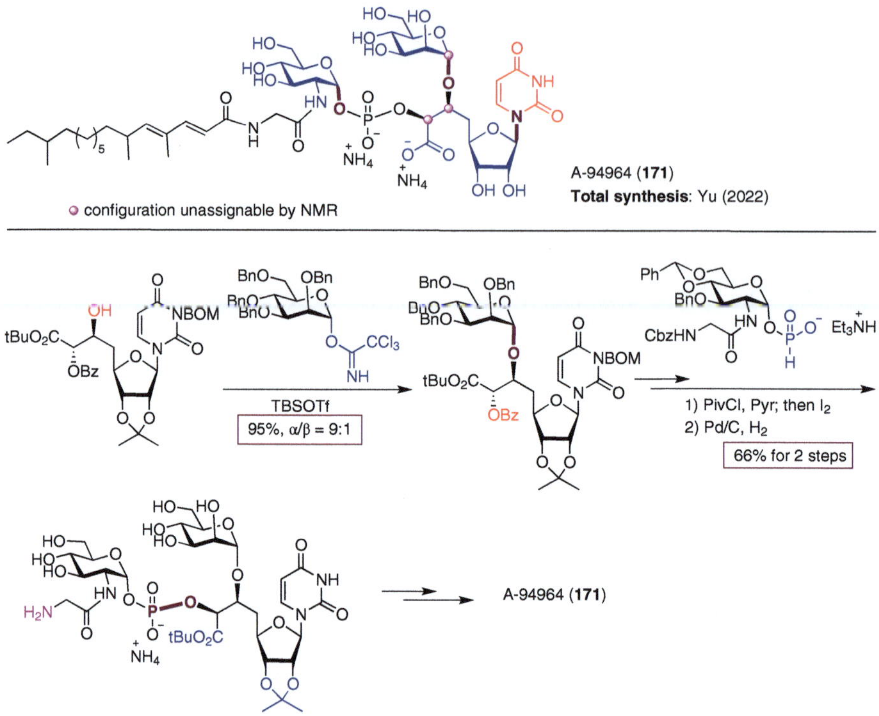


**Figure 6.16** A-94964 (**171**) and the glycosylation steps in the total synthesis by the Yu group.

α-D-*N*-acylglucosaminopyranosyl residue is linked via a phosphoric acid diester linker at C7′.

## Miharamycin B

Miharamycins A and B (**172, 173**) are peptidyl nucleoside antibiotics (PNAs) isolated from *Streptomyces miharaensis* by Niida and coworkers in 1967 (Figure 6.17) [76]. These compounds, along with amipurimycin (**111**), display potent antifungal, antibacterial, and antiviral activities. In particular, they can inhibit the growth of *Pyricularia oryzae*, a plant pathogenic fungus that causes rice blast disease, at 10–20 ppm concentrations. In 2019, the Yu group achieved the total synthesis of amipurimycin (**111**) and established the absolute configuration of miharamycins [42]. Structurally, miharamycins feature a unique nine-carbon pyranosyl amino acid core, decorated with a tetrahydrofuran ring, and a rare 2-aminopurine moiety. In 2022, Wang and coworkers reported the first asymmetric de novo synthesis of miharamycin B in 20 steps, including construction of the 3′-branched saccharide by Achmatowicz rearrangement, palladium-catalyzed *N*-glycosylation, stereocontrolled dihydroxylation, and cyclization of the tetrahydrofuran via a chemoselective *in situ* intramolecular $S_N2$ substitution [77].

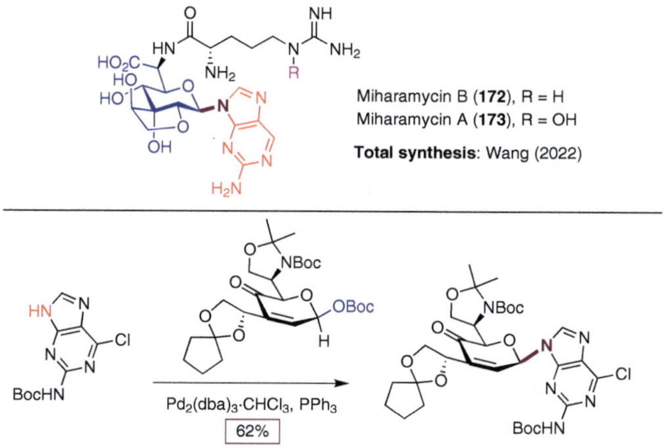


**Figure 6.17** Miharamycin B (**172**) and the glycosylation step in the total synthesis by the Wang group.

## References


1 Yamamoto, K. and Ichikawa, S. (2019). *J. Antibiot.* 72: 924–933.
2 Shiraishi, T. and Kuzuyama, T. (2019). *J. Antibiot.* 72: 913–923.
3 Osada, H. (2019). *J. Antibiot.* 72: 855–864.
4 Niu, G., Li, Z., Huang, P., and Tan, H. (2019). *J. Antibiot.* 72: 906–912.
5 Wiegmann, D., Koppermann, S., Wirth, M. et al. (2016). *Beilstein J. Org. Chem.* 12: 769–795.
6 Huang, R.-M., Chen, Y.-N., Zeng, Z. et al. (2014). *Mar. Drugs* 12: 5817.
7 Lamberth, C. (2005). *Heterocycles* 65: 667–695.
8 Isono, K., Asahi, K., and Suzuki, S. (1969). *J. Am. Chem. Soc.* 91: 7490–7505.
9 Kuzuhara, H., Ohrui, H., and Emoto, S. (1973). *Tetrahedron Lett.* 14: 5055–5058.
10 Ichikawa, S. (2008). *Chem. Pharm. Bull.* 56: 1059–1072.
11 More, J.D. (2007). *Org. Prep. Proced. Int.* 39: 107–133.
12 Ichikawa, S. and Matsuda, A. (2005). *Nucleos. Nucleot. Nucl.* 24: 319–329.
13 Pathak, T. (2002). *Chem. Rev.* 102: 1623–1668.
14 Knapp, S. (1995). *Chem. Rev.* 95: 1859–1876.
15 Takatsuki, A., Arima, K., and Tamura, G. (1971). *J. Antibiot.* 24: 215–223.
16 Tetsuo, S., Hiroaki, S., and Kazuhiro, M. (1983). *Chem. Lett.* 12: 819–822.
17 Suami, T., Sasai, H., Matsuno, K. et al. (1984). *Tetrahedron Lett.* 25: 4533–4536.
18 Suami, T., Sasai, H., Matsuno, K., and Suzuki, N. (1985). *Carbohydr. Res.* 143: 85–96.
19 Danishefsky, S. and Barbachyn, M. (1985). *J. Am. Chem. Soc.* 107: 7761–7762.
20 Danishefsky, S.J., DeNinno, S.L., Chen, S.H. et al. (1989). *J. Am. Chem. Soc.* 111: 5810–5818.
21 Myers, A.G., Gin, D.Y., and Widdowson, K.L. (1991). *J. Am. Chem. Soc.* 113: 9661–9663.

22 Myers, A.G., Gin, D.Y., and Rogers, D.H. (1993). *J. Am. Chem. Soc.* 115: 2036–2038.

23 Myers, A.G., Gin, D.Y., and Rogers, D.H. (1994). *J. Am. Chem. Soc.* 116: 4697–4718.

24 Li, J. and Yu, B. (2015). *Angew. Chem. Int. Ed.* 54: 6618–6621.

25 Yamamoto, K., Yakushiji, F., Matsumaru, T., and Ichikawa, S. (2018). *Org. Lett.* 20: 256–259.

26 Mitachi, K., Mingle, D., Effah, W. et al. (2022). *Angew. Chem. Int. Ed.* 61: e202203225.

27 Uchida, K., Ichikawa, T., Shimauchi, Y. et al. (1971). *J. Antibiot.* 24: 259–262.

28 Ikemoto, N. and Schreiber, S.L. (1990). *J. Am. Chem. Soc.* 112: 9657–9659.

29 Ikemoto, N. and Schreiber, S.L. (1992). *J. Am. Chem. Soc.* 114: 2524–2536.

30 Fujino, H., Fukuda, T., Nagatomo, M., and Inoue, M. (2020). *J. Am. Chem. Soc.* 142: 13227–13234.

31 Arai, M., Haneishi, T., Kitahara, N. et al. (1976). *J. Antibiot.* 29: 863.

32 Newcombe, N.J., Mahon, M.F., Molloy, K.C. et al. (1993). *J. Am. Chem. Soc.* 115: 6430–6431.

33 Ichikawa, S., Shuto, S., and Matsuda, A. (1999). *J. Am. Chem. Soc.* 121: 10270–10280.

34 Hager, D., Paulitz, C., Tiebes, J. et al. (2013). *J. Org. Chem.* 78: 10784–10801.

35 Kirst, H.A., Dorman, D.E., Occolowitz, J.L. et al. (1976). *Abstracts of Papers of 16th Interscience Conference on Antimicrobial Agents and Chemotherapy*, No. 61, Chicago, IL. Washington, DC: ASM.

36 Zhu, Q., Li, J., Ma, J. et al. (2021). *J. Antimicrob. Agents Chemother.* 56: 110–114.

37 Nie, S., Li, W., and Yu, B. (2014). *J. Am. Chem. Soc.* 136: 4157–4160.

38 Wang, J., Gao, J., Guo, T. et al. (2023). *Angew. Chem. Int. Ed.* 62: e202213810.

39 Iwasa, T., Kishi, T., Matsuura, K., and Wakae, O. (1977). *J. Antibiot.* 30: 1–10.

40 Harada, S. and Kishi, T. (1977). *J. Antibiot.* 30: 11–16.

41 Wang, S., Sun, J., Zhang, Q. et al. (2018). *Angew. Chem. Int. Ed.* 57: 2884–2888.

42 Wang, S., Zhang, Q., Zhao, Y. et al. (2019). *Angew. Chem. Int. Ed.* 58: 10558–10562.

43 Igarashi, M., Nakagawa, N., Doi, N. et al. (2003). *J. Antibiot.* 56: 580–583.

44 Hirano, S., Ichikawa, S., and Matsuda, A. (2005). *Angew. Chem. Int. Ed.* 44: 1854–1856.

45 Hirano, S., Ichikawa, S., and Matsuda, A. (2007). *J. Org. Chem.* 72: 9936–9946.

46 Gopinath, P., Wang, L., Abe, H. et al. (2014). *Org. Lett.* 16: 3364–3367.

47 Nakamura, H., Tsukano, C., Yoshida, T. et al. (2019). *J. Am. Chem. Soc.* 141: 8527–8540.

48 Chida, N., Koizumi, K., Kitada, Y. et al. (1994). *J. Chem. Soc., Chem. Commun.* 111–113.

49 Ghosh, A.K. and Wang, Y. (1999). *J. Org. Chem.* 64: 2789–2795.

50 Dondoni, A., Junquera, F., Merchant, F.L. et al. (1995). *J. Chem. Soc., Chem. Commun.* 2127–2128.

51 Dondoni, A., Franco, S., Junquera, F. et al. (1997). *J. Org. Chem.* 62: 5497–5507.

52 Fujino, H., Nagatomo, M., Paudel, A. et al. (2017). *Angew. Chem. Int. Ed.* 56: 11865–11869.
53 Isono, K., Crain, P.F., and McCloskey, J.A. (1975). *J. Am. Chem. Soc.* 97: 943–945.
54 Danishefsky, S. and Hungate, R. (1986). *J. Am. Chem. Soc.* 108: 2486–2487.
55 Kozaki, S., Sakanaka, O., Yasuda, T. et al. (1988). *J. Org. Chem.* 53: 281–286.
56 Hanessian, S., Kloss, J., and Sugawara, T. (1986). *J. Am. Chem. Soc.* 108: 2758–2759.
57 Goldin, S., Fisher, J., Kobayashi, K., et al. (1995). Fucosylated Guanosine Disulfates as Excitatory Amino Acid Antagonists. U.S. Patent No. 5,438,130, issued August 1.
58 McCormick, J., Li, Y., McCormick, K. et al. (1999). *J. Am. Chem. Soc.* 121: 5661–5665.
59 Benner, J.P., Boehlendorf, B.G.H., Kipps, M.R., et al. (2003). WO 03/062242, CAN 139: 132519.
60 Hanessian, S., Marcotte, S., Machaalani, R., and Huang, G. (2003). *Org. Lett.* 5: 4277–4280.
61 Yamaguchi, H., Sato, S., Yoshida, S. et al. (1986). *J. Antibiot.* 39, 1047–1053.
62 Knapp, S. and Nandan, S.R. (1994). *J. Org. Chem.* 59: 281–283.
63 Wang, Y., Siricilla, S., Aleiwi, B.A., and Kurosu, M. (2013). *Chem. Eur. J.* 19: 13847–13858.
64 Kurosu, M., Li, K., and Crick, D.C. (2009). *Org. Lett.* 11: 2393–2396.
65 He, H., Xu, L., Sun, R. et al. (2021). *Chem. Sci.* 12: 5143-5151.
66 McDonald, L.A., Barbieri, L.R., Carter, G.T. et al. (2002). *J. Am. Chem. Soc.* 124: 10260–10261.
67 Tanino, T., Ichikawa, S., Shiro, M., and Matsuda, A. (2010). *J. Org. Chem.* 75: 1366–1377.
68 Mitachi, K., Aleiwi, B.A., Schneider, C.M. et al. (2016). *J. Am. Chem. Soc.* 138: 12975–12980.
69 DeBoer, C., Caron, E.L., and Hinman, J.W. (1953). *J. Am. Chem. Soc.* 75: 499–500.
70 Stevens, C.L., Nemec, J., and Ransford, G.H. (1972). *J. Am. Chem. Soc.* 94: 3280–3281.
71 Sugimura, H. and Watanabe, K.-I. (2001). *Synth. Commun.* 31: 2313–2321.
72 Fu, J., Xu, P., and Yu, B. (2021). *Chin. J. Chem.* 39: 2679–2684.
73 Fu, J., Laval, S., and Yu, B. (2018). *J. Org. Chem.* 83: 7076–7084.
74 Murakami, R., Fujita, Y., Kizuka, M. et al. (2008). *J. Antibiot.* 61: 537–544.
75 Shao, X., Zheng, C., Xu, P. et al. (2022). *Angew. Chem. Int. Ed.* 61: e202200818.
76 Tsuruoka, T., Yumoto, H., Ezaki, N., and Niida, T. (1967). *Sci. Rep. Meiji Seika Kaisha* 9: 1–4.
77 Huang, W., Fan, S., Gao, J. et al. (2022). *Angew. Chem. Int. Ed.* 61: e202204907.

# 7

# Peptide Glycosides

Peptide glycosides, also referred to as glycopeptides, are peptides that incorporate carbohydrate moieties (glycans) covalently bonded to the side chains of the amino acid residues constituting the peptide [1–3]. These glycan attachments can be established through various types of glycosylations, including *O*-glycosylation, *N*-glycosylation, *S*-glycosylation, or *C*-glycosylations. Over the recent decades, the significance of glycosides located on cell surfaces (associated with membrane proteins or lipids) and those linked to proteins (glycoproteins) has been widely recognized in biological processes. The synthesis of glycopeptides not only serves as a valuable tool for researchers to investigate the role of glycans in nature but also yields products with potential therapeutic and biotechnological applications.

Glycopeptide antibiotics have emerged as significant natural products obtained from bacterial metabolites and are widely recognized for their clinical importance, which exhibits a broad spectrum of antibacterial activity against gram-positive pathogens while maintaining manageable side effects. One renowned glycopeptide is vancomycin (**1**, Figure 7.1) [4], which was approved for clinical use in 1958. Vancomycin displays potent activity against a range of gram-positive bacteria, with *in vitro* minimum inhibitory concentration (MIC) ranging from 0.25 to 10.0 μg/ml against *Staphylococcus aureus*. Another well-known naturally occurring glycopeptide is teicoplanin (e.g. teicoplanin $A_2$-2, **2**), which was discovered at the Lepitit Research Center (Milan, Italy) in 1978 [5]. Teicoplanin (Targocid) was first approved in Europe in 1988 and is currently available worldwide for the treatment of various infections. Due to the increased occurrence of resistance and the need for more effective treatments for gram-positive infections, second-generation semisynthetic lipoglycopeptide derivatives such as telavancin, dalbavancin, and oritavancin have been clinically developed, which offer broader spectra of activity and improved pharmacokinetic properties.

Glycopeptides have gained significant interest as attractive targets for drug design and development, while the addition of glycan moieties to peptides serves as a strategy for optimizing the bioavailability and metabolic stability of therapeutic peptides [6–8]. The incorporation of carbohydrates into the peptide sequence could introduce changes in their physiological properties, leading to improved biodistribution within tissues, which consequently facilitates membrane penetration and enhances

*Carbohydrate Chemistry in the Total Synthesis of Naturally Occurring Glycosides*,
First Edition. Biao Yu and Xiaoyu Yang.
© 2024 WILEY-VCH GmbH. Published 2024 by WILEY-VCH GmbH.

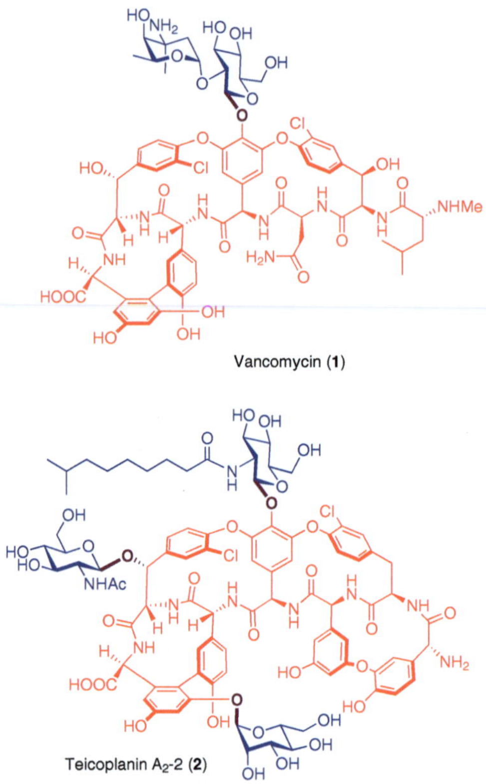


**Figure 7.1** Representative glycopeptide natural products.

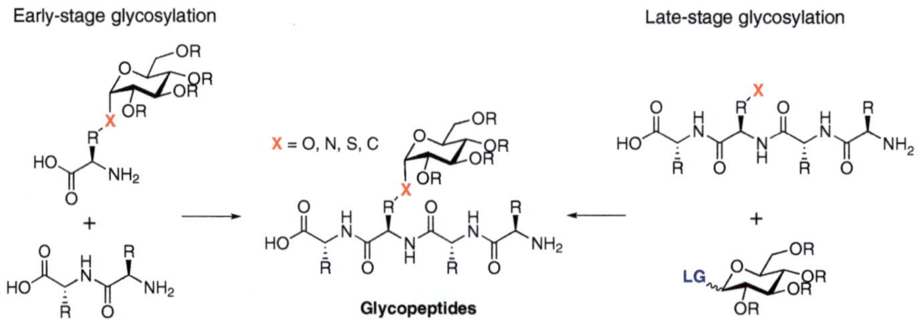


**Figure 7.2** Two major strategies for the synthesis of glycopeptide natural products.

metabolic stability. The synthesis of peptide glycosides is a complex process that necessitates precise control of regio- and stereoselectivity. Two major strategies have been used for the synthesis of glycopeptide natural products (Figure 7.2). The first strategy involves early-stage glycosylation, where the saccharide is introduced at an early stage onto the side chain of an amino acid, which then serves as a building block in subsequent peptide synthesis. The second strategy demands late-stage glycosylation, wherein the peptide is initially constructed, followed by glycosylation at the designated side chain of the peptide.

## Vancomycin

Vancomycin (**1**, Figure 7.3) was initially isolated from the fermentation broth of *Streptomyces orientalis* (later renamed *Nocardia orientalis* and eventually reclassified as *Amycolatopsis orientalis*) originating from a soil sample in the Borneo jungles in the early 1950s at Eli Lilly and received US Food and Drug Administration (FDA) approval in 1958 [4]. However, its structure was not elucidated until 25 years later through chemical degradation studies, NMR analysis, and pioneering X-ray crystallography. Vancomycin has been used as an intravenous treatment for various bacterial infections, including complicated skin infections, bloodstream infections, endocarditis, bone and joint infections, as well as meningitis caused by methicillin-resistant *S. aureus*. Structurally, vancomycin consists of a strained cyclic peptide unit with intricate rigid tricyclic heptapeptide structures, unique glycosylation patterns, and centers of axial or planar chirality (atropisomers). Since the determination of its structure, extensive synthetic efforts have been undertaken using both semisynthetic and total synthesis approaches [9]. These studies have served as the basis for ongoing investigations into the structure–function relationship, contributing to the understanding of their mechanism(s) of action and uncovering unforeseen roles for non-natural functionalities. These advancements have paved the way for the discovery of improved glycopeptide antibiotics based on vancomycin.

**Synthesis of the model aryl glycosides of vancomycin (1).** In 1992, Danishefsky and coworkers reported the synthesis of the aryloxyl carbohydrate segment of vancomycin (**1**, Scheme 7.1) [10]. The synthesis began with the treatment of 3,4,6-tri-*O*-benzyl-D-glucal (**3**) with 3,3-dimethyldioxirane (DMDO), yielding the 2α,3α-oxirane with high stereoselectivity, which was followed by addition of 2,6-di-methoxy phenol under alkaline conditions through a $S_N2$ type mechanism to give aryl 2α-hydroxyl-β-glycoside **4** in 67% yield with high stereoselectivity. The vancosamine-based glycal **5** was readily obtained by degradation of vancomycin followed by several transformation steps. By reacting a benzene solution of 2,6-dimethoxyphenyl β-glucoside **4** with an excess of glycal **5** in the presence

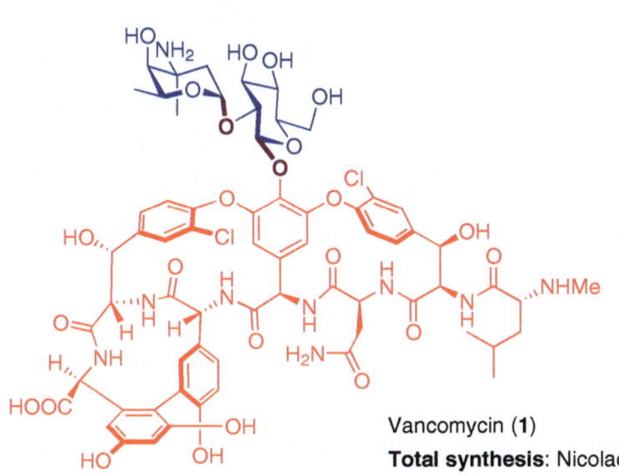

- Strained tricyclic heptapeptide
- Biaryl and diaryl ether linkage
- Unusual axial or planar chirality
- Rare glycosylation pattern

Vancomycin (**1**)
**Total synthesis**: Nicolaou (1999), Kahne (1999), Boger (2020)

**Figure 7.3** Vancomycin (**1**) and its structure features.

of camphorsulfonic acid (CSA), the desired disaccharide **6** was obtained with α-selectivity in 51% yield; 48% of the starting aryl glucoside was recovered, due to the hindered steric environment.

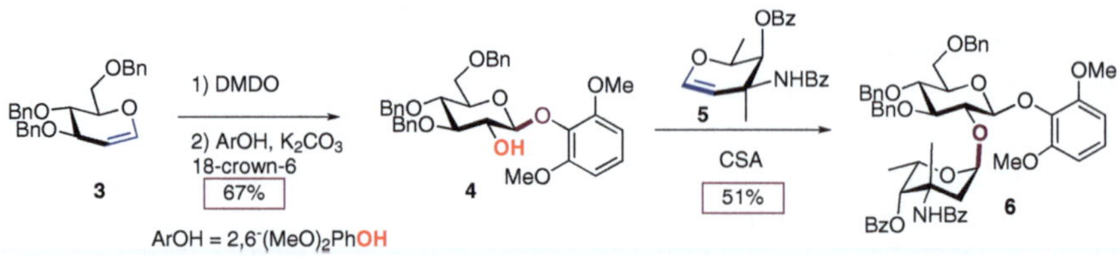


**Scheme 7.1** Synthesis of vancomycin disaccharide **6** by Danishefsky and coworkers.

**Total synthesis of vancomycin (1) by the Nicolaou group.** In 1999, Nicolaou and coworkers reported the first total synthesis of vancomycin (**1**, Scheme 7.2) [11–16]. The synthesis of the vancomycin aglycon began with a Suzuki coupling between aryl iodide **7** and boronic acid **8** enabled by $Pd(PPh_3)_4$ catalyst, which provided the axially chiral biaryl **9** in 82% yield with a 1 : 1 ratio of the atropisomers. After introducing the azido group, the coupling of carboxylic acid **10** with amine **11** gave access to amide **12**, which was subsequently coupled with amino acid **13** to yield amide **14**. The intramolecular Ullmann-type C–O coupling of **14**, facilitated by $CuBr{\cdot}SMe_2$, generated the cyclized product **15** with a 1 : 1 ratio of the atropisomers. With the product possessing the desired axial chirality, the intramolecular amide formation afforded the bicyclic product **16**. The condensation of **16** with tripeptide **17** readily produced heptapeptide **18**. Finally, another intramolecular Ullamnn-type C–O coupling of **18** resulted in the desired tricyclic product **19**, albeit favoring the undesired atropisomer. Fortunately, heating the undesired atropisomer at 140 °C yielded a 1 : 1 mixture of the two atropisomers, which could be readily separated by column chromatography.

The final stage of the total synthesis involved the installation of the disaccharide moiety and global deprotection (Scheme 7.3). Through several steps of functional group interconversion and protecting-group manipulation, the suitably protected vancomycin aglycone **20** was coupled with D-glucosyl trichloroacetimidate **21** in the presence of $BF_3{\cdot}Et_2O$, resulting in the formation of monosaccharide **22** in 82% yield with high β selectivity. Next, the C2 hydroxyl group of the glucose residue was selectively liberated, which was then coupled with an excess of vancosaminyl fluoride **23** in the presence of $BF_3{\cdot}Et_2O$. This provided the protected vancomycin derivative **24** with a yield of 84% and high α-selectivity, likely facilitated by the involvement of the C4 acetoxy group. Finally, sequential removal of the TBS groups using HF, the *O*-Ac groups using $K_2CO_3$, the *N*-Cbz groups using Raney Ni, and the methyl ester using LiOH treatment (65% for four steps) successfully accomplished the first synthesis of vancomycin (**1**), whose NMR and HPLC spectrum were identical to those of the authentic sample.

**Synthesis of vancomycin (1) from the aglycone by the Kahne group.** Kahne and coworkers reported their synthesis of vancomycin (**1**) starting from the aglycone, utilizing the sulfoxide glycosylation reaction (Scheme 7.4) [17, 18].

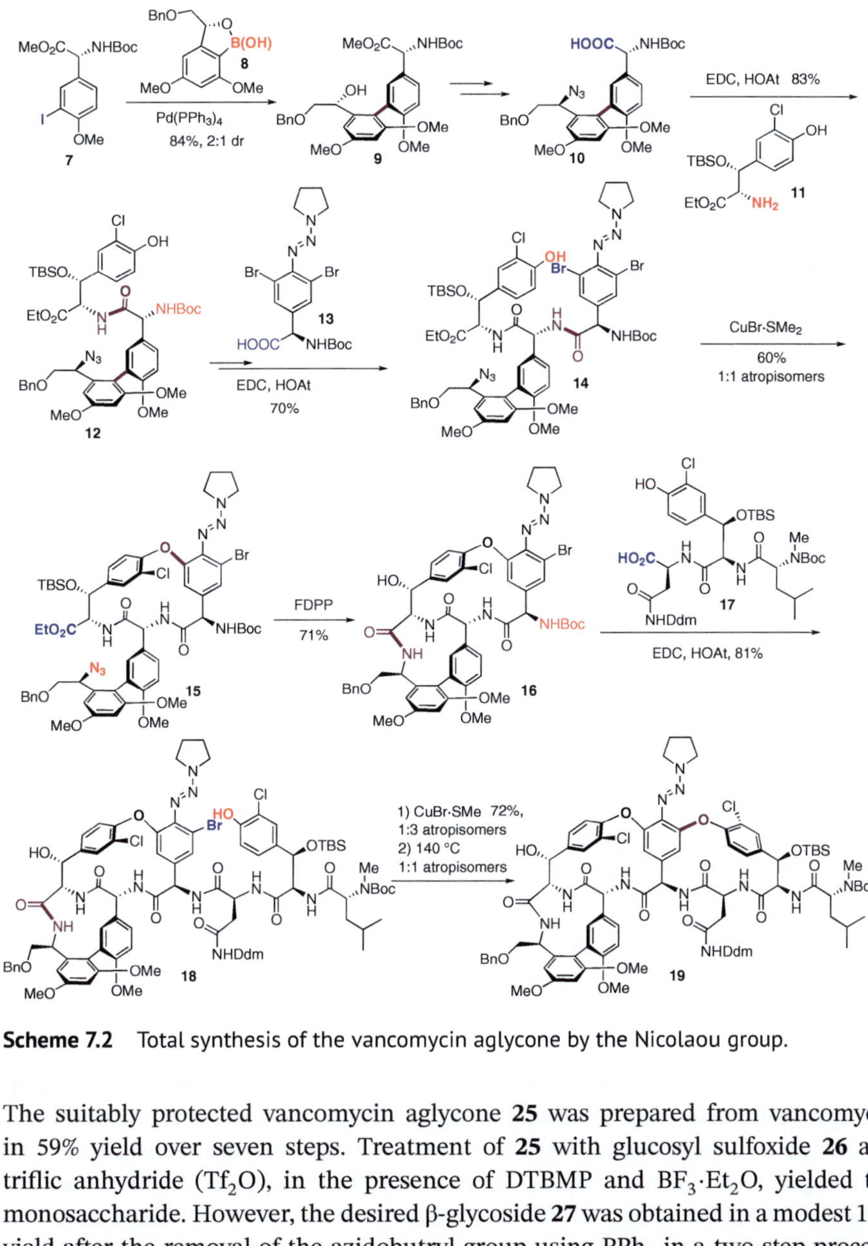

**Scheme 7.2** Total synthesis of the vancomycin aglycone by the Nicolaou group.

The suitably protected vancomycin aglycone **25** was prepared from vancomycin in 59% yield over seven steps. Treatment of **25** with glucosyl sulfoxide **26** and triflic anhydride ($Tf_2O$), in the presence of DTBMP and $BF_3 \cdot Et_2O$, yielded the monosaccharide. However, the desired β-glycoside **27** was obtained in a modest 13% yield after the removal of the azidobutryl group using $PPh_3$ in a two-step process, highlighting the limitation of this method. The coupling of **27** with vancosaminyl sulfoxide **28** under the promotion of $Tf_2O$ and $BF_3 \cdot Et_2O$ afforded disaccharide **29** in 60% yield. Finally, removal of the six acetyl groups with hydrazine and the four allyl groups and two Alloc groups with $Bu_3SnH/Pd$ accomplished the synthesis.

**Total synthesis of vancomycin (1) by the Boger group.** The Boger group has long dedicated their efforts to the total synthesis of vancomycin and analogs and the exploration of structure–activity relationships (SAR) [19–22]. In 2020, Boger and

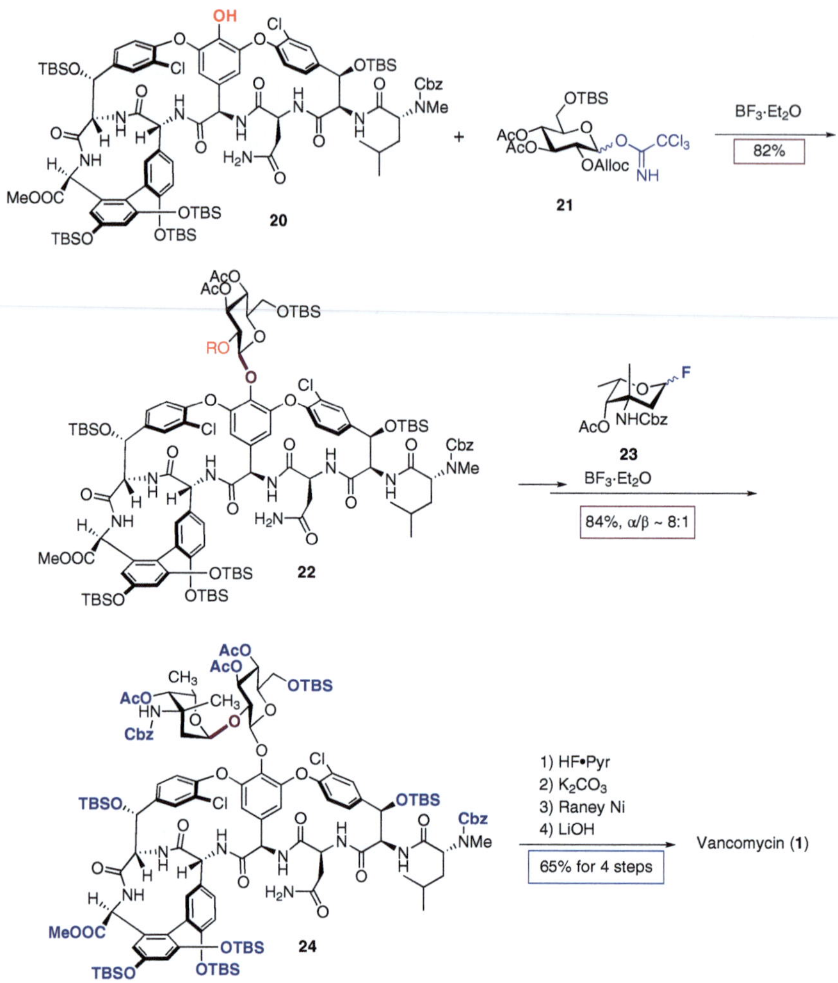

**Scheme 7.3** Total synthesis of vancomycin (**1**) by Nicolaou and coworkers.

coworkers reported the new-generation total synthesis of vancomycin in 19 steps (longest linear sequence, LLS), employing a kinetically controlled diastereoselective approach to incorporate all three elements of atropisomerism (Scheme 7.5) [23]. The asymmetric Suzuki coupling between aryl bromide **30** and boronic acid **31**, enabled by $Pd(OAc)_2$ and (*R*)-BINAP, afforded the axially chiral biaryl **32** in 93% yield with >20 : 1 dr. The subsequent intramolecular lactam formation afforded the cyclic product **33**, which was then condensed with amino acid **34** to generate amide **35**. Under basic conditions, the intramolecular $S_NAr$ cyclization of **35** produced the bicyclic product **36** in 84% yield with 8 : 1 diastereoselectivity. The coupling of **36** with tripeptide **37** readily afforded heptapeptide **38**, which, upon treatment of $Cs_2CO_3$, afforded the tricyclic product **39** in 82% yield with 14 : 1 diastereoselectivity. Finally, a series of functional group transformations and global deprotection on **39** afforded the vacomycin aglycone **40**. For the introduction of the protecting-group

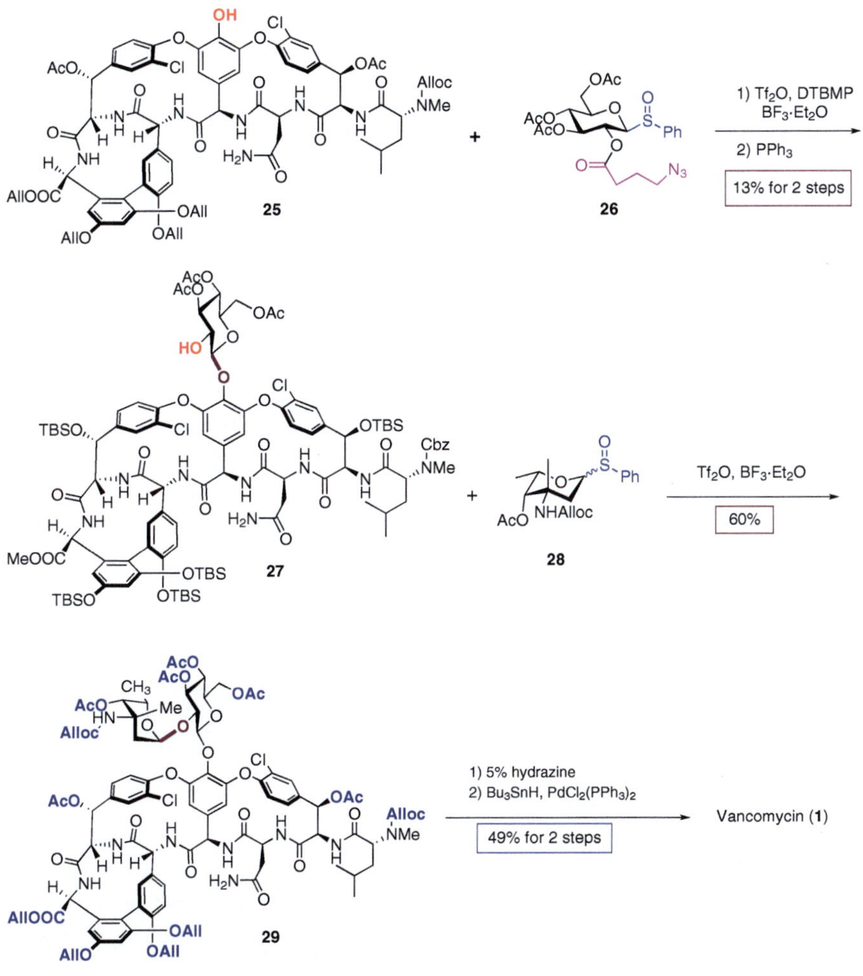


**Scheme 7.4** Synthesis of vancomycin (**1**) from the aglycone by Kahne and coworkers.

free disaccharide moiety, a more efficient enzymatic glycosylation was employed, catalyzed by the glycosyltransferases GtfE and GtfD, which achieved a remarkable 80% yield in just two steps and accomplished the total synthesis of vancomycin (**1**). This approach not only provided a pathway for the large-scale preparation of pocket-modified vancomycin analogs but also allowed for a direct investigation of the underlying mechanism of resistance to vancomycin.

## Bleomycin A2

The bleomycins, a class of glycopeptide antitumor antibiotics, were first isolated from the fermentation broths *Streptomyces verticellus* by Umezawa and coworkers in 1966 [24]. Bleomycin $A_2$ (**41**, Figure 7.4), distinguished by a structural variation in its cationic *C*-terminus, is the predominant component (making up 70%) of

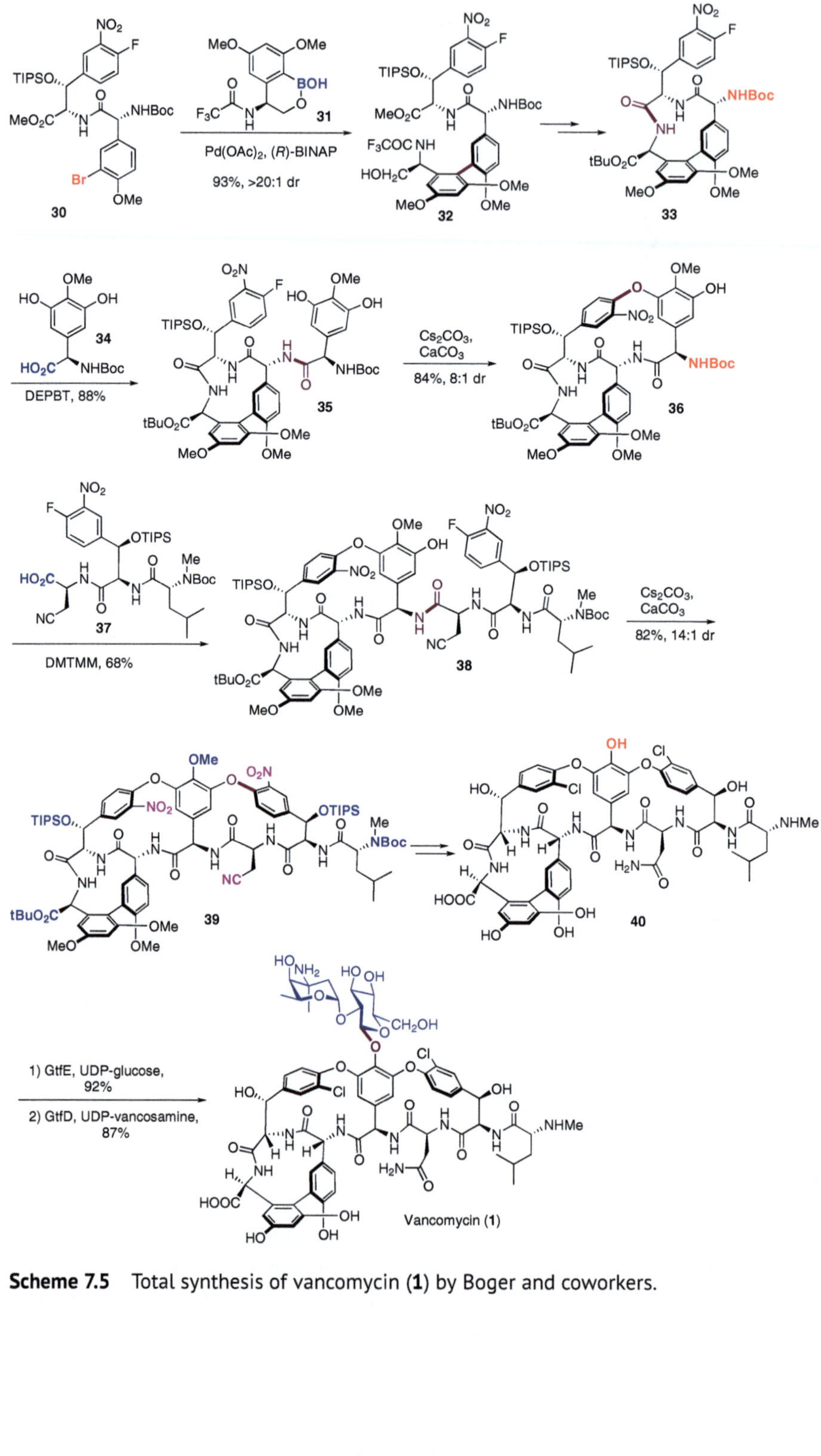

**Scheme 7.5** Total synthesis of vancomycin (**1**) by Boger and coworkers.

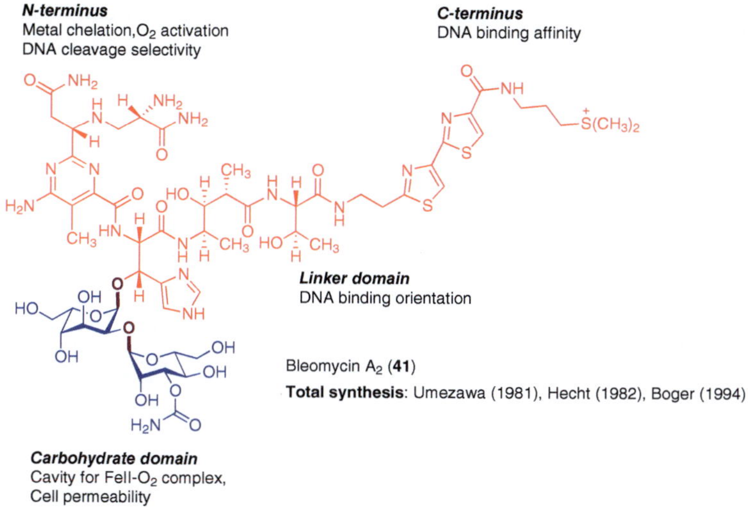


**Figure 7.4** Bleomycin $A_2$ (**41**).

the clinically used anticancer drug Blenoxane, which is utilized in the treatment of Hodgkin's lymphoma; carcinomas, of the skin, head, and neck; and testicular cancers. It is believed that bleomycin $A_2$ exerts its biological effects by binding to DNA and causing its degradation, a mechanism dependent on metal ions and oxygen. Specifically, it selectively cleaves double-stranded DNA at 5′-GC or 5′-GT sites through minor groove C4′-H atom abstraction and subsequent fragmentation of the deoxyribose backbone. Each structural component of bleomycin $A_2$ (**41**) plays a crucial role in its biological activities. Extensive early studies on bleomycin $A_2$ and its derivatives have elucidated the significance of the *N*-terminal pyrimidine as the domain responsible for metal binding. The *C*-terminus, encompassing the bithiazole and positively charged sulfonium salt, contributes significantly to the DNA binding affinity of bleomycin A2. In contrast, the impact of the carbohydrate domain has received less comprehensive examination despite its known enhancement of biological potency and efficacy. While the carbohydrate domain does not directly influence the selectivity of DNA cleavage, it does promote greater efficiency in the cleavage process. Additionally, it has been suggested that the carbohydrate domain facilitates the cellular uptake of bleomycin $A_2$.

**Total synthesis of bleomycin $A_2$ (41) by the Umezawa group.** In 1982, Umezawa and coworkers reported the first total synthesis of bleomycin $A_2$ (**41**, Scheme 7.6) [25–27]. Their synthesis began with the condensation of D-mannopyranosyl bromide **43** with 6-azido D-gluocofuranose **42** under the promotion of $Hg(CN)_2$, yielding disaccharide **44** in 52% yield with the correct α-configuration, which was attributed to the influence of the C2 acetate group. Steps of transformations, including the photolysis of the $-CH_2N_3$ group of **44** to an imine group, led to the reconstruction of the desired L-gluocopyranose scaffold in disaccharide **45**. Subsequent conversion of **45** into a suitably protected disaccharide bromide **46** enabled the coupling with pentapeptide **47** using $Hg(CN)_2$, thereby generating the targeted glycopeptide **48**. Amide formation between **48** and pyrimidoblamic acid **49** provided the protected form of bleomycin $A_2$ **50**. Finally, global deprotection of the eight *O*-Ac groups with NaOMe and one *N*-Boc group with TFA accomplished the synthesis of bleomycin $A_2$ (**41**). However, no yields were reported for the steps subsequent to the glycosylation. Notably, direct condensation of disaccharide bromide **46** with deglyco-bleomycin $A_2$ failed to produce the desired glycosylated product due to side reactions.

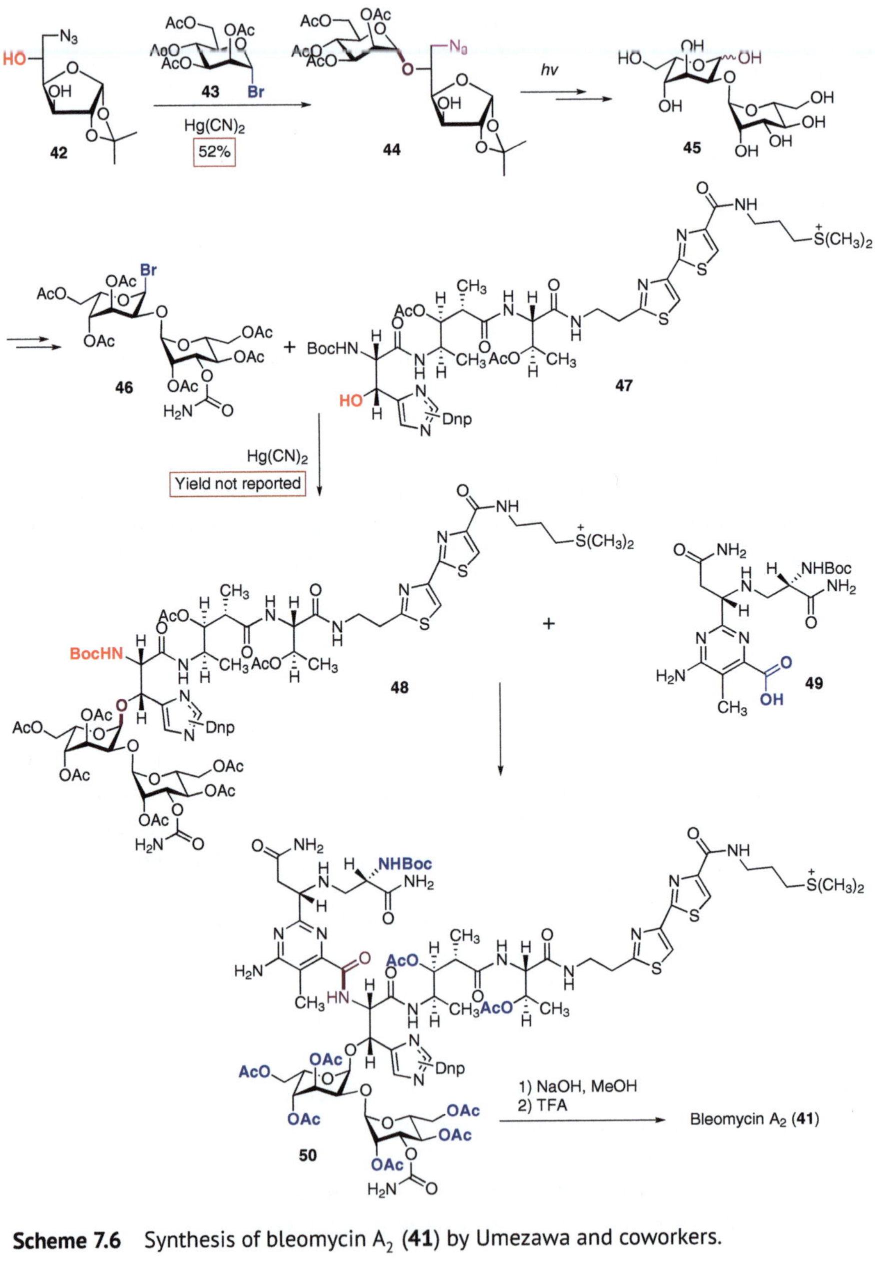


**Scheme 7.6** Synthesis of bleomycin $A_2$ (**41**) by Umezawa and coworkers.

**Total synthesis of bleomycin $A_2$ (41) by Hecht and coworkers.** In the same year of 1982, Hecht and coworkers reported their synthesis of bleomycin $A_2$ (**41**, Scheme 7.7) [28–30]. The synthesis commenced with the coupling of 1,6-anhydro-β-L-gluopyranose **51** with D-mannopyranosyl bromide **52** under the promotion of AgOTf, yielding the desired disaccharide **53** in 68% yield, which was subsequently converted into gluopyranosyl chloride **54** through two additional steps. Coupling of **54** with histidine derivative **55** with the assistance of AgOTf gave **56** in 25% yield, whose configuration was confirmed by analyzing the $^1$H NMR spectrum of its anomeric proton. Sequential amide formation between the disaccharide-containing histidine **56** and (*2S,3S,4R*)-4-amino-3-hydroxy-2-methylvalerate **57**, "tripeptide S" derivative **58**, and pyrimidoblamic acid **59** afforded the protected bleomycin demethyl $A_2$ **60**. Finally, removal of the *O*-Ac groups with NaOH and *N*-Boc group with TFA resulted in bleomycin demethyl $A_2$, which was readily converted to bleomycin $A_2$ (**41**) according to a known method.

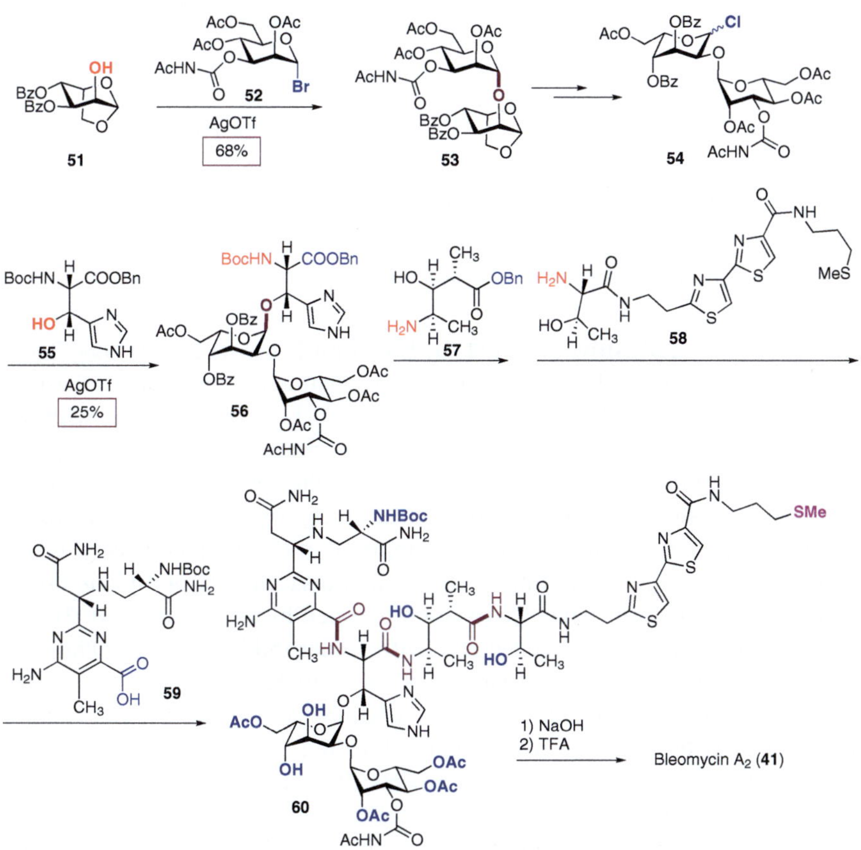


**Scheme 7.7** Total synthesis of bleomycin $A_2$ (**41**) by Hecht and coworkers.

In 1998, Hecht and coworkers reported an improved synthesis for the construction of the histidine disaccharide scaffold (Scheme 7.8) [31]. The process involved treating glycosyl acceptor **61** with D-mannopyranosyl trichloroacetimidate **62** in the presence of catalytic amount of $BF_3{\cdot}OEt_2$, which provided the 1,2-*trans* linked disaccharide **63** in 86% yield as a > 10 : 1 mixture of the α/β anomers. Accompanied by the selective removal of the anomeric acetyl group, the resulting disaccharide trichloroacetimidate **64** was coupled with histidine congener **65** using TMSOTf as a catalyst, resulting in the formation of disaccharide **66** in 44% yield, which represented the key intermediate for the total synthesis of bleomycin $A_2$ (**41**, Scheme 7.7).

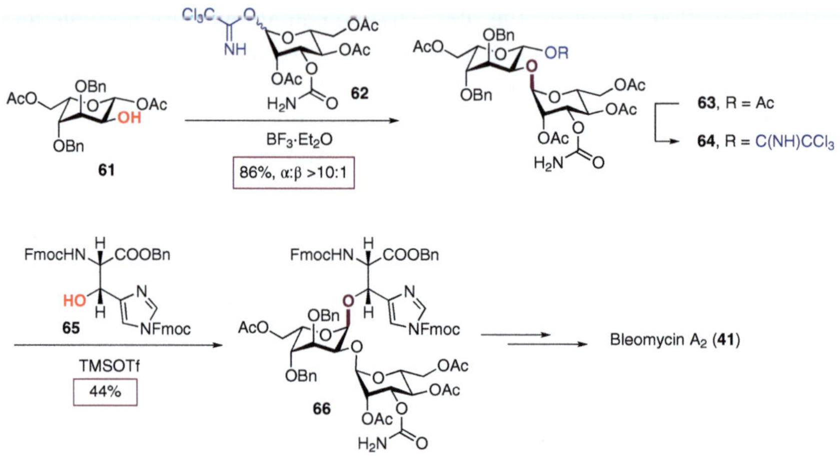


**Scheme 7.8** Improved synthesis of bleomycin $A_2$ (**41**) by Hecht and coworkers.

**Total synthesis of bleomycin $A_2$ (41) by Boger and coworkers.** In 1994, Boger and coworkers reported their comprehensive investigation into the total synthesis of bleomycin $A_2$ (**41**) and analogs (Scheme 7.9) [32–36]. The synthesis began with the coupling of C2-OH L-gluopyranose **67** and D-mannopyranosyl diphenylphosphate **68** under the promotion of TMSOTf, which provided the desired disaccharide **69** in 93% yield within a short reaction time (0 °C, 10 minutes). Switching the *O*-Bn groups into *O*-Ac groups was followed by selective removal of the anomeric acetyl group and phosphate formation, resulting in the desired disaccharide phosphate donor **70** in the β-anomeric form. The glycosylation between histidine derivative **71** and disaccharide phosphate **70** proceeded efficiently under mild conditions (TMSOTf, 25 °C, 30 minutes), affording the desired product **72** in 63% yield with >13 : 1 α/β selectivity. The authors attributed the high diastereoselectivity of this glycosylation to the low reactivity of the glycosyl acceptor, favoring the formation of the more stable α-anomer, as well as mechanistic aspects characteristic of a glycosyl phosphate donor, promoting inversion of stereochemistry at the reacting anomeric center. Following the removal of *O*-Ac groups on the disaccharide moiety, the sequential coupling of **73** with the free amine of tetrapeptide S **74** and pyrimidoblamic acid **75** led to the formation of product **76**. Finally, through the deprotection of the *N*-Boc and *N*-trityl group using TFA treatment, the total synthesis of bleomycin $A_2$ (**41**) was achieved.

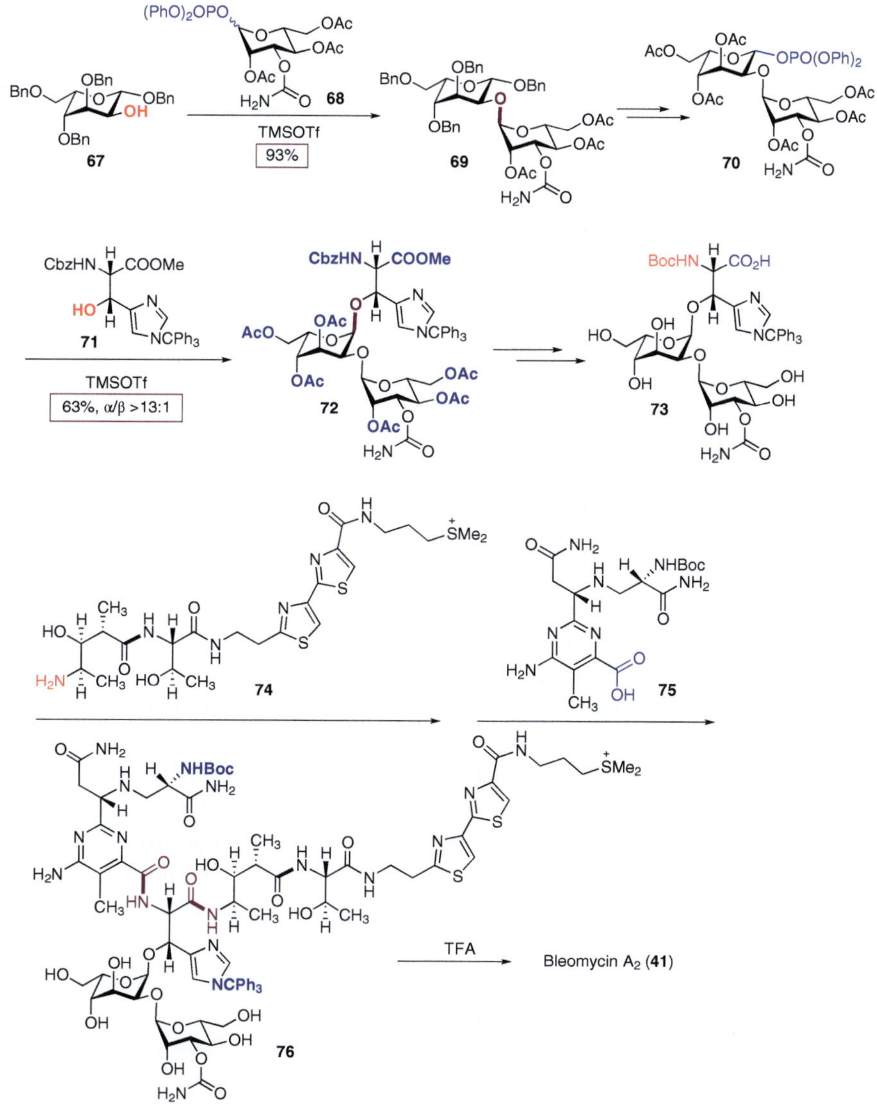

**Scheme 7.9** Total synthesis of bleomycin $A_2$ (**41**) by Boger and coworkers.

## Mannopeptimycin

Mannopeptimycins (MMPs), such as mannopeptimycin α (**77**) and β (**78**), represent a novel class of glycopeptides, which were originally discovered as an antibiotic complex from the culture broth of *Streptomyces hygroscopicus* LL-AC98 in the 1950s (Figure 7.5) [37]. However, their structures were only elucidated in 2002 through nuclear magnetic resonance (NMR) and chemical degradation studies. The MPPs consist of a cyclic hexapeptide core comprised of alternating L- and D-α-amino acids (αAAs). Notably, these αAA units include unique features such

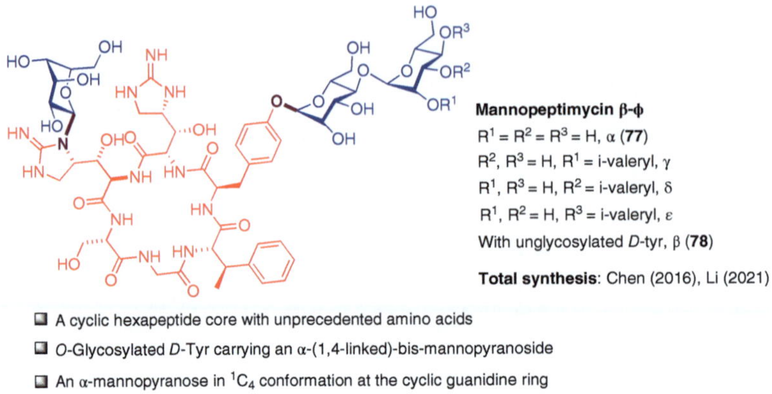


**Figure 7.5** Mannopeptimycins and their structural features.

as a pair of unprecedented β-hydroxyenduracididine, an L-β-methylated Phe, and an *O*-glycosylated D-Tyr carrying an α-(1,4)-linked-bis-mannopyranosside. More strikingly, it has been proposed that the D-βhEnd unit bears an α-mannopyranose in $^{1}C_{4}$ conformation at the most hindered N atom on the cyclic guanidine ring. This *N*-glycosylated guanidine motif is unparalleled in any other natural product. The mannopeptimycins exhibited potent antibiotic activity against a wide range of gram-positive multidrug-resistant pathogens, including methicillin-resistant *S. aureus* (MRSA) and vancomycin-resistant *Enterococci* (VRE).

**Synthesis of tyrosine bis-manno-disaccharide by the O'Doherty group.** In 2006, O'Doherty and coworkers reported the synthesis of the C-4′ acylated 1,4-α,α-manno,manno-disaccharide fragment of mannopeptimycin α from D-tyrosine using a palladium-catalyzed glycosylation strategy (Scheme 7.10) [38]. Their syntheses began with the coupling of protected D-tyrosine **79** with pyranone

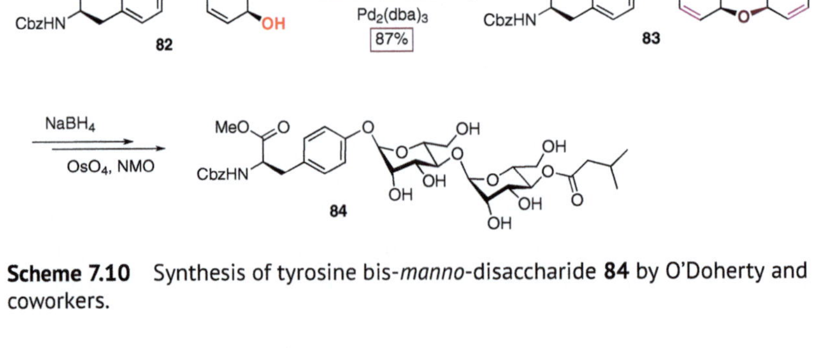


**Scheme 7.10** Synthesis of tyrosine bis-*manno*-disaccharide **84** by O'Doherty and coworkers.

**80** under the catalysis of $Pd_2(dba)_3·CHCl_3$, which afforded the glycosylated product **81** with complete α-selectivity in 92% yield. A diastereoselective 1,2-reduction of enone **81** using $NaBH_4$ afforded allylic alcohol **82** as a single diastereomer, which was followed by an iterative glycosylation between C-4 alcohol and pyranone **80** to give the 1,4-linked-α-bis pyranone **83** in good yield and nearly complete stereoselectivity. The sequential stereoselective reduction of the ketone group, introduction of the isovalerate group, dihydroxylations, and removal of the TBS-ethers protecting groups led to tyrosine α-1,4-linked-bis-manno-disaccharide **84**.

**Total synthesis of mammopeptimycin α and β by the Chen group.** After the seminal work on total synthesis and stereochemistry revision of mannopeptimycin aglycone by Fuse et al. in 2014 [39], the total synthesis of mannopeptimycins α (**77**) and β (**78**) was ultimately accomplished by Chen and coworkers in 2016, incorporating fully elaborated *N*- and *O*-linked sugars [40]. The primary challenge in the synthesis of MPP lay in the preparation of a suitable *N*-α-mannosyl-D-βhEnd unit (Scheme 7.11). Difficulties arose due to the limited compatibility with Lewis acid-promoted conditions, steric hindrance surrounding the *N*-glycosylation site on the cyclic guanidine ring, and the delicate structure of the D-βhEnd substrate. Extensive screening of various glycosyl donors, such as glycosyl trichloroacetimidate, glycosyl sulfide, and glycosyl bromide, revealed that the gold(I)-catalyzed glycosylation with glycosyl *ortho*-alkynylbenzoate donor was the most efficient for achieving the desired *N*-α-mannosylation. The unique π acid activation mode offered by this method, orthogonal to the Lewis basic guanidine NH, facilitated efficient *N*-mannosylation compared to other glycosylation approaches. The coupling of *N*-mannosylation of D-βhEnd **85** with *ortho*-alkynylbenzoate **86**, under the catalysis of $PPh_3AuNTf_2$ (20 mol%) in toluene at 65 °C, proceeded efficiently to give the desired product **87** in 86% isolated yield and complete α stereoselectivity on a gram scale. Subsequent amide coupling with serine derivative **88** and L-βhEnd **89** readily furnished tripeptide **90**, wherein the *N*-linked mannose residue remained intact throughout the process.

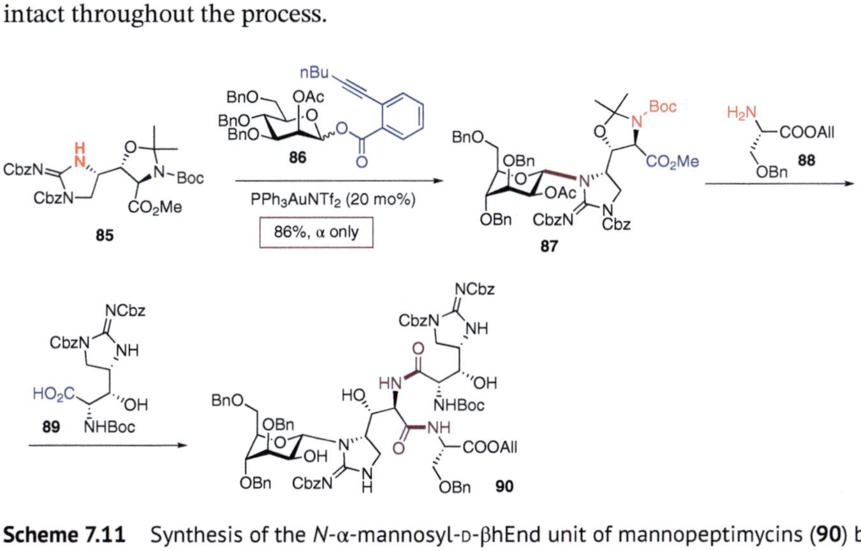


**Scheme 7.11** Synthesis of the *N*-α-mannosyl-D-βhEnd unit of mannopeptimycins (**90**) by Chen and coworkers.

On the other hand, $BF_3{\cdot}OEt_2$-promoted glycosylation between D-mannsoyl acceptor **91** and D-mannsoyl trichloroacetimidate donor **92** successfully yielded the 1,4-linked di-mannose **93** with excellent yield and α selectivity (Scheme 7.12). After the conversion of **93** into the corresponding trichloroacetimidate **94**, the $BF_3{\cdot}OEt_2$ promoted *O*-glycosylation of **94** with tyrosine **95,** resulting in the desired glycosylated product **96** in excellent yield and α-selectivity. Subsequent amide

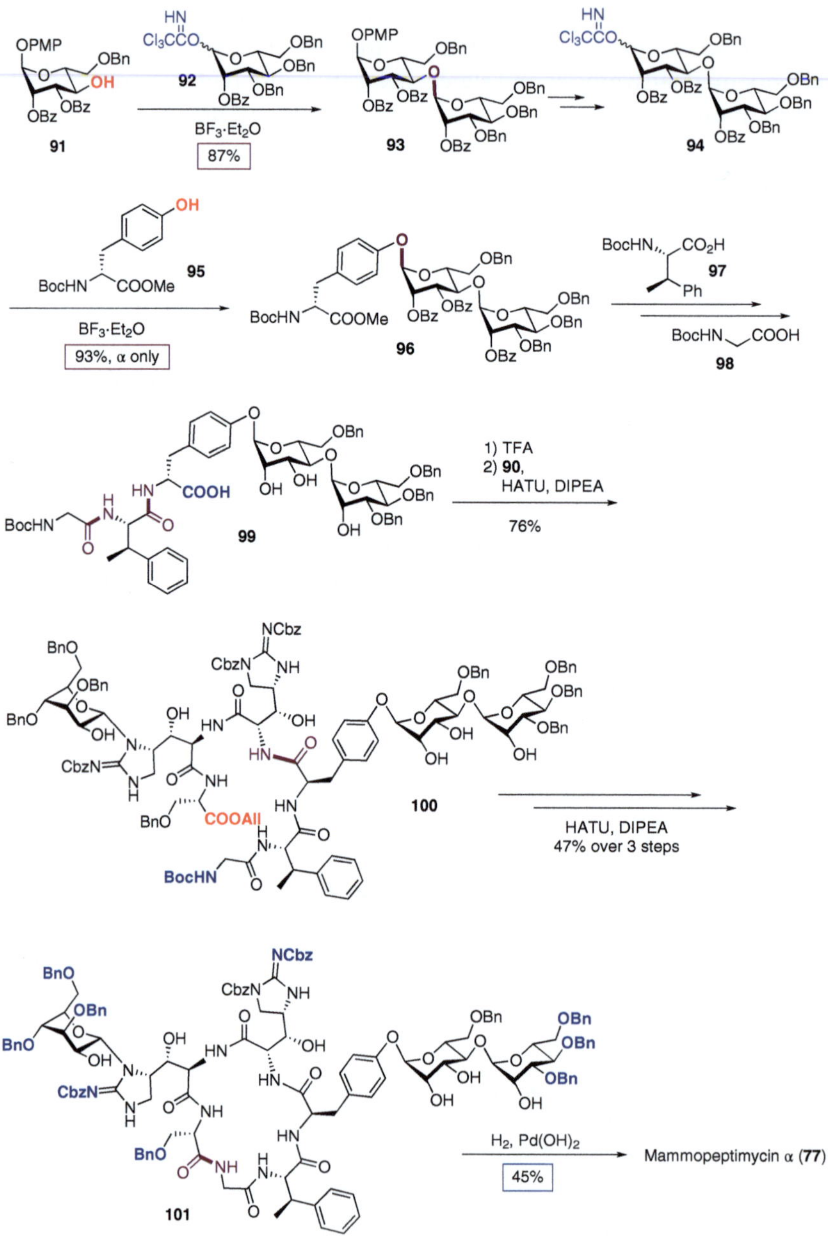


**Scheme 7.12** Total synthesis of mammopeptimycin α (**77**) by Chen and coworkers.

coupling of **96** with Boc-βMePhe-OH **97** and Boc-Gly-OH **98** readily furnished tripeptide Boc-Gly-βMePheD-Tyr(di-Man)-OH **99**. Upon removal of the *N*-Boc group, HATU-mediated amide coupling of **99** with *N*-α-mannosyl-D-βhEnd **90** gave the linear hexapeptide **100** in 76% yield. Following the removal of the *C*-terminus allyl group and the *N*-terminus Boc group, HATU-mediated macrolactamization was conducted, leading to the cyclized hexapeptide **101** in 47% yield (for three steps). Finally, the global deprotection of the *O*-Bn and *N*-Cbz groups through catalytic hydrogenation accomplished the synthesis of mannopeptimycin α (**77**) in 45% yield. Employing a similar strategy, Chen and colleagues also accomplished the total synthesis of mannopeptimycin β (**78**). Their synthesis provided definitive evidence for the structural determination of these highly complex glycopeptide natural products.

**Total synthesis of mammopeptimycin β (78) by Li and coworkers.** In 2021, the Li group reported a streamlined solution phase–solid phase synthesis and chemical ligation-mediated peptide cyclization for the total synthesis of mannopeptimycin β (**78**) (Scheme 7.13) [41]. Their syntheses began with the ester coupling of the 2′-hydroxycinnamoyl AM resin **102** with D-BocTyr(Bn)-OH **103**, followed by sequential amide coupling with (*2S,3R*)-β-methyl phenylalanine **104**, the *N*-mannosyl-containing tripeptide **105** and L-βhEnd building block **106**, yielding

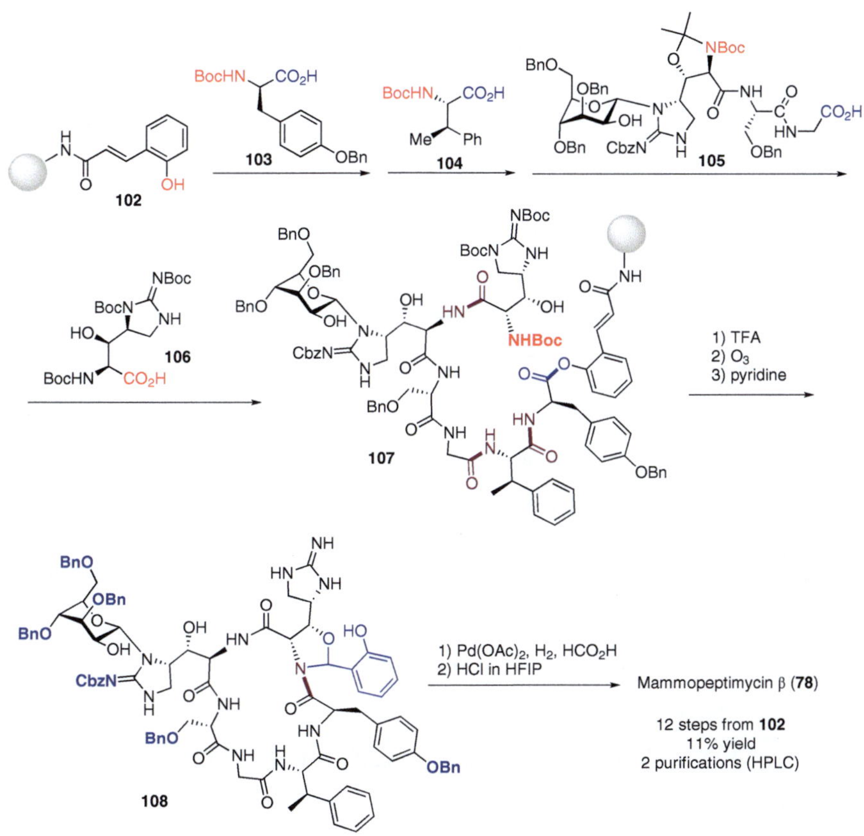


**Scheme 7.13** Total synthesis of mammopeptimycin β (**78**) by Li and coworkers.

hexapeptide **107**. Notably, the *N*-mannosyl moiety in tripeptide **105** was installed using the same Au(I)-catalyzed glycosylation approach as described in Chen's work. Treatment of **107** with TFA led to the removal of the Boc protecting groups, which was followed by ozonolysis to liberate the peptide *C*-terminal salicylaldehyde ester, serving as a precursor for cyclization. Encouragingly, the β-hydroxyenduracididine ligation [42] proceeded smoothly ($c = 2$ mM), granting access to the cyclized *N*-acyl oxazolidine intermediate **108**. Finally, the removal of five benzyl groups and one Cbz group, along with acidolysis of the *N*-acyl oxazolidine intermediate, successfully accomplished the synthesis of mammopeptimycin β (**78**) in 11% yield, with 12 steps from the resin and only two purifications by HPLC.

## Syndecan

Heparan sulfate proteoglycans (HSPGs) are essential components found on the cell membrane and extracellular matrix, which comprise three parts: heparan

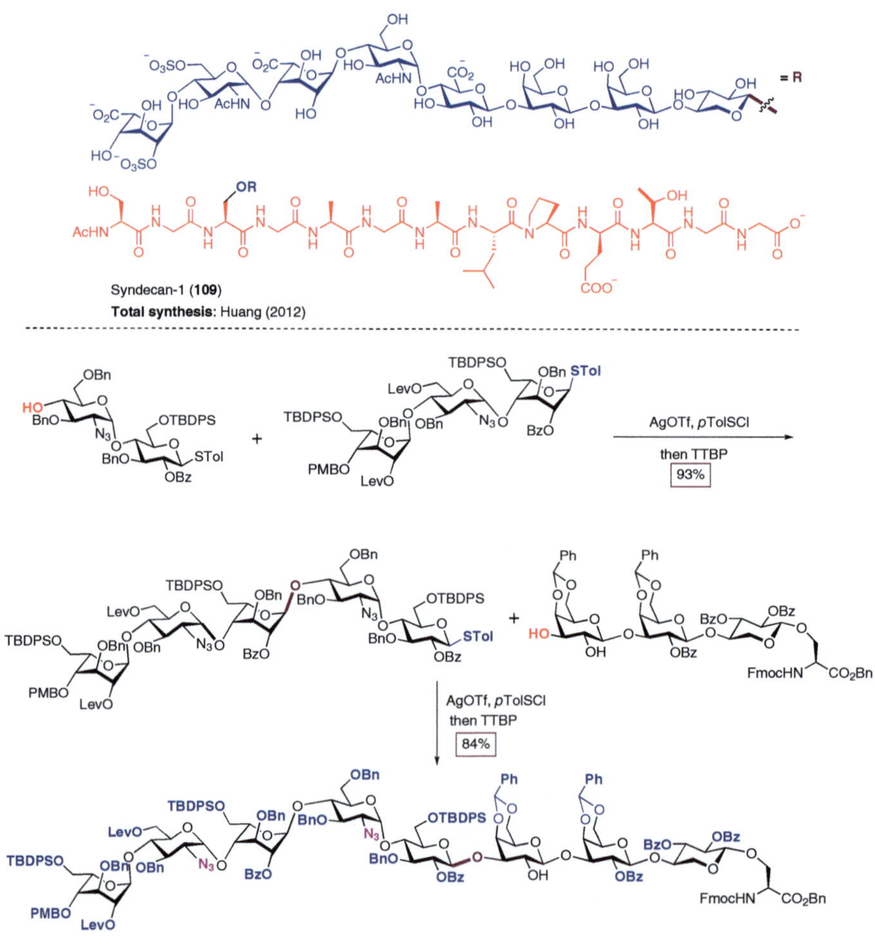


**Figure 7.6** Syndecan-1 (**109**) and the glycosylation steps in the synthesis.

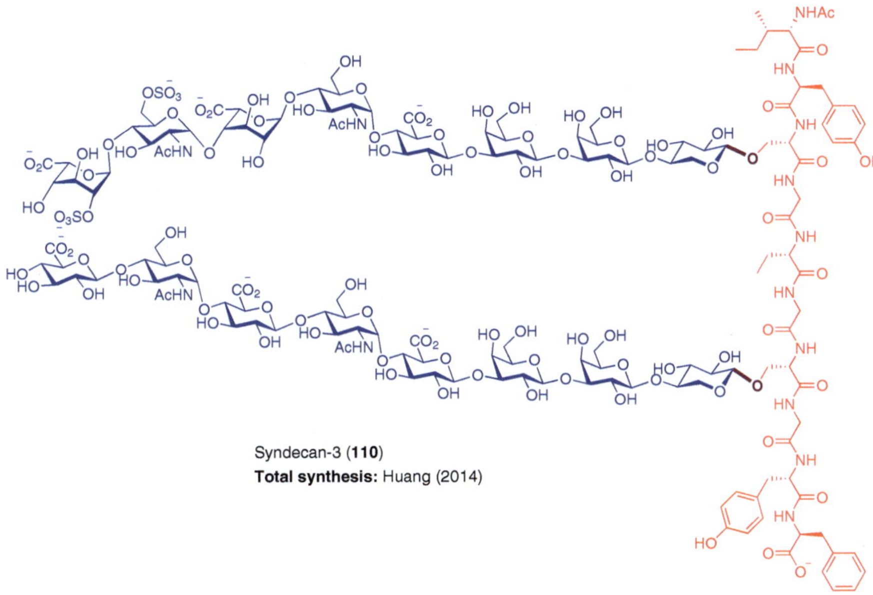


**Figure 7.7** Syndecan-3 (**110**).

sulfate (HS), a tetrasaccharide linker, and the core protein. Among them, human syndecan-1 (**109**, Figure 7.6) holds significance as an HSPG [43]. Syndecan-1 and its family molecules can function as receptors for various adhesion molecules and growth factors, such as fibroblast growth factor, integrin, Wnt proteins, and herpes simplex virus glycoprotein D. On a structural level, syndecan-1 (**109**) exhibits typical features of HSPGs, including the complete tetrasaccharide linker with glucosamine (GlcN) attached to both glucuronic acid (GlcA) and iduronic acid (IdoA), 2-O sulfation, 6-O sulfation, and N acetylation. In 2012, Huang and colleagues reported the first total synthesis of syndecan-1 (**109**) using a 3+2+3 glycosylation approach with the preactivation protocol of glycosyl sulfide donors and a promoter system involving *p*-TolSCl/AgOTf [44].

In 2014, the Huang group reported the total synthesis of a more complex glycosaminoglycan family glycopeptide, syndecan-3 (**110**), which consists of two different octasaccharide chains (Figure 7.7) [45].

## Lipoglycopeptide Arylomycin

The arylomycins, initially isolated in 2002 from a strain of Streptomyces, consist of three interrelated compound series, which feature a conserved *C*-terminal tripeptide macrocycle attached to an *N*-terminal lipopeptide [46]. In contrast to other arylomycins, the macrocycles of the lipoglycopeptide arylomycin (**111**, Figure 7.8) exhibit glycosylation with a deoxy-α-mannose substituent, and the absolute configuration of the deoxy-mannose remained unassigned. The arylomycins

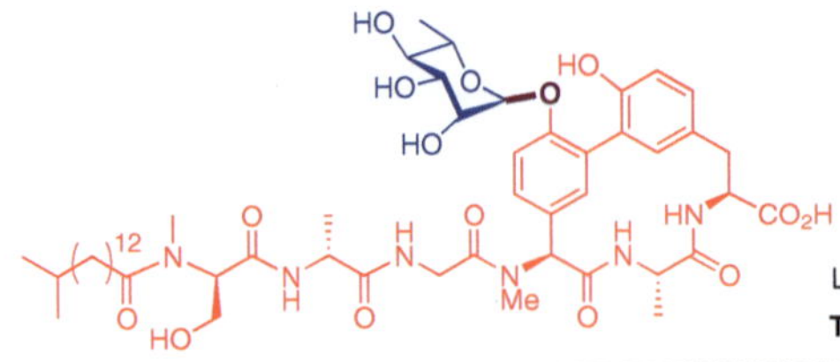

Lipoglycopeptide arylomycin (**111**)
**Total synthesis**: Romesberg (2011)

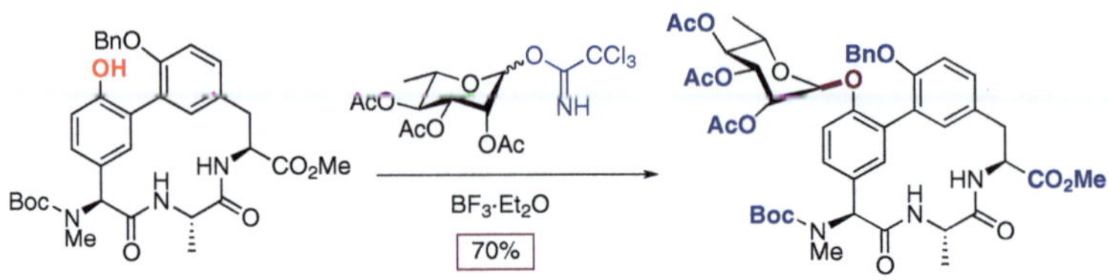


**Figure 7.8** Lipoglycopeptide arylomycin (**111**) and the glycosylation step in the synthesis.

exhibited moderate activity against various bacteria, with MICs ranging from 8 to >64 μM against *Streptococcus pneumoniae*, from 32 to >64 μM against *S. aureus*, and an MIC of 64 μM against *Haemophilus influenzae*. In 2011, Romesberg and coworkers reported the total synthesis of arylomycin derivatives bearing each possible deoxy-α-mannose enantiomer, which enabled to assign the absolute stereochemistry of the sugar in the natural product as L-deoxy-mannose [47].

## References


**1** Nicolaou, K.C., Boddy, C.N.C., Bräse, S., and Winssinger, N. (1999). *Angew. Chem. Int. Ed.* 38: 2096–2152.

**2** Butler, M.S., Hansford, K.A., Blaskovich, M.A.T. et al. (2014). *J. Antibiot.* 67: 631–644.

**3** Kahne, D., Leimkuhler, C., Lu, W., and Walsh, C. (2005). *Chem. Rev.* 105: 425–448.

**4** McCormick, M.H., Stark, W.M., Pittenger, G.E. et al. (1955–1956). *Antibiot. Annu.* 3: 606.

**5** Parenti, F., Beretta, G., Berti, M., and Arioli, V. (1978). *J. Antibiot.* 31: 276–283.

**6** Ashford, P.-A. and Bew, S.P. (2012). *Chem. Soc. Rev.* 41: 957–978.

**7** Cheng, S., Wantuch, P.L., Kizer, M.E. et al. (2019). *Org. Biomol. Chem.* 17: 2646–2650.

**8** Rodríguez-Mayor, A.V., Peralta-Camacho, G.J., Cárdenas-Martínez, K.J., and García-Castañeda, J.E. (2020). *Curr. Org. Chem.* 24: 2475–2497.

**9** Okano, A., Isley, N.A., and Boger, D.L. (2017). *Chem. Rev.* 117: 11952–11993.

**10** Dushin, R.G. and Danishefsky, S.J. (1992). *J. Am. Chem. Soc.* 114: 3471–3475.

**11** Nicolaou, K.C., Jain, N.F., Natarajan, S. et al. (1998). *Angew. Chem. Int. Ed.* 37: 2714–2716.

**12** Nicolaou, K.C., Mitchell, H.J., van Delft, F.L. et al. (1998). *Angew. Chem. Int. Ed.* 37: 1871–1874.

**13** Nicolaou, K.C., Natarajan, S., Li, H. et al. (1998). *Angew. Chem. Int. Ed.* 37: 2708–2714.

**14** Nicolaou, K.C., Takayanagi, M., Jain, N.F. et al. (1998). *Angew. Chem. Int. Ed.* 37: 2717–2719.

**15** Nicolaou, K.C., Mitchell, H.J., Jain, N.F. et al. (1999). *Chem. Eur. J.* 5: 2648–2667.

**16** Nicolaou, K.C., Mitchell, H.J., Jain, N.F. et al. (1999). *Angew. Chem. Int. Ed.* 38: 240–244.

**17** Ge, M., Thompson, C., and Kahne, D. (1998). *J. Am. Chem. Soc.* 120: 11014–11015.

**18** Thompson, C., Ge, M., and Kahne, D. (1999). *J. Am. Chem. Soc.* 121: 1237–1244.

**19** Okano, A., Nakayama, A., Schammel, A.W., and Boger, D.L. (2014). *J. Am. Chem. Soc.* 136: 13522–13525.

**20** Okano, A., Nakayama, A., Wu, K. et al. (2015). *J. Am. Chem. Soc.* 137: 3693–3704.

**21** Nakayama, A., Okano, A., Feng, Y. et al. (2014). *Org. Lett.* 16: 3572–3575.

**22** Boger, D.L., Miyazaki, S., Kim, S.H. et al. (1999). *J. Am. Chem. Soc.* 121: 10004–10011.

**23** Moore, M.J., Qu, S., Tan, C. et al. (2020). *J. Am. Chem. Soc.* 142: 16039–16050.

**24** Umezawa, H., Maeda, K., Takeuchi, T., and Okami, Y. (1966). *J. Antibiot. Ser. A* 19: 200–209.

**25** Tsuchiya, T., Miyake, T., Kageyama, S. et al. (1981). *Tetrahedron Lett.* 22: 1413–1416.

**26** Takita, T., Umezawa, Y., Saito, S.-i. et al. (1982). *Tetrahedron Lett.* 23: 521–524.

**27** Ohno, M. and Otsuka, M. (1990). *Recent Progress in the Chemical Synthesis of Antibiotics* (ed. G. Lukacs and M. Ohno), 387–414. Berlin, Heidelberg: Springer.

**28** Pozsgay, V., Ohgi, T., and Hecht, S.M. (1981). *J. Org. Chem.* 46: 3761–3763.

**29** Aoyagi, Y., Katano, K., Suguna, H. et al. (1982). *J. Am. Chem. Soc.* 104: 5537–5538.

**30** Ohgi, T. and Hecht, S.M. (1981). *J. Org. Chem.* 46: 1232–1234.

**31** Katano, K., An, H., Aoyagi, Y. et al. (1998). *J. Am. Chem. Soc.* 120: 11285–11296.

**32** Boger, D.L., Colletti, S.L., Honda, T., and Menezes, R.F. (1994). *J. Am. Chem. Soc.* 116: 5607–5618.

**33** Boger, D.L. and Honda, T. (1994). *J. Am. Chem. Soc.* 116: 5647–5656.

**34** Boger, D.L., Honda, T., and Dang, Q. (1994). *J. Am. Chem. Soc.* 116: 5619–5630.

**35** Boger, D.L., Honda, T., Menezes, R.F., and Colletti, S.L. (1994). *J. Am. Chem. Soc.* 116: 5631–5646.

**36** Boger, D.L. and Cai, H. (1999). *Angew. Chem. Int. Ed.* 38: 448–476.

**37** De Voe, S.E. and Kunstmann, M.P. (1970). Antibiotic AC98 and production. U.S. Patent 3495004, 10 February 1970.

**38** Babu, R.S., Guppi, S.R., and O'Doherty, G.A. (2006). *Org. Lett.* 8: 1605–1608.

**39** Fuse, S., Koinuma, H., Kimbara, A. et al. (2014). *J. Am. Chem. Soc.* 136: 12011–12017.

**40** Wang, B., Liu, Y., Jiao, R. et al. (2016). *J. Am. Chem. Soc.* 138: 3926–3932.

**41** Wang, J., Lin, D.A., Liu, M. et al. (2021). *J. Am. Chem. Soc.* 143: 12784–12790.

**42** Lam, H.Y., Zhang, Y., Liu, H. et al. (2013). *J. Am. Chem. Soc.* 135: 6272–6279.

**43** Yang, Y., MacLeod, V., Dai, Y. et al. (2007). *Blood* 110: 2041–2048.

**44** Yang, B., Yoshida, K., Yin, Z. et al. (2012). *Angew. Chem. Int. Ed.* 51: 10185–10189.

**45** Yoshida, K., Yang, B., Yang, W. et al. (2014). *Angew. Chem. Int. Ed.* 53: 9051–9058.

**46** Holtzel, A., Schmid, D.G., Nicholson, G.J. et al. (2002). *J. Antibiot.* 55: 571–577.

**47** Liu, J., Luo, C., Smith, P.A. et al. (2011). *J. Am. Chem. Soc.* 133: 17869–17877.

# 8

# Resin Glycosides

Resin glycosides, also known as glycoresins, constitute a diverse group of amphiphilic glycolipids, which consist of various acylated oligosaccharide bearing a mono- or di-hydroxy C14 or C16 fatty acid [1, 2]. Typically, the acid chain folds back to form a macrolactone ring of varied size, spanning two or more sugar units (Figure 8.1). Resin glycosides are mainly isolated from plants of the *Convolvulaceae* (morning glory family). Initial investigations into resin glycosides date back to the mid-nineteenth century when they were recognized as essential components of traditional folk medicines. These compounds contain several deoxy sugars such as D-quinovose, D-fucose, and L-rhamnose, along with homologous optically active 11-hydroxy fatty acids, namely jalapinolic acid (11(*S*)-hydroxyhexadecanoic acid) and convolvulinolic acid (11(*S*)-hydroxytetradecanoica acid), in addition to various short-chain carboxylic acids. Due to their complex structures, resin glycosides were initially regarded as heterogeneous polymers linked through glycosidic and esteric linkages. Certain resin glycosides exhibit unique bioactivities, including cytotoxic effects against human breast cancer cell lines, antimicrobial properties, and plant growth-regulating effects. For instance, calonyctins (**1a** and **1b**), extracted from the leaves of Yue–Guang–Hua (*Calonyction aculeatum* L. House), acted as a plant growth regulator to promote the tuber production of sweet potato and increased the crop yields of beans and wheat [3]. Tricolorin A (**2**), isolated in 1993 from *Ipomoea tricolor* cav. (*convolvulaceae*), is traditionally used in Mexican agriculture to inhibit the growth of invasive weeds [4]. This 19-membered tetrasaccharide macrolactone demonstrated significant cytotoxic activity against cultured P-388 and human breast cancer cells ($ED_{50} = 2.2\,\mu g/ml$).

Based on the methods for the ring closure step, the synthesis of resin glycosides can be classified into three strategies (Figure 8.2) [5–8]. The first approach involves macrolactonization between the carboxylic acid moiety and the saccharide alcohol. The second strategy utilizes ring-closing metathesis/hydrogenation of diene moieties to achieve the ring closure. The third strategy entails intramolecular glycosylation to construct the macrocyclic ring.

*Carbohydrate Chemistry in the Total Synthesis of Naturally Occurring Glycosides*, First Edition. Biao Yu and Xiaoyu Yang.
© 2024 WILEY-VCH GmbH. Published 2024 by WILEY-VCH GmbH.

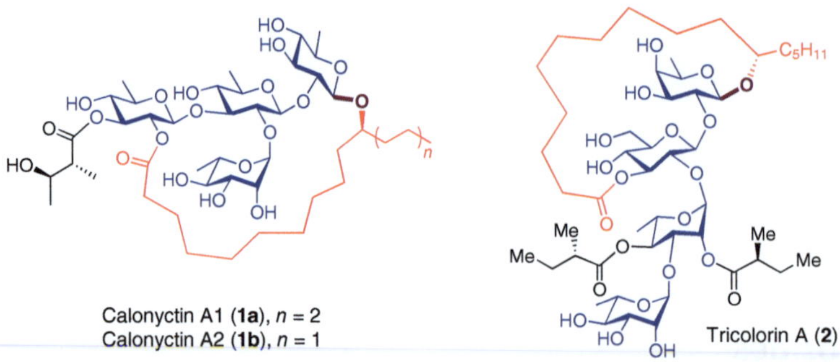


**Figure 8.1** Representative resin glycosides: calonyctin A (**1**) and tricolorin A (**2**).

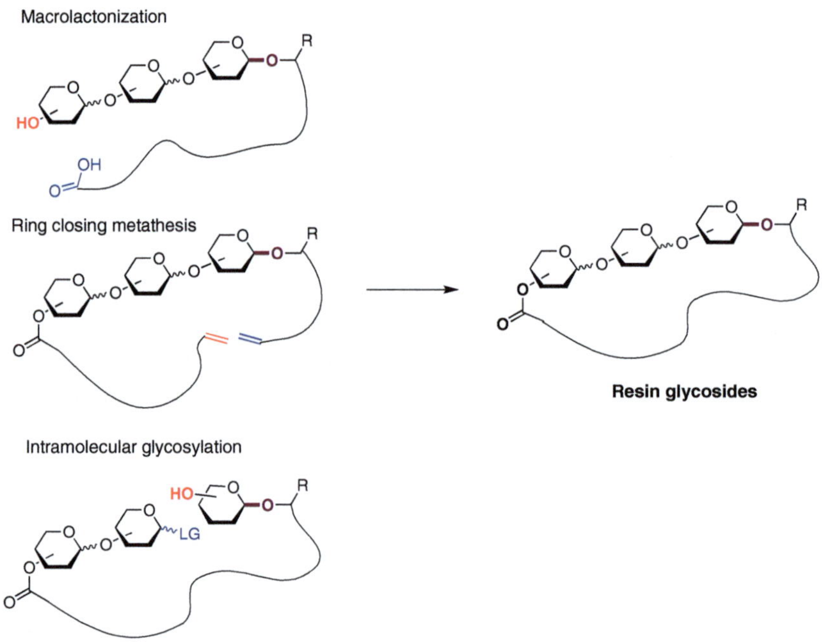


**Figure 8.2** Three general strategies utilized in the total synthesis of resin glycosides.

## Calonyctin A

Calonyctin A was isolated from the leaves of Yue–Guang–Hua (*C. aculeatum* L. House) in 1993 by Xin and coworkers [3]. Notably, calonyctin A possesses a significant promoting effect on the growth and yield of various crops, such as peanuts, beans, and wheat. Structurally, it is a mixture of homologous glycolipids consisting of a shared tetrasaccharide backbone and 11-hydroxy fatty acids of different length, referred to as calonyctin A1 (**1a**, Figure 8.3) and A2 (**1b**), respectively. The most remarkable feature lies in the presence of a 22-membered macrolide ring. However, the configurations of 3-hydroxy-2-methylbutyric acid and the 11-hydroxy

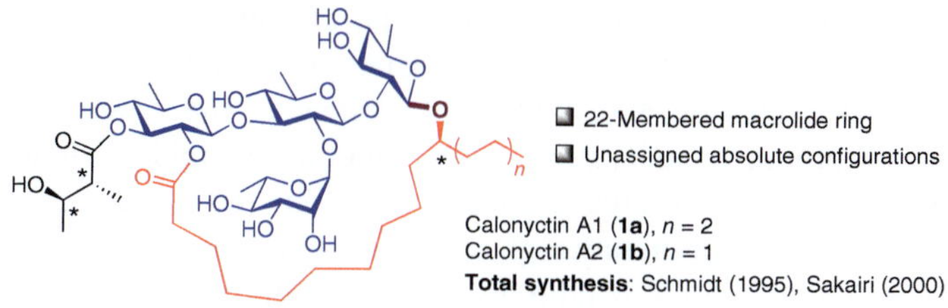


**Figure 8.3** Calonyctin A (**1**) and its structural features.

fatty acids moieties were not assigned initially. In addition, it was assumed that only the common *D*-isomer of quinovose (*D*-Qui) occurred in calonyctin A.

**Total synthesis of calonyctin A1 by Schmidt and coworkers.** In 1995, Schmidt and coworkers reported the first total synthesis of calomyctin A1 (**1a**) and the assignment of its absolute configurations (Scheme 8.1) [9, 10]. The

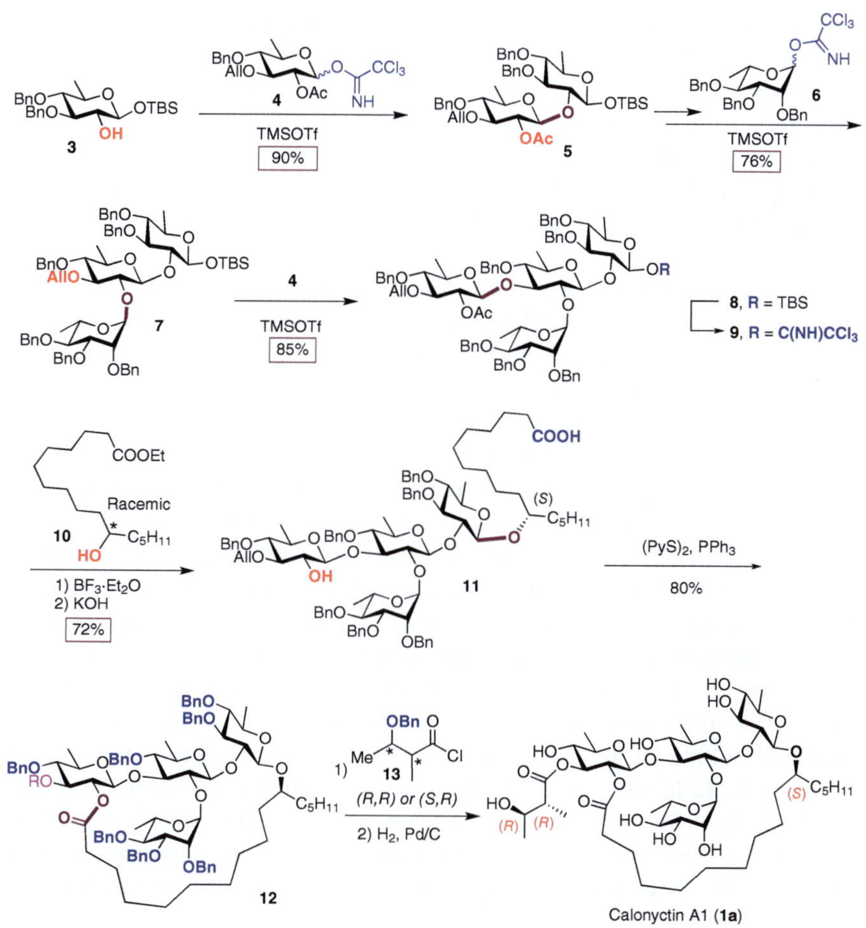

**Scheme 8.1** Total synthesis and configuration determination of calonyctin A1 (**1a**) by Schmidt and coworkers.

synthesis of the tetrasaccharide segment began with the coupling of quinovose **3** and quinovosyl trichloroacetimidate **4** under the catalysis of TMSOTf, which provided the desired β-disaccharide **5** in 90% yield. Following the removal of the C2′-OAc protecting group, condensation with L-rhamnosyl imidate donor **6** enabled by TMSOTf gave the desired trisaccharide **7** in 76% yield. Upon removal of the C4′-OAll group, the iterative glycosylation step using quinovosyl trichloroacetimidate **4** under similar conditions afforded tetrasaccharide **8** in 85% yield, which was further converted into the corresponding tetrasaccharide imidate **9**. Considering the unknown absolute configuration of the fatty acid moiety, a racemic jalapinolic ester **10** was glycosylated with **9**, under the catalysis of $BF_3 \cdot Et_2O$, which resulted in the formation of a 1 : 1 mixture of two diastereomeric β-glycosides in 80% yield. The (*S*) configuration of the natural jalapinolic acid was assigned by hydrogenolytic debenzylation and acidic hydrolysis of **11**, followed by a comparison of the optical rotations with the literature data. The desired (*S*)-isomer of **11** was cyclized using 2-pyridyl disulfide and $Ph_3P$ (the Corey–Nicolaou protocol) under high dilution conditions (0.4 mM in toluene), leading to the formation of macrolactone **12** in 80% yield. After the removal of the *O*-allyl protecting group, the resulting alcohol was acylated with (*R,R*) or (*S,S*) 3-benzyloxy-2-methylbutyric acid chloride **13**. Final subjection to hydrogenolytic debenzylation yielded calonyctin A1 (**1a**) and its epimer. By comparing the optical rotations and NMR data with those of the authentic natural compound, the (*R,R*) configuration was confidently assigned to the 3-hydroxy-2-methylbutyric acid side chain.

**Total synthesis of calonyctin A2 by Sakairi and coworkers.** In 2000, Sakairi and coworkers reported the total synthesis of calonyctin A2 (**1b**) (Scheme 8.2) [11, 12]. The synthesis began with the preparation of disaccharide sulfide donor **15** from naturally occurring β-laminaribiose, involving several steps of functional group interconversions and protecting group modifications. Coupling of 11-hydroxy fatty acid **16** with glucosyl sulfide **17** under the action of NIS-TfOH afforded the desired glycosylated product **18** after subsequent removal of the C2-OAc group. Glycosylation of **18** with disaccharide sulfide **15** under similar conditions provided the desired β-trisaccharide **19** in 71% yield. Upon removal of the *O*-Bz groups, the macrolactonization of **19**, enabled by $Cl_3C_6H_2COCl$ under high dilution conditions in toluene (17 mM), afforded the expected macrocycle **20** in 60% yield. Remarkably, despite the presence of two hydroxyl groups at the 2b- and 2c-positions, the desired 22-membered macrolide **20** was obtained as a single product. However, the following glycosylation of **20** with L-rhamnosyl sulfide **21** in the presence of NIS-TfOH only gave the expected tetrasaccharide **22** in 15% yield, which could be ascribed to the steric hindrance imposed by the hydroxyl group at the 2b-position. Finally, installation of the (2*R*,3*R*)-3-benzyloxy-2-methylbutyric acid moiety and global deprotection of all the *O*-Bn groups through catalytic hydrogenation accomplished the synthesis of calonyctin A2 (**1b**).

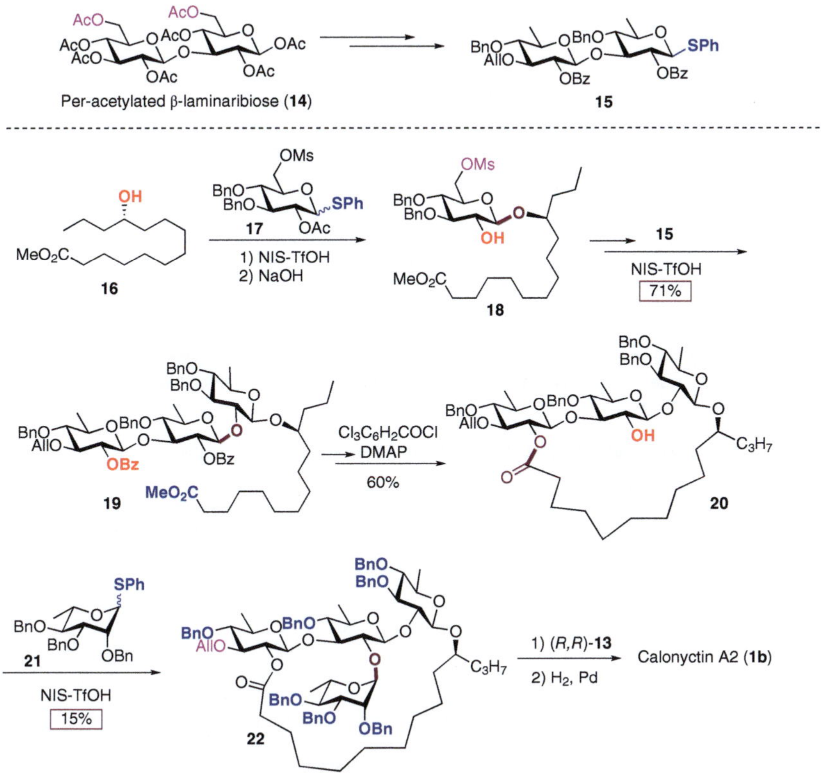


**Scheme 8.2** Total synthesis of calonyctin A2 (**1b**) by Sakairi and coworkers.

# Tricolorin

The tricolorin family of resin glycosides constitutes the allelochemical principle of *I. tricolor* Cav., a plant widely employed in traditional Mexico agriculture as a cover crop for the protection of sugar cane. Structurally, tricolorins feature a macrolide entity formed by (11*S*)-hydroxyhexadecanoic acid, which spans two sugar units in tricolorin A (**2**) [4] or three sugar units in tricolorin F (**23**) [13], leading to the formation of 19-membered or 21-membered macrolactone, respectively (Figure 8.4). Among its congeners, tricolorin A (**2**) exhibited significant cytotoxic properties against cultured P-388 and human breast cancer cell lines, with $ED_{50}$ = 2.2 μg/ml. Studies have shown that the bioactivities of the tricolorin family are closely associated with their macrocyclic structure, as the glycosidic acids derived from saponification were found inactive.

**Total synthesis of tricolorin A by Hui and coworkers.** In 1996, Yu, Hui, and coworkers reported the first total synthesis of tricolorin A (**2**) utilizing a total of 45 steps, where the longest linear sequence involved 20 steps, resulting in an overall yield of 0.65% starting from D-(+)-mannitol (Scheme 8.3) [14, 15]. Their

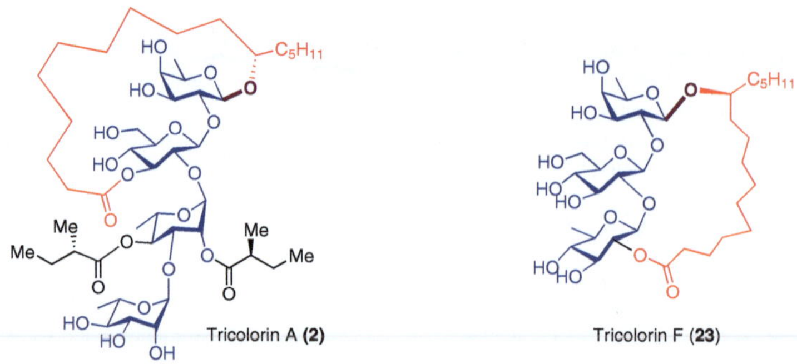


- A disaccharide containing 19-membered macrolactone
  **Total synthesis**: Yu and Hui (1997), Heathcock (1997), Fürstner (1998), Wan (2020)
- A trisaccharide containing 21-membered macrolactone
  **Total synthesis**: Heathcock (2004), Sakairi (2009)

**Figure 8.4** Tricolorin A (**2**) and tricolorin F (**23**) and the structural features.

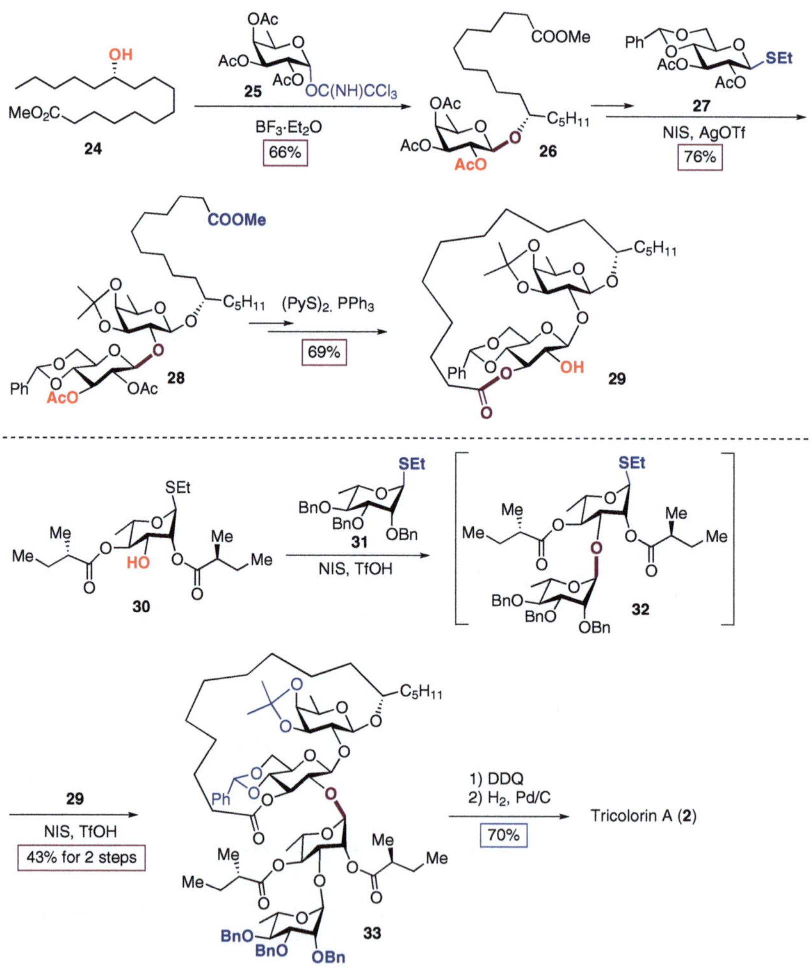


**Scheme 8.3** Total synthesis of tricolorin A (**2**) by Yu, Hui, and coworkers.

synthetic route featured regioselective macrolactonization of a disaccharide to form a 19-membered ring, and a one-pot glycosylation strategy for constructing a macrolactone tetrasaccharide. The synthesis began with the coupling of methyl 11(*S*)-jalapinolate **24** with glycosyl imidate **25** under the promotion of $BF_3 \cdot Et_2O$, which afforded the desired β-glycoside **26** in 66% yield. Subsequent steps involved manipulations of protecting groups, followed by the coupling of the 2′-OH group with thioglycoside **27** in the presence of *N*-iodosuccinimide (NIS) and AgOTf, leading to disaccharide **28** in a good yield of 76%. Upon removal of the *O*-Ac groups and release of the carboxylic acid group, the regioselective macrolactonization of **28** was achieved using the Corey–Nicolaou protocol ($(PyS)_2$, $PPh_3$), which gave the desired 3″-OH macrolactonized product **29** in 69% yield, along with the 2″-OH macrolactonized byproduct in 11% yield. One the other hand, the glycosylation of thioglycoside donors **30** and **31** in the presence of NIS/TfOH, employing the "armed-disarmed" glycosidation approach [16], yielded disaccharide sulfide intermediate **32**, which was subsequently coupled with macrolactone disaccharide **29** in one-pot to give the desired tetrasaccharide **33** in 43% yield (for two steps). Ultimately, the 3′,4′-*O*-isopropylidene and 4″,6″-*O*-benzylidene groups on compound **33** were deprotected using DDQ oxidation, followed by the removal of the benzyl groups through catalytic hydrogenation, resulting in the synthesis of tricolorin A (**2**) in 70% yield for the two-step process.

**Total synthesis of tricolorin A (2) by Heathcock and coworkers.** In the same year, Heathcock and coworkers reported another approach to the synthesis of tricolorin A (**2**) with the longest linear sequence of 14 steps and an overall yield of 6% (Scheme 8.4) [17, 18]. The synthesis began with the coupling of C3-OH rhamnoside **34** with rhamnosyl trichloroacetimidate **35** in the presence of $BF_3 \cdot Et_2O$, resulting in the desired α-disaccharide **36** in 81% yield, which was then converted into the corresponding imidate donor **37** in two steps. Simultaneously, glycosylation of methyl 11(*S*)-jalapinolate **24** with imidate donor **38** under the catalysis of TMSOTf afforded β-glycoside **39** in 76% yield. Following the removal of the C2-OPiv protecting group, subsequent glycosylation with glucosyl imidate **40** under the promotion of AgOTf provided β-disaccharide **41** in 62% yield. Upon removal of the *O*-Ac groups, the regioselective condensation with disaccharide trichloroacetimidate **37**, under the promotion of TMSOTf, gave the desired tetrasaccharide **42** in 54% yield. Saponification of the four ester groups in **42** were performed to give the acid triol, which was subjected to macrolactonization using the Yonemitsu protocol [19] (2,4,6-trichlorobenzoyl chloride, DMAP), yielding the desired lactone **43** in 61% yield, following the introduction of the (*S*)-2-methylbutyric acid chain in one-pot. Finally, global deprotection of the ketal and the Bn protecting groups by catalytic hydrogenation under acidic conditions accomplished the synthesis of tricolorin A (**2**) in 77% yield.

**Formal total synthesis of tricolorin A (2) by Fürstner and coworkers.** In 1998, Fürstner and coworkers reported a formal total synthesis of tricolorin A (**2**) utilizing ring-closing olefin metathesis (RCM) as the key step (Scheme 8.5) [20, 21]. The synthesis commenced with the coupling of glycosyl imidate **40** and fucose derivative **44** in the presence of catalytic amounts of $BF_3 \cdot Et_2O$, which provided the desired

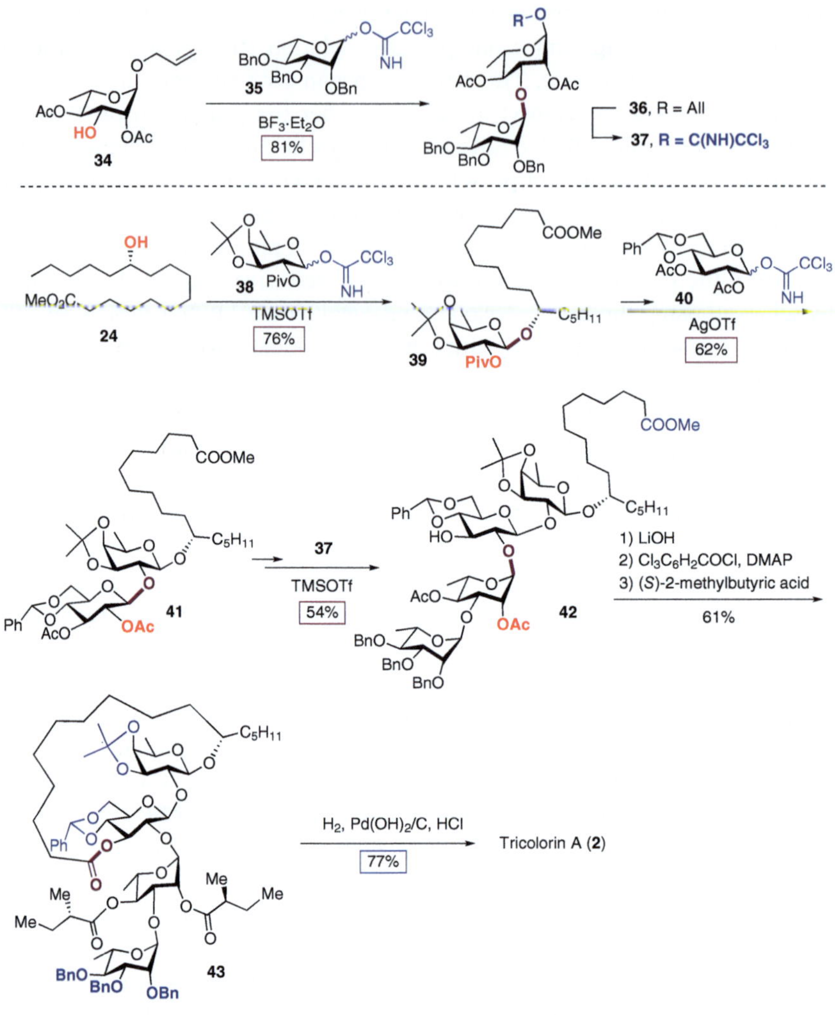

Scheme 8.4 Total synthesis of tricolorin A (**2**) by Heathcock and coworkers.

β-configurated disaccharide **45** in 82% yield. Subsequent deprotection of the two *O*-Ac groups was followed by the regioselective acylation of the resulting diol with 6-heptenoic acid (**46**), providing the *O*-3-substituted ester **47** as the only product in 71% yield. The diene **47** smoothly cyclized to form the desired 19-membered ring under the catalysis of ruthenium carbene catalyst **48** (5 mol%), which was followed by hydrogenation of the crude cycloalkene (*E*/*Z* mixture) to give disaccharide macrolactone **29** in 77% yield for two steps.

**Total synthesis of tricolorin A (2) by Wan and coworkers.** In 2020, Wan and coworkers reported their total synthesis of tricolorin A (**2**) based on an interrupted Pummerer reaction-mediated (IPRm) glycosylation [22], featuring the longest linear sequence of 16 steps and 12.5% overall yield (Scheme 8.6) [23]. The synthesis commenced with the coupling of *S*-2-[(propan-2-yl) sulfanyl]benzyl (SPTB) glycoside

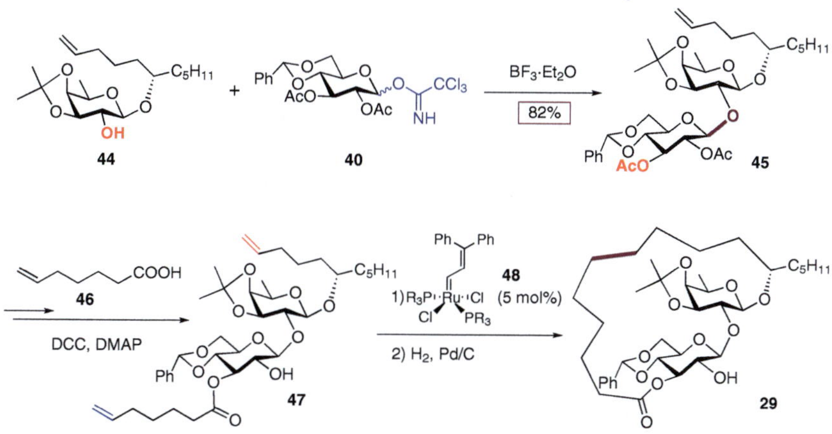


**Scheme 8.5** Formal total synthesis of tricolorin A (**2**) by Fürstner and coworkers.

acceptor **49** with *O*-2-[(propan-2-yl)sulfinyl]benzyl (OPSB) glycosyl donor **50** under the action of $Tf_2O$ and DTBMP to afford the expected disaccharide in 88% yield, which was then converted into the corresponding *S*-2-[(propan-2-yl)sulfinyl]benzyl (SPSB) disaccharide donor **51**. Simultaneously, glycosylation between methyl 11-(*S*)-jalapinolate **24** with SPSB glycosyl donor **52**, in the presence of $Tf_2O$, afforded glycoside **53** in 83% yield. Sequential manipulations of protecting groups were followed by glycosylation with SPSB glucosyl donor **54** under similar conditions, yielding β-disaccharide **55** in 95% yield. After release of the diol and the acid group, the macrolactonization was performed using $(PyS)_2$ and $PPh_3$, which afforded the desired 19-membered macrocycle **56** in 74% yield, along with the C2′-isomer in 14% yield. The subsequent coupling of macrocycle **56** with SPSB disaccharide donor **51** provided the tetrasaccharide **57** with only moderate diastereoselectivity (97%, $\alpha/\beta = 3.6:1$). Interestingly, the authors discovered that using the corresponding OPSB disaccharide donor analogs of **51** could produce tetrasaccharide **57** in a satisfying yield of 92%, with only a trace amount of the β-anomer. Finally, successive removal of the ketal protecting groups under acidic conditions and the Bn groups via catalytic hydrogenation accomplished the total synthesis of tricolorin A (**2**).

**Total synthesis of tricolorin F (23) by Heathcock and coworkers.** In 2004, Heathcock and coworkers reported the total synthesis of tricolorin F (**23**) (Scheme 8.7) [24]. The synthesis began with the coupling of the previously prepared jalapinolic acid glycoside **39** with glucosyl imidate **58** in the presence of $BF_3 \cdot Et_2O$, which afforded the desired β-disaccharide **59** in 70% yield. Subsequent removal of the C2-*O*-Ac protecting group was followed by glycosylation with quinovosyl imidate **60** under analogous conditions, yielding β-trisaccharide **61** in 60% yield. Upon release of the alcohol and acid group, the macrolactonization of **61** was performed in the presence of 2,4,6-$Cl_3C_6H_2COCl$, resulting in the formation of the desired 21-membered macrocycle **62** in 60% yield. Finally, removal of the ketal and Bn groups by hydrogenolysis under acidic conditions accomplished the total synthesis of tricolorin F (**23**).

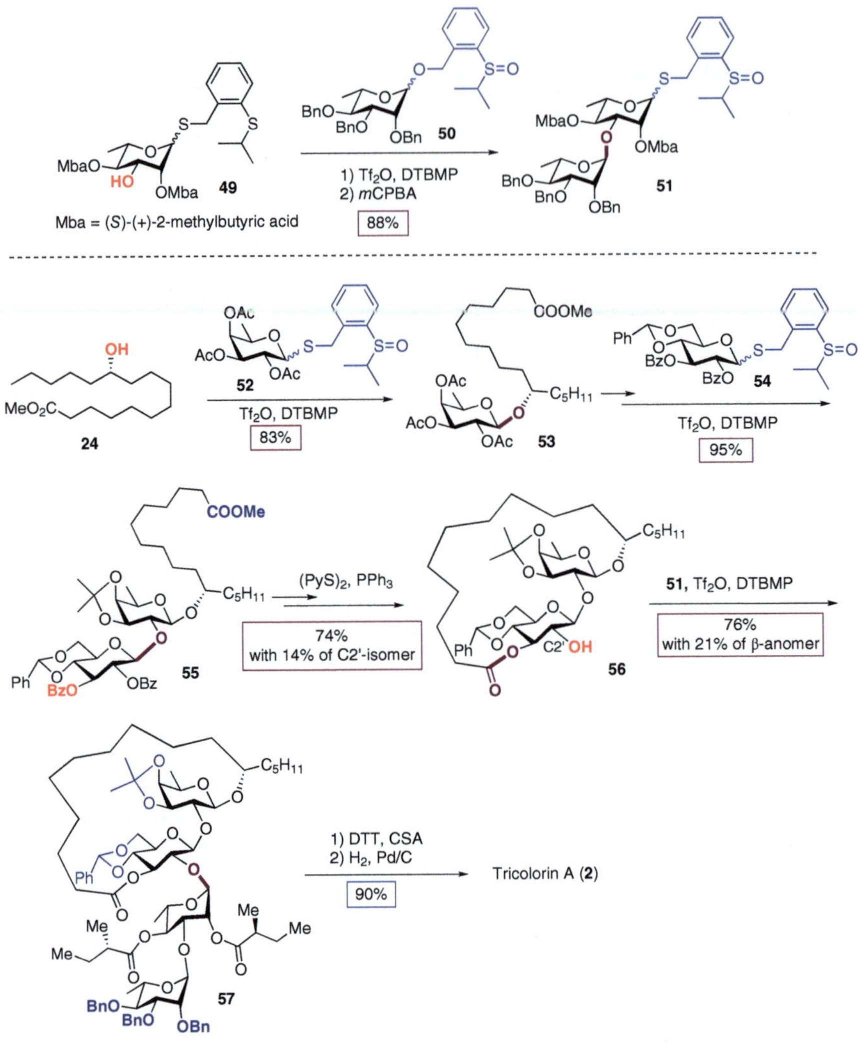

Scheme 8.6 Total synthesis of tricolorin A (**2**) by Wan and coworkers.

**Total synthesis of tricolorin F (23) by Sakairi and coworkers.** In 2009, Sakairi and coworkers reported their total synthesis of tricolorin F (**23**) utilizing the intramolecular glycosylation strategy (Scheme 8.8) [25]. The synthesis commenced with the glycosylation between the known D-fucoside acceptor **39** and thioglycoside donor **63** under the promotion of MeOTf, which gave the desired β-disaccharide **64** in 83% yield. Upon hydrolysis of the ester group, the resulting carboxylic acid was coupled with the C2-OH of thioquinovoside **65**, affording β-trisaccharide **66** in 72% yield. The pivotal macrocyclization step involved the intramolecular glycosylation of **66**, which was succeeded in the presence of MeOTf at a substrate concentration of 1.8 mM, providing the 21-membered derivative **67** in 70% yield. Finally, global deprotection on **67** using the previously reported two-step procedure accomplished the synthesis of tricolorin F (**23**) in nearly quantitative yield.

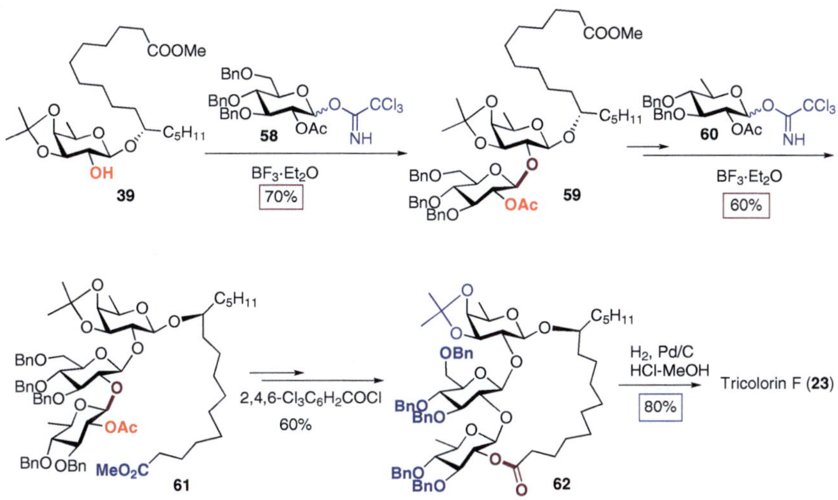


**Scheme 8.7** Total synthesis of tricolorin F (**23**) by Heathcock and coworkers.

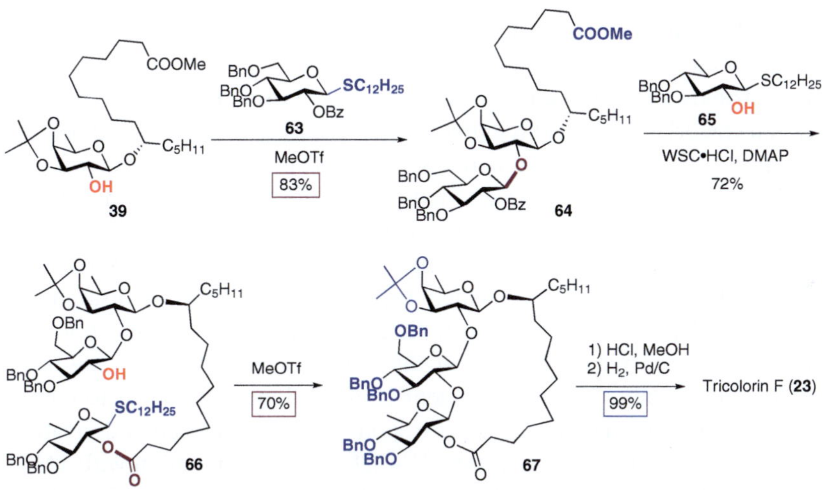


**Scheme 8.8** Total synthesis of tricolorin F (**23**) by Sakairi and coworkers.

## Ipomoeassin

Ipomoeassins (ipomoeassin A to F, **68–73**, Figure 8.5) were isolated from the morning glory *Ipomoea squamosal* collected in the Suriname rainforest during a biodiversity program in 2005 and 2007 [26, 27]. These compounds quickly captured the attention of synthetic chemists due to their remarkable biological activities. Ipomoeassins D (**71**) and F (**73**) exhibited exceptionally potent cytotoxicity, with single-digit nanomolar $IC_{50}$ values against several cancer cell lines. More importantly, ipomoeassin A (**68**), the most abundant member of the family, displayed potent and selective growth inhibition against different cancer cells in

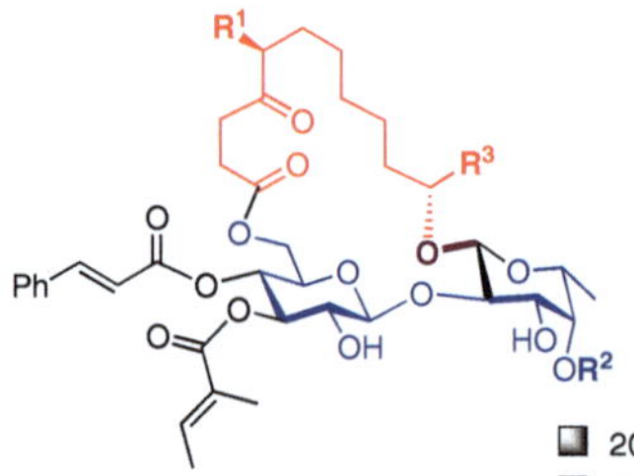


Ipomoeassin A (**68**), $R^1$ = H, $R^2$ = Ac, $R^3$ = $C_3H_7$
Ipomoeassin B (**69**), $R^1$ = H, $R^2$ = H, $R^3$ = $C_3H_7$
Ipomoeassin C (**70**), $R^1$ = OH, $R^2$ = Ac, $R^3$ = $C_3H_7$
Ipomoeassin D (**71**), $R^1$ = OAc, $R^2$ = Ac, $R^3$ = $C_3H_7$
Ipomoeassin E (**72**), $R^1$ = OAc, $R^2$ = H, $R^3$ = $C_3H_7$
Ipomoeassin F (**73**), $R^1$ = H, $R^2$ = Ac, $R^3$ = $C_5H_{11}$

- 20-Membered macrolactone
- Presence of oxygenation on the linking acid chain

**Total synthesis**: Fürstner (2007), Postema (2009), Shi (2015)

**Figure 8.5** Ipomoeassin family resin glycosides and their structural features.

the NCI 60-cell line screen. A COMPARE analysis revealed that its activity pattern differed from that of other known anticancer agents, indicating the potential of ipomoeassins as promising leads for developing a novel class of antineoplastic agents. Structurally, the presence of oxygenation on the linking acid chain and the shorter tail at the fucoside anomeric center are uncommon compared to other resin glycosides. Furthermore, the inclusion of cinnamate and tiglate functions is relatively rare in resin glycosides, while the reduced or hydrated forms of the tiglate (2-methylbutyrate or 3-hydroxy-2-methylbutyric acid) are more commonly encountered motifs.

**Total synthesis of ipomoeassins by Fürstner and coworkers.** In 2007, Fürstner and coworkers reported the first total synthesis of ipomoeassin B (**69**) and E (**72**) using the RCM strategy (Scheme 8.9) [28]. The synthesis began with the silver-mediated glycosylation between fucosyl bromide **75** and chiral alcohol **74**, resulting in β-glycoside **76** in 84% yield. Subsequent protecting group manipulations were followed by the coupling between free C2-OH of **76** and glucosyl imidate donor **77**, enabled by $BF_3 \cdot Et_2O$, which furnished the desired β-disaccharide **78** in 75% yield. Upon removal of the *O*-Ac groups, selective esterification with tiglic acid **79** occurred regioselectively at the more nucleophilic *O*-3″ site of the disaccharide **78**. Despite that many reagent combinations are known to convert sugar 4,6-*O*-benzylidene acetal into the corresponding 4-*O*-benzyl ether derivatives, the authors discovered that only the use of $NaBH_3CN$ in combination with TMSCl furnished a mixture of the corresponding PMB-ether derivatives in high yield, surprisingly favoring the 6″-*O*-PMB-ether **80** with a 4 : 1 isomeric ratio. Based on this unexpected outcome, the synthetic strategy was revised. The authors proposed protecting the cinnamate moiety with a *C*-silyl substituent (**81**) since, as a trisubstituted alkene, it would neither interfere with the RCM reaction nor be affected during alkene hydrogenation. Consequently, alcohol **80** was esterified with cinnamic acid surrogate **81** under Yamaguchi conditions to give product **82**. After oxidative cleavage of the PMB group, condensation with 4-oxo-8-nonenoic acid ester segment (**83a** or **83b**) generated diene **84**. Treatment of diene **84** with catalytic amounts of ruthenium carbene catalyst afforded the macrocycle as an *E*/*Z* mixture, which was subsequently hydrogenated using $RhCl(PPh_3)_3$ to afford macrocycle **85** without affecting the cinnamate and tiglate groups. Finally, removal of the *C*-silyl group and the *O*-TBS groups with TASF, followed by cleavage of the remaining isopropylidene acetal with TFA, accomplished the synthesis of ipomoeassin B (**69**) and ipomoeassin E (**72**). In 2009, Fürstner and coworkers reported the divergent total synthesis and biological evaluation of the cytotoxicity of ipomoeassin A–F and analogs using similar synthetic routes [29].

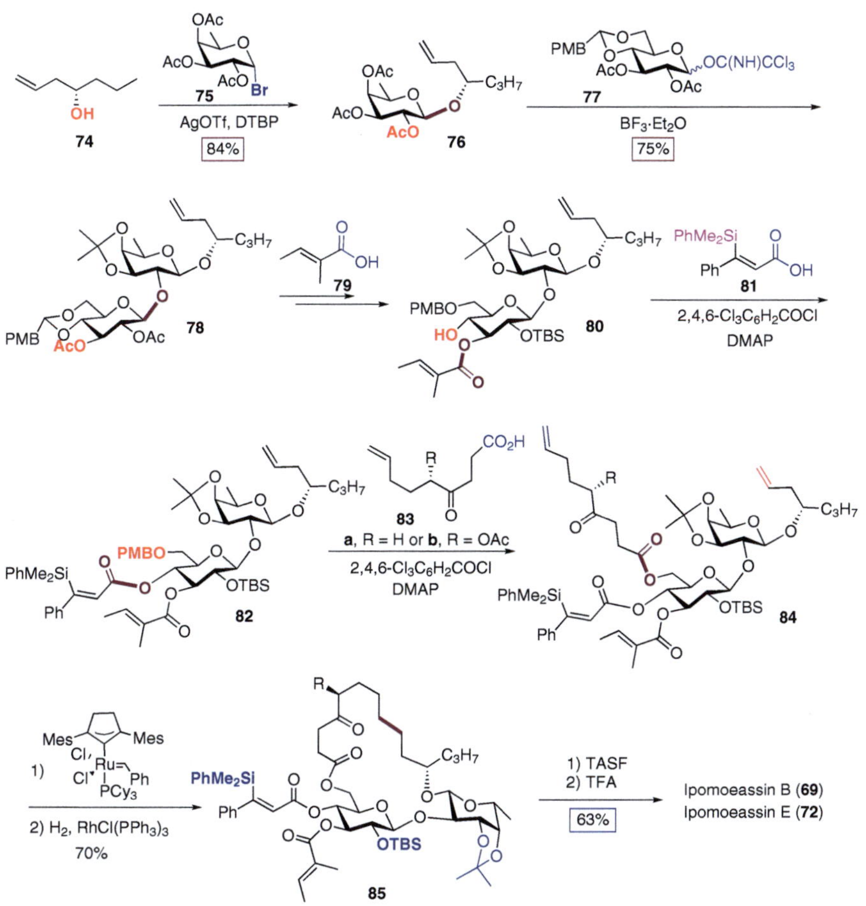

**Scheme 8.9** Total synthesis of ipomoeassin B (**69**) and E (**72**) by Fürstner and coworkers.

**Total synthesis of ipomoeassin F (73) by Postema and coworkers.** In 2009, Postema et al. reported the first total synthesis of ipomoeassin F (**73**) through the RCM-based strategy (Scheme 8.10) [30]. Their syntheses began with the coupling of chiral alcohol **86** with the known fucosyl bromide **75** with $Ag(CO_3)_2$ and $AgClO_4$ as catalyst, which gave β-glycoside **87** in 64% yield. After several steps of protecting group manipulation, the C2-OH of **87** was coupled with glucosyl imidate **88** in the presence of $BF_3{\cdot}Et_2O$, which afforded the expected β-disaccharide **89** in 73% yield after ketal group removal. The regioselective condensation of C6′-OH with acid **83a** afforded the desired ester **90**, whose diene moiety underwent the RCM reaction as Fürstner's approach and subsequent hydrogenation, yielding the expected macrolactone **91** in 70% yield. Successive condensation of **91** with cinnamic acid and tiglic acid **79** at the C4′ and C3′-positions, respectively, furnished the suitably protected ipomoeassin F **92**. Finally, the selective removal of the two α-chloroacetates (CA) groups with DABCO, in the presence of OAc group, concluded the total synthesis of ipomoeassin F (**73**).

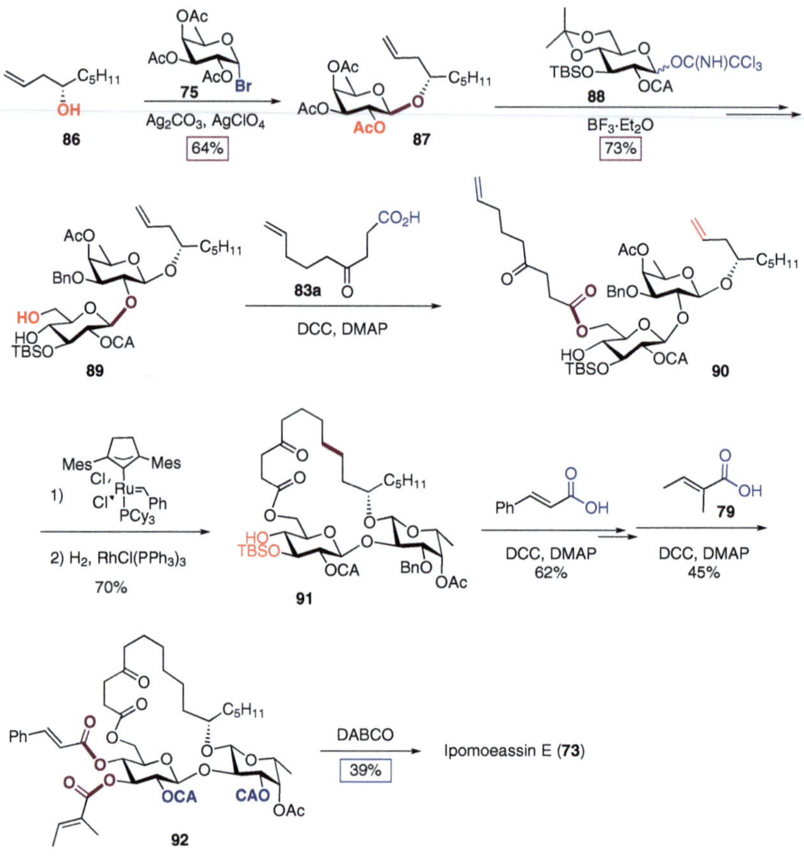

**Scheme 8.10** Total synthesis of ipomoeassin F (**73**) by Postema and coworkers.

**Total synthesis of ipomoeassin F (73) by Shi and coworkers.** In 2015, Shi and coworkers reported a gram-scale synthesis of ipomoeassin F (**73**) with the longest linear sequence of 17 steps and in 3.8% overall yield (Scheme 8.11) [31, 32]. The coupling of chiral alcohol **86** with galactosyl imidate **93** in the presence of TMSOTf afforded β-glycoside **94** in an impressive yield of 93%. Leveraging the established reactivity order for secondary hydroxyl groups in the galacto configuration (3-OH > 2-OH > 4-OH), selective silylation on 3-OH was performed after the removal of all O–Ac groups, leading to the formation of diol **95**. Regioselective Schmidt glycosylation at the 2-OH of **95** with glucosyl imidate **96**, in the presence of TMSOTf, gave disaccharide **97** in 93% yield. Following the removal of isopropylidene group, the regioselective esterification of the 6-OH of glucose residue with carboxylic acid **83a** gave diene **98**, serving as the RCM precursor. Treatment of diene **98** with Hoveyda–Grubbs catalyst (II) gave the macrolide as a mixture of *E*/*Z* isomers, which were subjected to hydrogenation, resulting in the desired macrolide **99** in 70% yield for two steps. Further condensation of **99** with cinnamic acid provided the protected ipomoeassin F **100**, which was deprotected with TBAF and AcOH to complete the synthesis of ipomoeassin F (**73**).

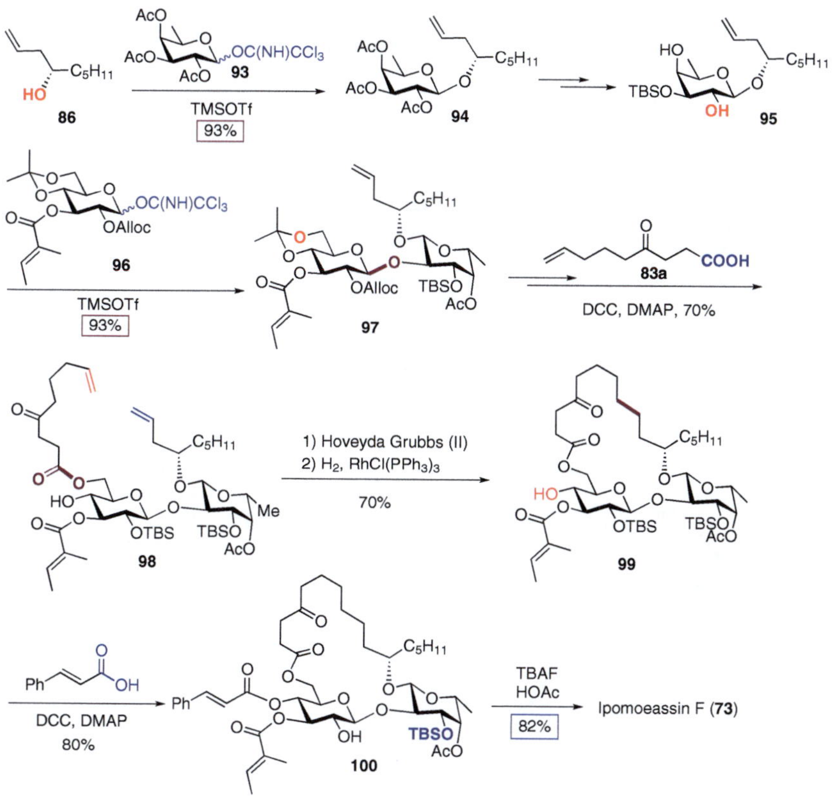


**Scheme 8.11** Total synthesis of ipomoeassin F (**73**) by Shi and coworkers.

In 2020, the Shi group reported the synthesis of two novel analogs of ipomoeassin F (**73**) with varying ring sizes [33]. These analogs demonstrated enhanced cytotoxicity and *in vitro* protein translocation inhibition activities. The improved outcomes observed with the ring-expanded analogs indicated that the binding pocket of Sec61α could accommodate additional structural modifications, particularly in the fatty acid portion.

## Woodrosin I

Woodrosin I (**101**, Figure 8.6) was initially isolated from the stems of *Ipomoea tuberosa* L by Miyahara and coworkers in 1993 [34]. Structurally, woodrosin I

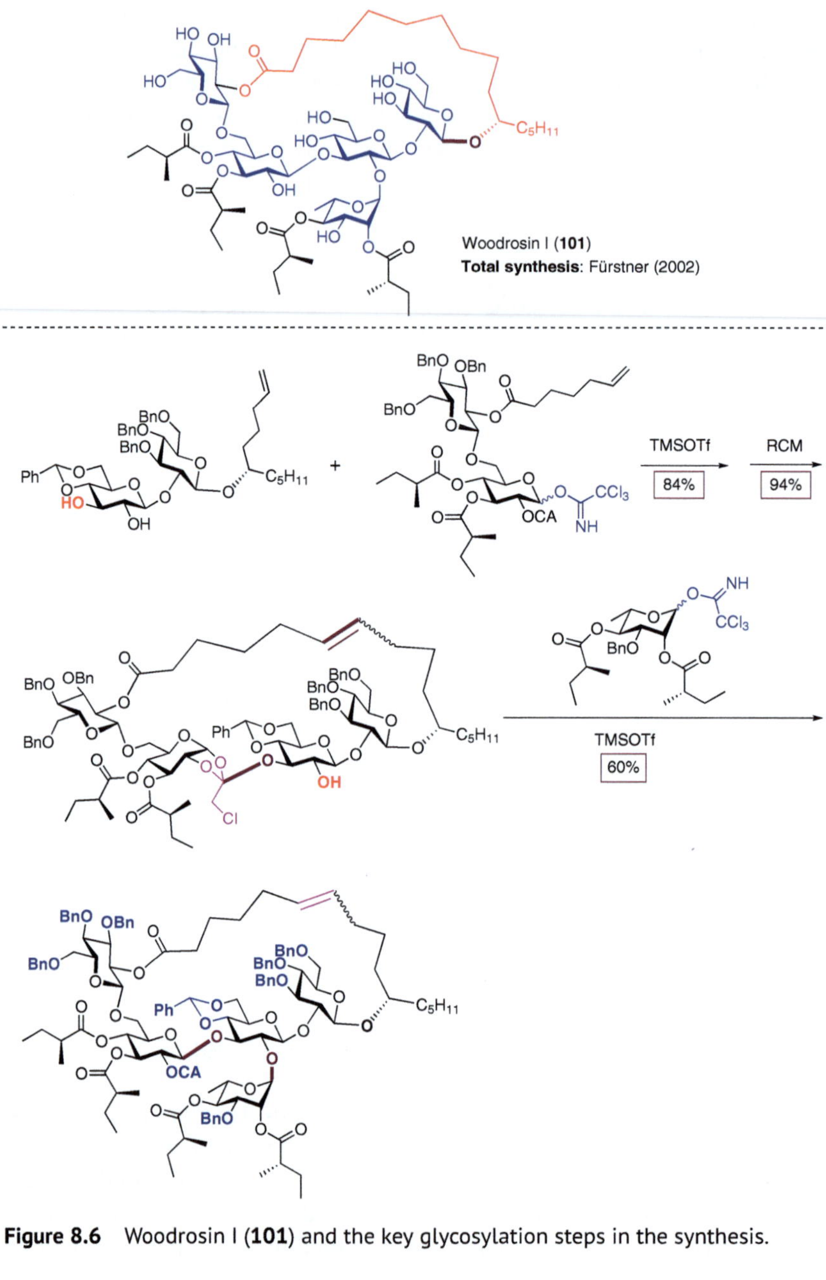


**Figure 8.6** Woodrosin I (**101**) and the key glycosylation steps in the synthesis.

consists of a lactone moiety that spans four glucose units, forming a 27-membered macrocycle. From a synthetic point of view, the distinctive acylation pattern at its periphery poses restrictions on the selection of temporary protecting groups. Additionally, the branched pentasaccharide backbone gives rise to significant steric hindrance at vicinal positions. In 2002, Fürstner et al. reported the first total

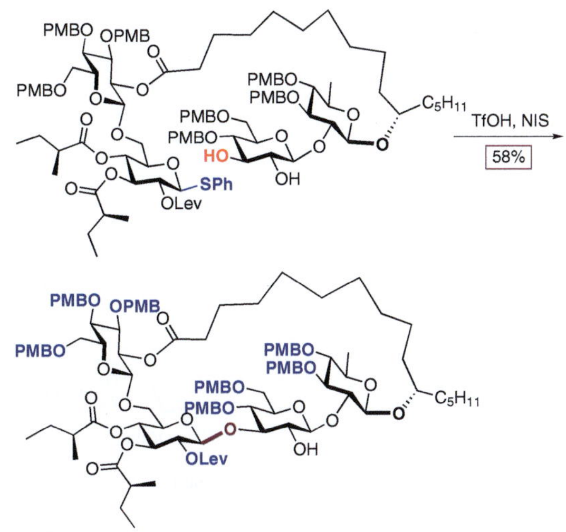


**Figure 8.7** Synthesis of the macrolide core of calysolin IX using an intramolecular glycosylation by Mlynarski and coworkers.

synthesis of woodrosin I (**101**), employing glycosylations with glycosyl imidate donors and RCM reactions as the key steps [35–37].

Recently, Mlynarski and coworkers reported the utility of intramolecular glycosylation for the synthesis of the 27-membered macrocyclic ring of a structurally related resin glycoside calysolin IX (Figure 8.7) [38].

## Batatosides L

Batatoside-type resin glycosides were isolated from the tuber of *Ipomoea batatas* (L.) Lam. (Convolvulaceae) by Kong and coworkers in 2008 [39]. Among these congeners, batatosides L (**102**) demonstrated significant cytotoxic activity against laryngeal carcinoma (Hep-2) cells, with $ED_{50}$ values at 3.5 μg/ml (Figure 8.8). In 2010, Yang and coworkers reported the total synthesis of batatoside L (**102**) employing a convergent approach, which involved the Corey–Nicolaou macrolactonization and glycosylation with a disaccharide imidate donor as the key steps [40].

## Batatin VI

Batatin VI (proposed as **103**, Figure 8.9) was isolated from the tuberous roots of sweet potato (*I. batatas*) by Pereda-Miranda et al. in 2011 [41]. Batatin VI (**103**) represents the first known example of an ester-type dimer of heterotetrasaccharide operculinic acid C. The ester linkage between the macrocycle-containing unit A and the acyclic unit B is positioned at C-3 of rhamnose moiety c within unit A. In 2013, Yang and coworkers reported the first total synthesis of the proposed structure of batatin VI (**103**) via a convergent [5+3] glycosidic coupling approach. However, the analysis data of the synthesized compound did not match that of the natural batatin VI, thus necessitating a revision of its proposed structure [42].

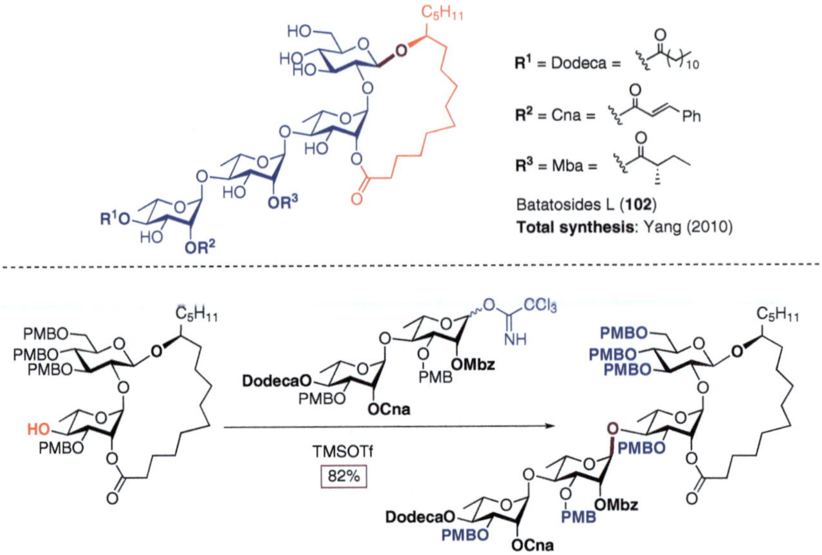


**Figure 8.8** Batatoside L (**102**) and the glycosylation step in the synthesis.

## Merremoside D

The merremoside family, including merremoside D (proposed as **104**, Figure 8.10), was initially isolated from the fresh tuber of *Merremia mammosa* (Lour.) Hall. f. (convolvulaceae) by Kitagawa et al. in 1996 [43]. Structurally, these resin glycosides are characterized by a 20- or 21-membered macrolactone composed of a bisrhamnose disaccharide bridged at C-1 and either C-2′ or C-3′ by jalapinolic acid. Despite their interesting biological activities, a comprehensive investigation regarding their structure–activity relationship (SAR) remains unexplored. In 2014, O'Doherty and coworkers reported the first synthesis of the proposed structure of merremoside D (**104**) in 22 longest linear steps, which features the de novo asymmetric synthesis of all 21 stereocenters in the final product through asymmetric catalysis [44]. In 2020, Wan and coworkers reported a novel one-pot relay glycosylation protocol, involving the activation of SPSB glycosides with $Tf_2O$, to construct the first glycosidic linkage. Simultaneously, a cyclic thiosulfonium ion is generated *in situ*, which serves as a relay activator for the subsequent activation of the thioglycoside at a higher temperature, leading to the formation of the second glycosidic linkage. The expedient total synthesis of merremoside D (**104**) was realized by utilizing this one-pot relay glycosylation approach twice, establishing all four glycosidic bonds [45].

## Murucoidins

Murucoidins V (**105**), IV (**106**), and congeners were isolated from the flowers of *Ipomoea murucoides* by Pereda-Miranda in 2006 (Figure 8.11) [46]. Murucoidin V

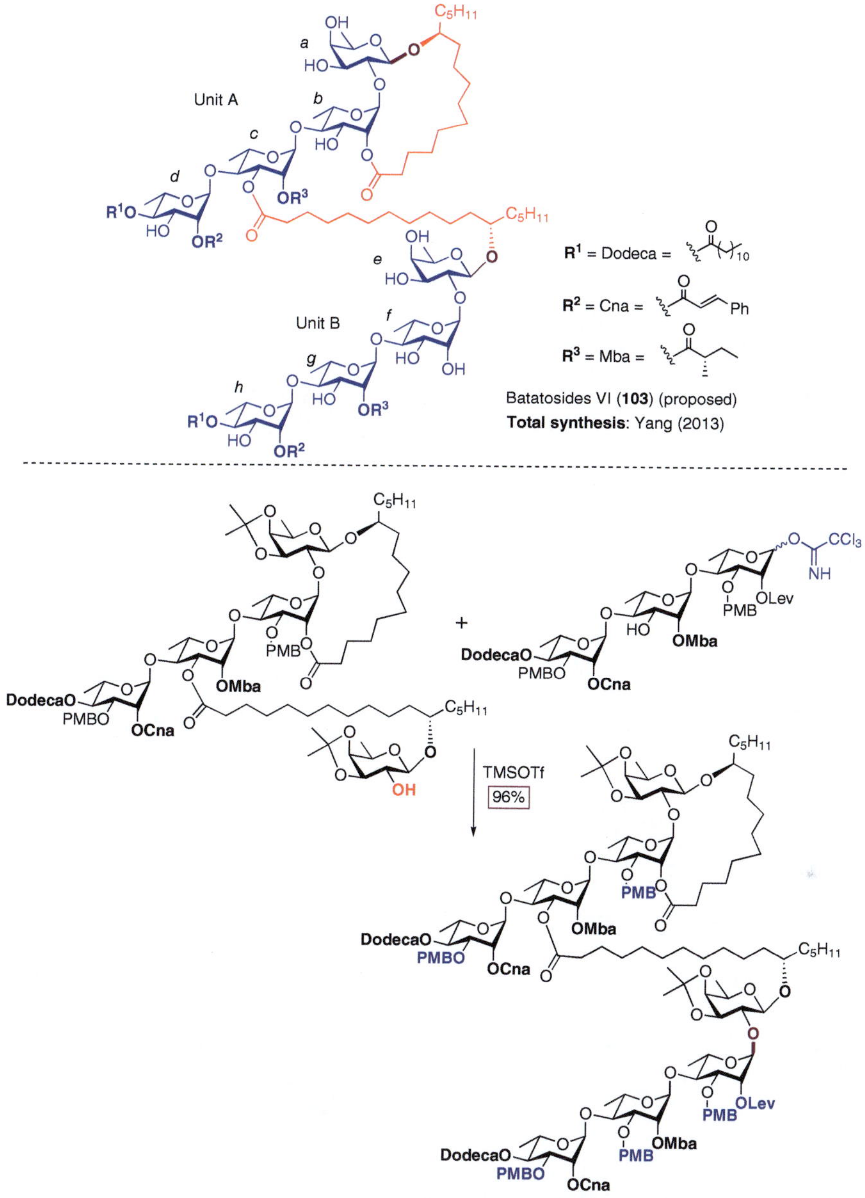


**Figure 8.9** Proposed structure of batatin VI (**103**) and the glycosylation step in its synthesis.

(**105**) exhibited significant multidrug-resistant (MDR) reversal activity, which was able to enhance the susceptibility of vinblastine by 255-fold through the inhibition of the efflux pump of vinblastine-resistant human breast carcinoma cells (MCF-7/Vin). Structurally, murucoidins IV (**106**) and V (**105**) comprise a hydrophilic oligosaccharide chain containing rare deoxy sugars and a hydrophobic

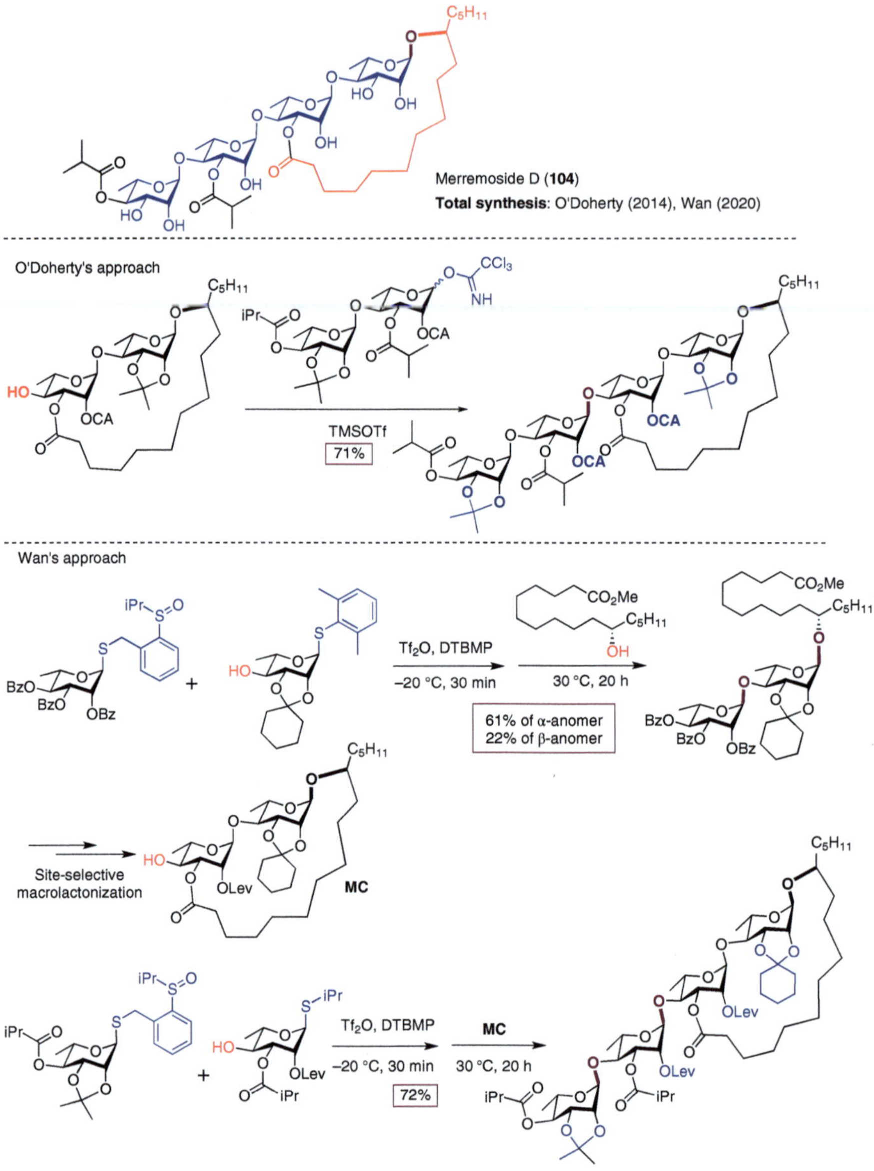


**Figure 8.10** Merremoside D (**104**) and the glycosylation steps in its synthesis.

hydroxy fatty acid aglycone chain. The only difference lies in the size of their macrolactone rings, with murucoidin V (**105**) possessing an 18-membered ring and murucoidin IV (**106**) featuring a 19-membered ring. In 2019, Wan and coworkers reported the first total syntheses of murucoidins IV (**106**) and V (**105**) utilizing a key [3+2] glycosidic coupling by the interrupted Pummerer reaction-mediated (IPRm) glycosylations [47].

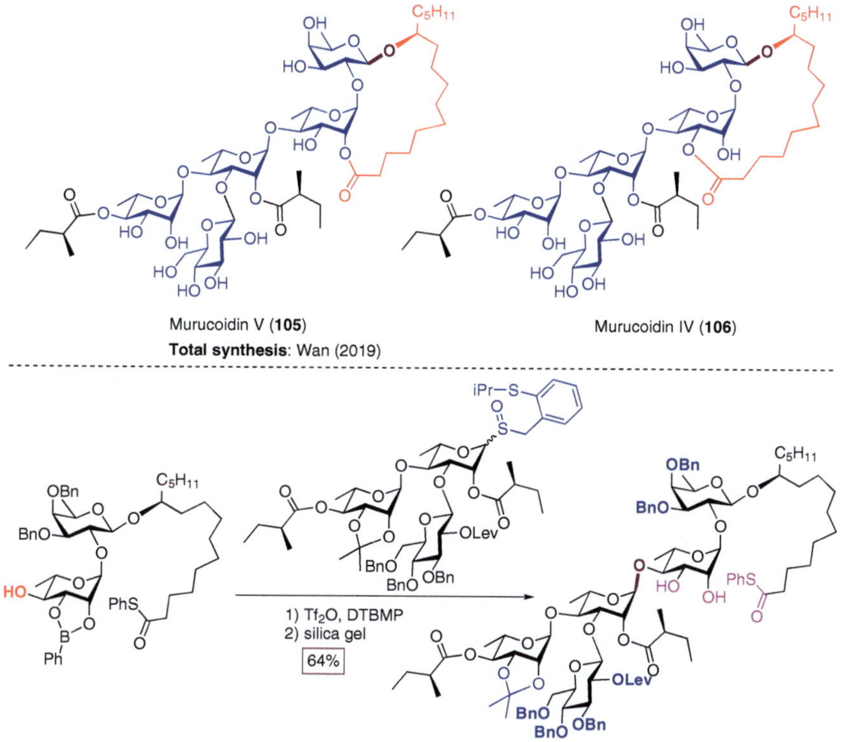


**Figure 8.11** Murucoidins IV (**105**) and V (**106**) and the glycosylation step in the syntheses.

## References


**1** Ono, M. (2017). *J. Nat. Med.-Tokyo* 71: 591–604.

**2** Maharani, R., Fajar, M., and Supratman, U. (2022). *Molecules* 27: 8161.

**3** Fang, Y., Chai, W., Chen, S. et al. (1993). *Carbohydr. Res.* 245: 259–270.

**4** Pereda-Miranda, R., Mata, R., Anaya, A.L. et al. (1993). *J. Nat. Prod.* 56: 571–582.

**5** Furukawa, J.S. (2001). *Trends Glycosci. Glycotechnol.* 12: 1–10.

**6** Fürstner, A. (2004). *Eur. J. Org. Chem.* 2004: 943–958.

**7** Xie, J. and Bogliotti, N. (2014). *Chem. Rev.* 114: 7678–7739.

**8** Wang, W., Li, Y., He, Y. et al. (2022). *Front. Chem.* 10: 1036954.

**9** Jiang, Z.-H., Geyer, A., and Schmidt, R.R. (1995). *Angew. Chem. Ed. Int. Engl.* 34: 2520–2524.

**10** Jiang, Z.-H. and Schmidt, R.R. (1994). *Liebigs Ann. Chem.* 1994: 645–651.

**11** Sakairi, N., Okazaki, Y., Furukawa, J.-i. et al. (1998). *Bull. Chem. Soc. Jpn.* 71: 679–683.

**12** Furukawa, J.-I., Kobayashi, S., Nomizu, M. et al. (2000). *Tetrahedron Lett.* 41: 3453–3457.

**13** Bah, M. and Pereda-Miranda, R. (1997). *Tetrahedron* 53: 9007–9022.

**14** Lu, S.-F., O'Yang, Q., Guo, Z.-W. et al. (1997). *Angew. Chem. Ed. Int. Engl.* 36: 2344–2346.

**15** Lu, S.-F., O'Yang, Q., Guo, Z.-W. et al. (1997). *J. Org. Chem.* 62: 8400–8405.
**16** Mootoo, D.R., Konradsson, P., Udodong, U., and Fraser-Reid, B. (1988). *J. Am. Chem. Soc.* 110: 5583–5584.
**17** Larson, D.P. and Heathcock, C.H. (1996). *J. Org. Chem.* 61: 5208–5209.
**18** Larson, D.P. and Heathcock, C.H. (1997). *J. Org. Chem.* 62: 8406–8418.
**19** Hikota, M., Sakurai, Y., Horita, K., and Yonemitsu, O. (1990). *Tetrahedron Lett.* 31: 6367–6370.
**20** Fürstner, A. and Müller, T. (1999). *J. Am. Chem. Soc.* 121: 7814–7821.
**21** Fürstner, A. and Müller, T. (1998). *J. Org. Chem.* 63: 424–425.
**22** Shu, P., Xiao, X., Zhao, Y. et al. (2015). *Angew. Chem. Int. Ed.* 54: 14432–14436.
**23** Sun, J., Fang, J., Xiao, X. et al. (2020). *Org. Biomol. Chem.* 18: 3818–3822.
**24** Brito-Arias, M., Pereda-Miranda, R., and Heathcock, C.H. (2004). *J. Org. Chem.* 69: 4567–4570.
**25** Son, S.H., Yanagiya, N., Furukawa, J., and Sakairi, N. (2009). *Synlett* 18: 2957–2960.
**26** Cao, S., Guza, R.C., Wisse, J.H. et al. (2005). *J. Nat. Prod.* 68: 487–492.
**27** Cao, S., Norris, A., Wisse, J.H. et al. (2007). *Nat. Prod. Res.* 21: 872–876.
**28** Fürstner, A. and Nagano, T. (2007). *J. Am. Chem. Soc.* 129: 1906–1907.
**29** Nagano, T., Pospíšil, J., Chollet, G. et al. (2009). *Chem. Eur. J.* 15: 9697–9706.
**30** Postema, M.H.D., TenDyke, K., Cutter, J. et al. (2009). *Org. Lett.* 11: 1417–1420.
**31** Zong, G., Barber, E., Aljewari, H. et al. (2015). *J. Org. Chem.* 80: 9279–9291.
**32** Zong, G., Whisenhunt, L., Hu, Z., and Shi, W.Q. (2017). *J. Org. Chem.* 82: 4977–4985.
**33** Zong, G., Hu, Z., Duah, K.B. et al. (2020). *J. Org. Chem.* 85: 16226–16235.
**34** Ono, M., Nakagawa, K., Kawasaki, T., and Miyahara, K. (1993). *Chem. Pharm. Bull.* 41: 1925–1932.
**35** Fürstner, A., Jeanjean, F., and Razon, P. (2002). *Angew. Chem. Int. Ed.* 41: 2097–2101.
**36** Fürstner, A., Jeanjean, F., Razon, P. et al. (2003). *Chem. Eur. J.* 9: 320–326.
**37** Fürstner, A., Jeanjean, F., Razon, P. et al. (2003). *Chem. Eur. J.* 9: 307–319.
**38** Nawój, M., Grobelny, A., and Mlynarski, J. (2020). *Eur. J. Org. Chem.* 2020: 47–51.
**39** Yin, Y., Li, Y., and Kong, L. (2008). *J. Agric. Food. Chem.* 56: 2363–2368.
**40** Xie, L., Zhu, S.-Y., Shen, X.-Q. et al. (2010). *J. Org. Chem.* 75: 5764–5767.
**41** Rosas-Ramírez, D., Escalante-Sánchez, E., and Pereda-Miranda, R. (2011). *Phytochemistry* 72: 773–780.
**42** Zhu, S.-Y., Huang, J.-S., Zheng, S.-S. et al. (2013). *Org. Lett.* 15: 4154–4157.
**43** Kitagawa, I., Baek, N.I., Kawashima, K. et al. (1996). *Chem. Pharm. Bull.* 44: 1680–1692.
**44** Sharif, E.U., Wang, H.-Y.L., Akhmedov, N.G., and O'Doherty, G.A. (2014). *Org. Lett.* 16: 492–495.
**45** Xiao, X., Zeng, J., Fang, J. et al. (2020). *J. Am. Chem. Soc.* 142: 5498–5503.
**46** Chérigo, L. and Pereda-Miranda, R. (2006). *J. Nat. Prod.* 69: 595–599.
**47** Fang, J., Zeng, J., Sun, J. et al. (2019). *Org. Lett.* 21: 6213–6216.

# 9

# Steroid Glycosides

Steroid glycosides, often named steroidal saponins, are predominantly found in monocotyledonous angiosperms and include four major types of steroid skeletons, i.e. tetracyclic cholestane, lactone-containing cardenolide, pentacyclic furostane, and hexacyclic spirostane (Figure 9.1). The name "saponin" derives from the Latin word "*sapo*," meaning soap, because these compounds can function as surfactants, producing soapy foams when shaken with water. This is due to the amphiphilic nature of these molecules, which contain hydrophilic sugar chains and lipophilic sapogenin. In nature, plants employ saponins to defend themselves against pathogens, pests, and herbivores. Therefore, saponins commonly show antimicrobial, fungicidal, insecticidal, molluscicidal, and antiparasitic activities. Steroid saponins also demonstrate various biological and pharmacological properties, such as hemolytic, cytotoxic, immunomodulatory, anti-inflammatory, and antitumor activities, and play pivotal roles as essential constituents in traditional folk medicines, particularly within the realm of traditional Chinese medicine.

The isolation of pure saponins from crude plant extracts, especially in substantial quantities, is tedious and challenging due to the heterogeneity of these complex molecules. Chemical synthesis of steroid glycosides necessitates the collaborative integration of two distinct research domains of organic chemistry: carbohydrate chemistry and steroid chemistry. Some steroid aglycones, such as diosgenin and cholesterol, can be obtained in abundant amounts by direct extraction from natural sources. Many others require synthesis from readily accessible precursors. In recent decades, remarkable advancements have been made in the synthesis of steroid glycoside natural products and their derivatives [1–4], which facilitated the structure-activity-relationships (SARs) and mechanistic studies on their biological properties.

## Cholestane Type Steroid Glycosides

### OSW-1

OSW-1 (**1**, Figure 9.2), a cholestan disaccharide, was first discovered in 1992 from the bulbs of *Ornithogalum saundersid*, a perennial lily plant in southern

*Carbohydrate Chemistry in the Total Synthesis of Naturally Occurring Glycosides*,
First Edition. Biao Yu and Xiaoyu Yang.
© 2024 WILEY-VCH GmbH. Published 2024 by WILEY-VCH GmbH.

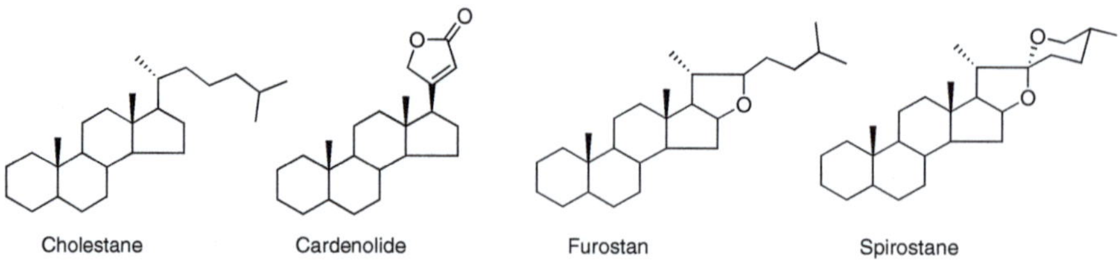


**Figure 9.1** Representative skeletons of the steroidal aglycones.

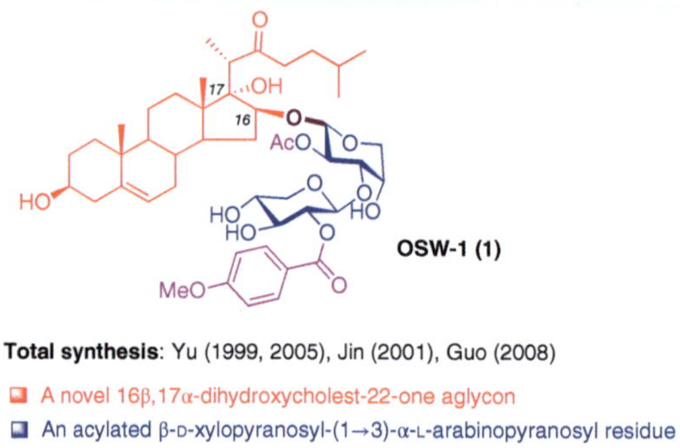


**Figure 9.2** OSW-1 (**1**) and its structural features.

Africa [5]. It exhibited remarkable cytotoxicity, with $IC_{50}$ value between 0.1 and 0.7 nM against a variety of malignant tumor cells, which are 10- to 100-fold more potent than the clinically utilized anticancer agents, including mitomycin C, adriamycin, cisplatin, camptothecin, and taxol [6]. The exceptionally potent cytotoxicity made OSW-1 a promising lead for development of novel antitumor drugs [7]. Structurally, OSW-1 (**1**) features a unique 16β,17α-dihydroxycholest-22-one aglycone and an acylated β-D-xylopyranosyl-(1 → 3)-α-L-arabinopyranosyl residue linked to the 16-hydroxyl group. The SAR studies have underscored the critical role of the acyl groups on the disaccharide segment for the antitumor activity of this molecule.

**Total synthesis of OSW-1 (1) by Yu, Hui, and coworkers.** In 1999, Yu, Hui, and coworkers reported the first total synthesis of OSW-1 (**1**) from commercial available dehydroisoandrosterone (**6**), L-arabinose, and D-xylose, with the longest linear sequence of 14 steps and in 6% overall yield (Scheme 9.1) [8]. Given that the equatorial 3-OH is more reactive than the axial 4-OH in a pyranoside, a regioselective glycosylation between arabinoside diol **2** and xylosyl trichloroacetimidate **3** was carried out. Under the catalysis of $BF_3 \cdot Et_2O$, this reaction yielded the desired C-3 glycosylated disaccharide **4** in 70% yield, after protection of the remaining 4-OH with TES group. The C-4 glycosylated product was also isolated in 21% yield as the major byproduct. Subsequent removal of the anomeric benzyl group on **4**, followed by treatment with $CCl_3CN$/DBU, provided disaccharide imidate donor **5**. On the other hand, the preparation of 16β-OH aglycone was achieved in a total of 12 steps from dehydroisoandrosterone (**6**). After the Wittig reaction on compound **6**, an ene reaction with paraformaldehyde was carried out, resulting in the formation of a homoallylic alcohol, which was subsequently oxidized to yield the corresponding aldehyde **7**. Addition of **7** with (3-methylbutyl)magnesium bromide (**8**) installed the side chain, giving product **9** after subsequent transformations. Dihydroxylation of the 16,17-alkene moiety on **9** using $OsO_4$ afforded the *cis*-diol, which was then converted into the corresponding *trans*-diol **10** after the oxidation–reduction manipulation. With both the disaccharide donor and aglycone in hand, the glycosylation of the 16β,17α-diol **10** with disaccharide imidate **5** was performed under the promotion of TMSOTf, which proceeded smoothly to provide the fully protected OSW-1 **11** in a satisfactory yield (69%). Finally, all of the protecting groups (one TBS, three TES, and one ethylene glycol ketal) were cleanly removed in a single step by employing $Pd(MeCN)_2Cl_2$ as catalyst (79%), thus accomplishing the synthesis of OSW-1 (**1**).

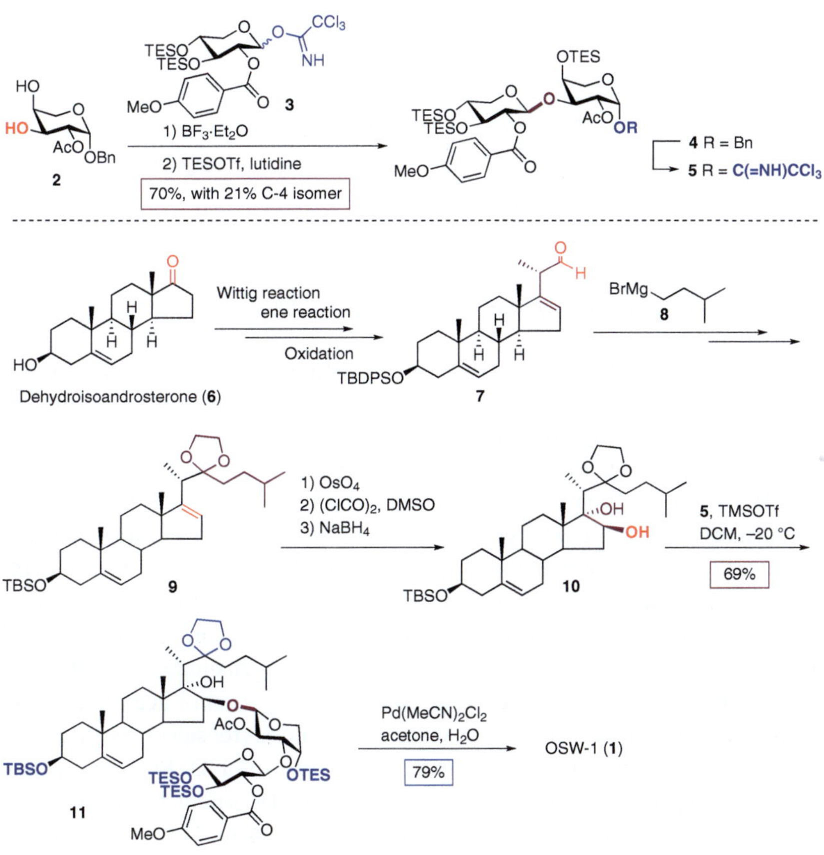


**Scheme 9.1** Total synthesis of OSW-1 (**1**) by Yu, Hui, and coworkers.

In 2005, Yu and coworkers reported an alternative approach to the synthesis of OSW-1 (**1**, Scheme 9.2) [9]. The key step in the construction of the aglycone involved a stereoselective aldol reaction between 16α-hydroxy-5-androsten-17-one **12** and propionate under basic conditions, which yielded 16α,17α-dihydroxycholest-22-one **13**.

After installation of the isoamyl ketone side chain, inversion of the 16-hydroxyl configuration and selective protection of the C22-hydroxy group led to the suitable protected aglycone **14**, which was then coupled with disaccharide imidate **5**, producing the desired β-glycosylated product **15** in 61% yield. Finally, oxidation of the C-22–OH group to ketone and global deprotection of the silyl ether protection group with $Pd(MeCN)_2Cl_2$ completed the synthesis of OSW-1 (**1**) in 65% (for three steps).

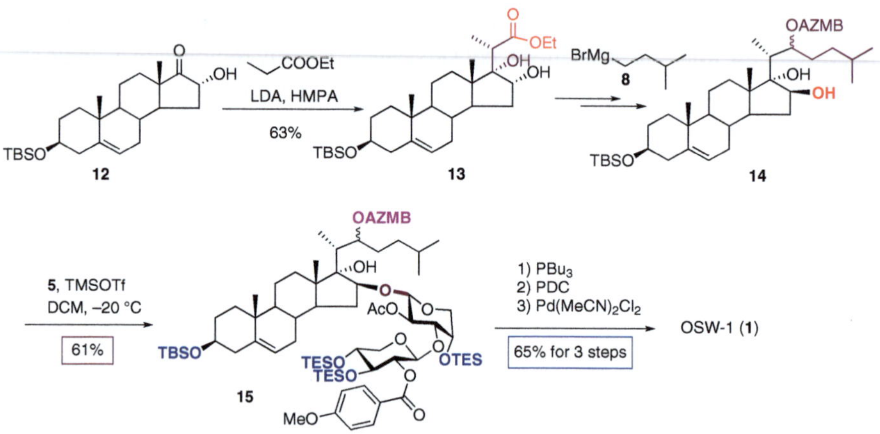


**Scheme 9.2** Alternative approach to the total synthesis of OSW-1 (**1**) by Yu and coworkers.

**Total synthesis of OSW-1 (1) by Jin and coworkers.** In 2001, Jin and coworkers reported their total synthesis of OSW-1(**1**) in 10 linear operations from the commercially available dehydroisoandrosterone (**6**), achieving an overall yield of 28% (Scheme 9.3) [10, 11]. The disaccharide unit was constructed by coupling of thioglycoside **16** with xylosyl imidate **17** under the promotion $BF_3 \cdot Et_2O$, which afforded β-disaccharide **18** in 93% yield. Hydrolysis of the anomeric sulfide group and subsequent formation of trichloroacetimidate provided disaccharide donor **19**. Simultaneously, starting from dehydroisoandrosterone (**6**), the consecutive Wittig reaction, allylic oxidation, and Swern oxidation afforded enone **20**. The crucial conjugate addition of α-alkoxy vinyl cuprate **21** with enone **20** enabled the efficient and stereoselective installation of the side chain, leading to enol acetate **22**. The subsequent α-hydroxylation with the Davis reagent and stereoselective reduction of the ketone group using $LiAlH_4$ gave the expected *trans*-diol **23**. Further, the key glycosylation of aglycone **23** with disaccharide imidate **19**, under the catalysis of TMSOTf, furnished the desired β-glycosylated product **24** in 71% yield. Finally, removal of all the protecting groups by sequential treatment of **24** with DDQ and $Pd(MeCN)_2Cl_2$ in a single operation completed the synthesis of OSW-1 (**1**) in 81% yield.

**Total synthesis of OSW-1 (1) by Guo and coworkers.** In 2008, Guo and coworkers reported their total synthesis of OSW-1 (**1**) in 10 linear steps with an overall yield of 6.4%, starting from (+)-dehydroisoandrosterone (Scheme 9.4) [12]. The disaccharide segment was constructed through the coupling of glycosyl acceptor **25**

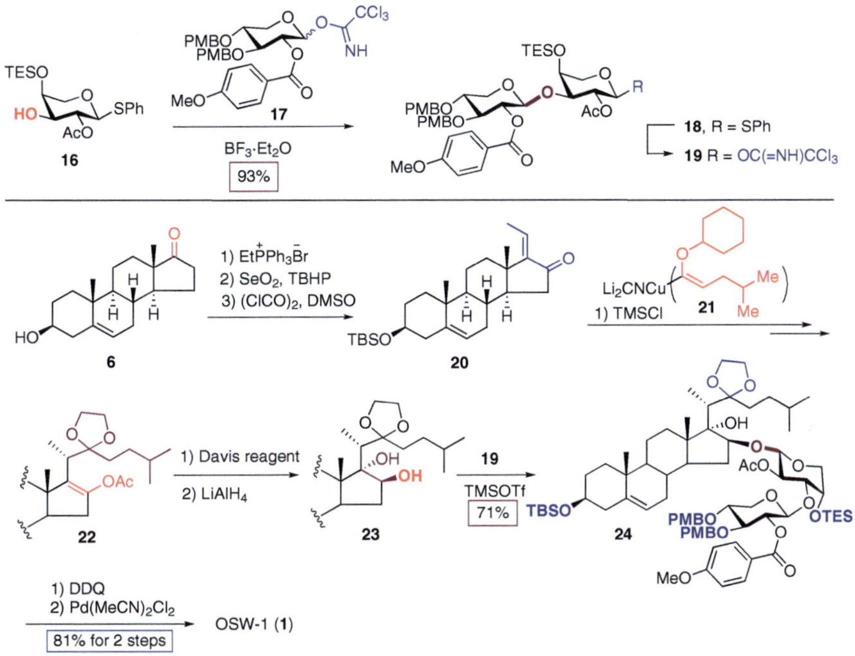

**Scheme 9.3** Total synthesis of OSW-1 (**1**) by Jin and coworkers.

with xylosyl imidate **26** under the promotion of $BF_3 \cdot Et_2O$, generating β-disaccharide **27** in 93% yield. A similar two-step procedure was employed to prepare disaccharide imidate donor **28**. The addition of propanenitrile onto dehydroisoandrosterone (**6**) under basic conditions afforded compound **29**. Further elongation of the side chain by refluxing of **29** with (3-methylbutyl)magnesium bromide afforded enone **30**, which upon treatment of NaOH could be isomerized to product **31** with the correct C21-methyl configuration. The alkene moiety was converted to a *trans*-diol by using Yu's approach in three steps. The coupling between aglycone **32** and disaccharide imidate **28** proceeded smoothly with TMSOTf as the catalyst, giving the desired β-disaccharide **33** in 58% yield. Finally, cleavage of the TBS and ethylene glycol ketal group by $HOAc/H_2O$ and removal of all the PMB groups by DDQ accomplished the synthesis in 72% yield for two steps.

## Periploside A

Periploside A (**34**, Figure 9.3), a representative member of the pregnane glycosides family, was initially isolated from *Periploca sepium* and *Periploca forrestii* (Asclepiadaceae) in 1987 [13]. This glycoside possesses immunosuppressive activities comparable to rapamycin and cyclosporin A, as it effectively inhibited ConA-induced T cell proliferation, while exhibiting favorable selectivity with low toxicity ($IC_{50}$ = 0.64 μM and $CC_{50}$ = 10.1 μM). The significant immunosuppressive effect of periploside A (**34**) has been further validated in mice models, affirming

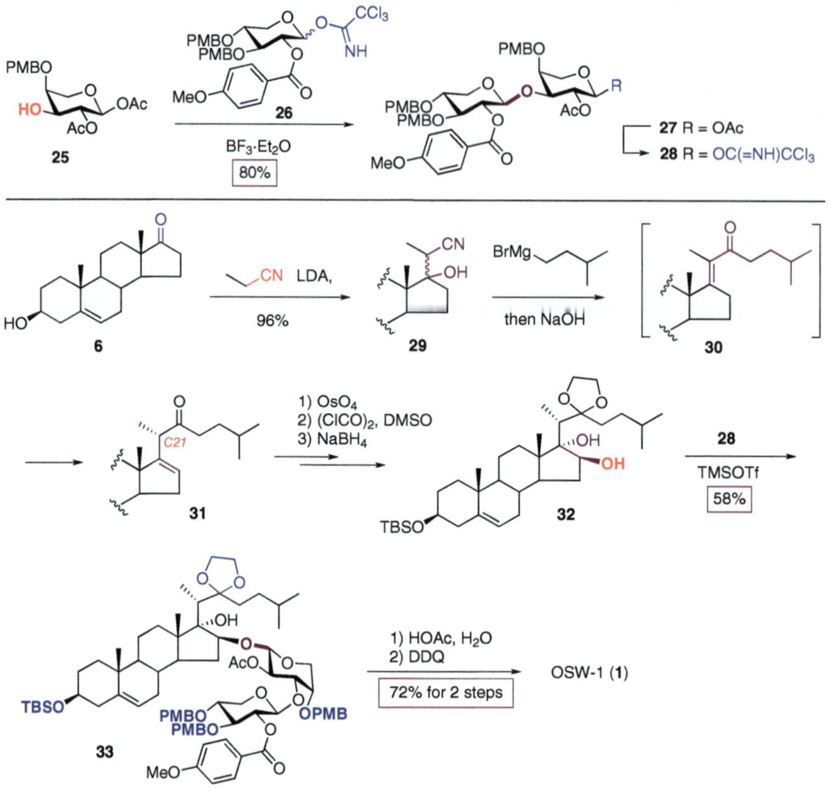

**Scheme 9.4** Total synthesis of OSW-1 (**1**) by Guo and coworkers.

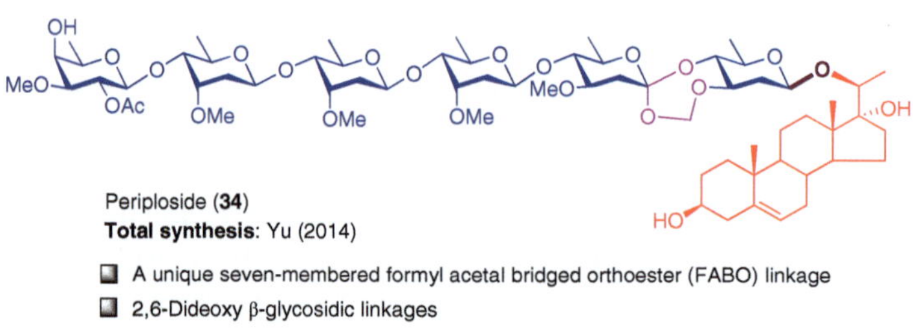

**Figure 9.3** Periploside A (**34**) and the structural features.

the traditional use of the plant in Chinese medicine for rheumatoid arthritis. Structurally, periploside A (**34**) features a unique seven-membered formyl acetal bridged orthoester (FABO) linkage between two sugar units, along with several 2,6-di-deoxy β-glycosidic linkages.

**Total synthesis of periploside A (34) by Yu and coworkers.** In 2014, Yu and coworkers achieved the total synthesis of peroploside A (**34**) in a total 76 steps, with the longest linear sequence of 29 steps and 9.2% overall yield [14]. The synthesis of the tetrasaccharide unit of peroploside A was described in

Scheme 9.5. The glycosylation reaction between digtoxoside accepter **35** and digtoxosyl *ortho*-cyclopropylethynylbenzoate **36** in the presence of $Ph_3PAuNTf_2$ (0.1 equiv.) afforded β-glycoside **37** as the sole anomer in an excellent 99% yield, in which the presence of the 3,4-*O*-TBDPS groups on donor **36** played a critical role in achieving the excellent β-selectivity. Following several steps of protecting group manipulations, the iterative coupling of the C4′-OH of **37** with **36** under the gold(I)-catalyzed conditions gave trisaccharide **38** in 99% yield. After cleavage of the TBDPS ethers, the selective protection of the C4″-OH group was followed by *O*-methylation of all the axial –OH groups, leading to alcohol **39**. Subsequently, coupling with digitalosyl *ortho*-alkynylbenzoate **40** in the presence of $Ph_3PAuNTf_2$ yielded the coupled tetrasaccharide **41** in a high yield of 93%. Notably, despite the presence of a 2-*O*-acetyl group in donor **39**, the resultant tetrasaccharide **41** existed as anomeric mixtures (β/α = 4 : 1). Removal of the anomeric MP group in tetrasaccharide **41** was followed by condensation with *ortho*-alkynylbenzoic acid, generating the corresponding tetrasaccharide donor **42**.

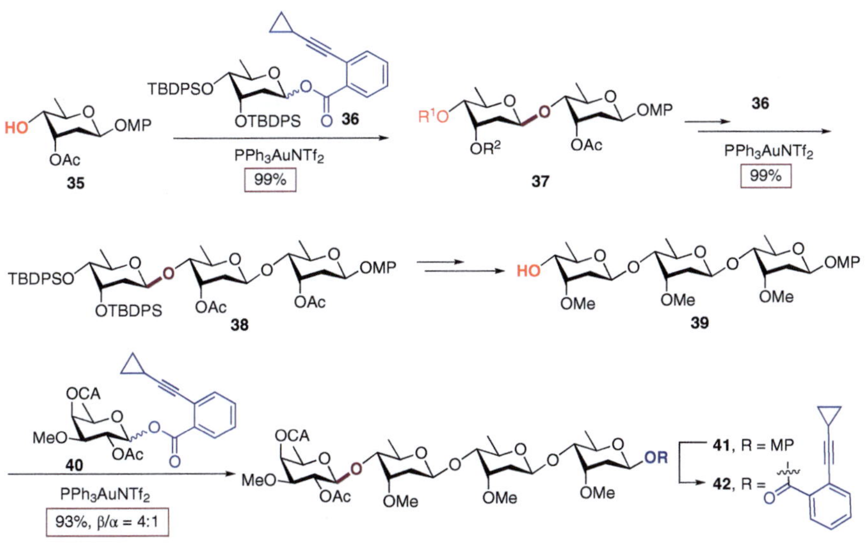

**Scheme 9.5** Synthesis of tetrasaccharide donor **42**.

On the other hand, 2-phenylseleno-D-mannopyranoside **44** was *in situ* converted into the corresponding glycosyl fluoride with diethylamino sulfur trifluoride (DAST), which was then coupled with sugar acceptor **43** under the action of $SnCl_2$ to give α-disaccharide **45** in 85% yield (Scheme 9.6). Subsequent selenoxide formation and syn-elimination, employing the modified Sinay conditions, gave the ketene acetal, which was condensed with $EtSCH_2OH$ to furnish the desired orthoester **46** as a single α-isomer. To achieve the key cyclization reaction, various conditions were screened for glycosidation of thioglycosides, and the Crich conditions ($Tf_2O$/BSP/TTBP, −114 °C) successfully yielded the desired seven-membered FABO product **47b** in a satisfactory 64% yield, along with the five-membered orthoester **47a** isolated in 18% yield. The FABO disaccharide **47b** was then subjected to selective cleavage of the anomeric MP group and further converted into the trifluoroacetimidate **48**. Subsequently, coupling of pregnane diol **49** with the FABO disaccharide donor **48**, under the promotion of TBSOTf, led to the corresponding disaccharide **50** in high yield (87%) and high β selectivity (β/α = 6.1 : 1). Following the removal of the bromide, iodide, and the 4′-*O*-benzoyl group, the subsequent coupling with tetrasaccharide donor **42** in the presence of $PPh_3AuOTf$ afforded hexasaccharide **51** in 80% yield with a β/α ratio of 2 : 1. Finally, the terminal *O*-CA and *O*-TBS groups were successively removed with thiourea and HF-pyridine, thus accomplishing the total synthesis of periploside A (**34**).

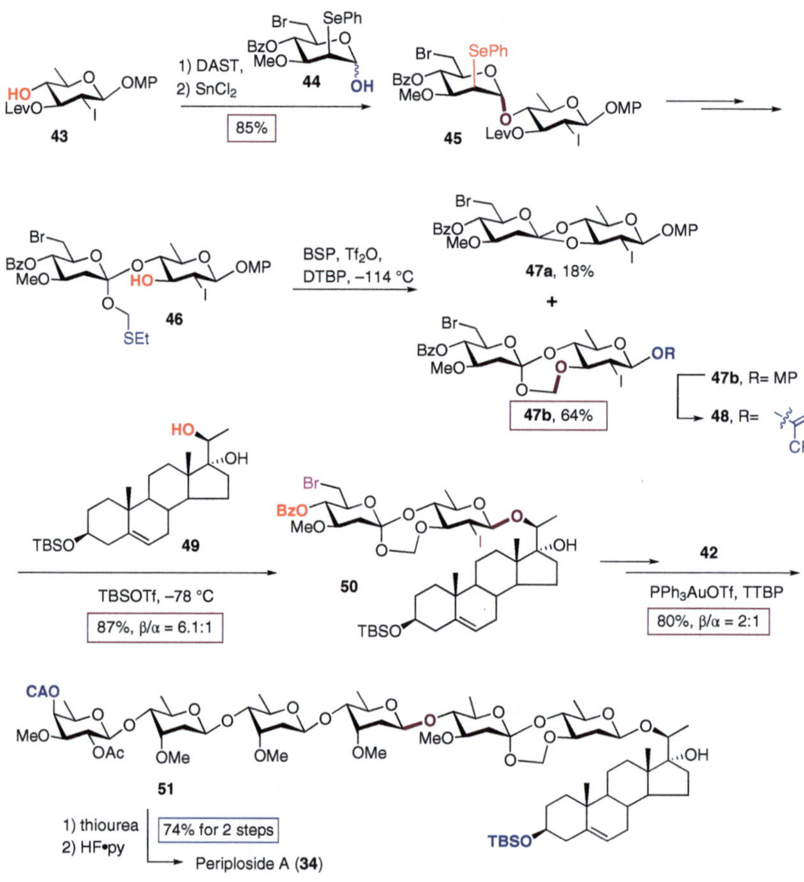


**Scheme 9.6** Total synthesis of periploside A (**34**) by Yu and coworkers.

## Luzonicosides and Sepositoside

Luzonicoside A (**52**, Figure 9.4) and sepositoside (**54**), two major cyclic starfish glycosides, were initially discovered from *Echinaster luzonicus* and *Echinaster*

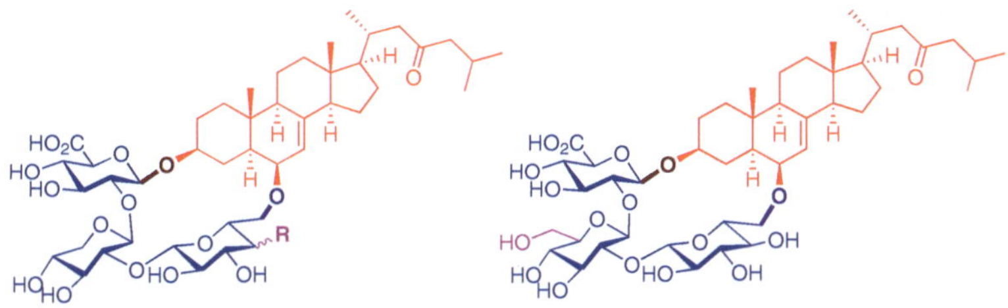

Luzonicoside A (**52**), R = axial OH
Luzonicoside D (**53**), R = equatorial OH
**Total synthesis**: Yu (2022)

Sepositoside A (**54**)

- Unprecedented 16-membered ring imposed by congested steroid and trisaccharide
- A sterically demanding and vulnerable C6β ether linkage

**Figure 9.4** Starfish cyclic steroid glycosides luzonicoside A (**52**) and D (**53**) and sepositoside A (**54**) and the structural features.

*sepositus* in early 1980s by Minale and coworkers [15]. Luzonicoside D (**53**) was discovered by Kicha et al. from *E. luzonicus* collected from a different region in 2015 [16]. Among these cyclic starfish glycosides, luzonicoside A (**52**) exhibited the highest activity in stimulating lysosomal activity, elevating intracellular ROS levels, and up-regulating NO synthesis in RAW 264.7 murine macrophages, all at non-toxic concentrations ranging from 0.01 to 0.1 μM [17]. Structurally, these steroidal cyclic glycosides feature a 16-membered ring formed by bridging C3 and C6 of a 3β,6β-dihydroxy-$\Delta^7$ steroid with a trisaccharide moiety through a glycosidic and ether linkage. The central synthetic challenge associated with these products lies in the construction of the conformationally rigid 16-members macrocycles, including formation of the sterically demanding and vulnerable C6β ether linkage.

**Total synthesis of luzonicoside A, D, and sepositoside A by Yu and coworkers.** The challenge in the construction of the strained 16-membered ring in starfish cyclic steroid glycosides was evidenced by the successful construction of simplified 16-membered rings through Au(I)-catalyzed intramolecular addition of sugar alcohol to a steroidal 6,7-epoxide, whereas the similar macroetherification on the authentic steroid trisaccharide was not successful [18–20]. Until 2022, Yu and coworkers realized a collective total synthesis of luzonicosides A (**52**), D (**53**), and sepositoside A (**54**) (Scheme 9.7) [21]. Due to the previously unsuccessful attempts to form the C6β ether linkage, the authors proposed a de novo protocol for the construction of the C6β-hexopyranosyl moiety. The synthesis began with the ring-closing metathesis (RCM) reaction between steroidal 6β-*O*-allyl derivative **55** and L-arabinose-derived olefine **56**, which afforded the coupled olefine **57** on a multi-gram scale. Following Sharpless asymmetric dihydroxylation, a series of transformations were carried out to construct the C6-galactopyranosyl *ortho*-hexynylbenzoate moiety **58**. Subsequent glycosylation of **58** with β-D-glucuronic acid imidate donor **59** under the catalysis of $TrBAr^F{}_4$ (10 mol%) afforded steroid bisglycoside **60** in 59% yield based on the recovered starting material (brsm). It was observed that the gold(I)-catalyzed intramolecular glycosylation failed to yield the desired cyclic glycoside after the introduction of the arabinopyranoside moiety, leading to the formation of decomposition derivatives. Consequently, a less sterically demanding olefinic pyranose imidate **61** was employed as the counterpart in the glycosylation with **60**, resulting in trisaccharide **62** in 74% yield. Satisfyingly, the ring-closing glycosylation of **62** in the presence of $PPh_3AuNTf_2$ gave rise to the desired cyclic trisaccharide **63** in a moderate yield of 33%. The final stage of the synthesis involved the elaboration of the arabinopyanoside unit and the removal of all protecting groups, achieving the total synthesis of luzonicoside A (**52**) in 38% yield over five steps. Similarly, modification of the synthetic route enabled the successful synthesis of luzonicoside D (**53**) and sepositoside A (**54**).

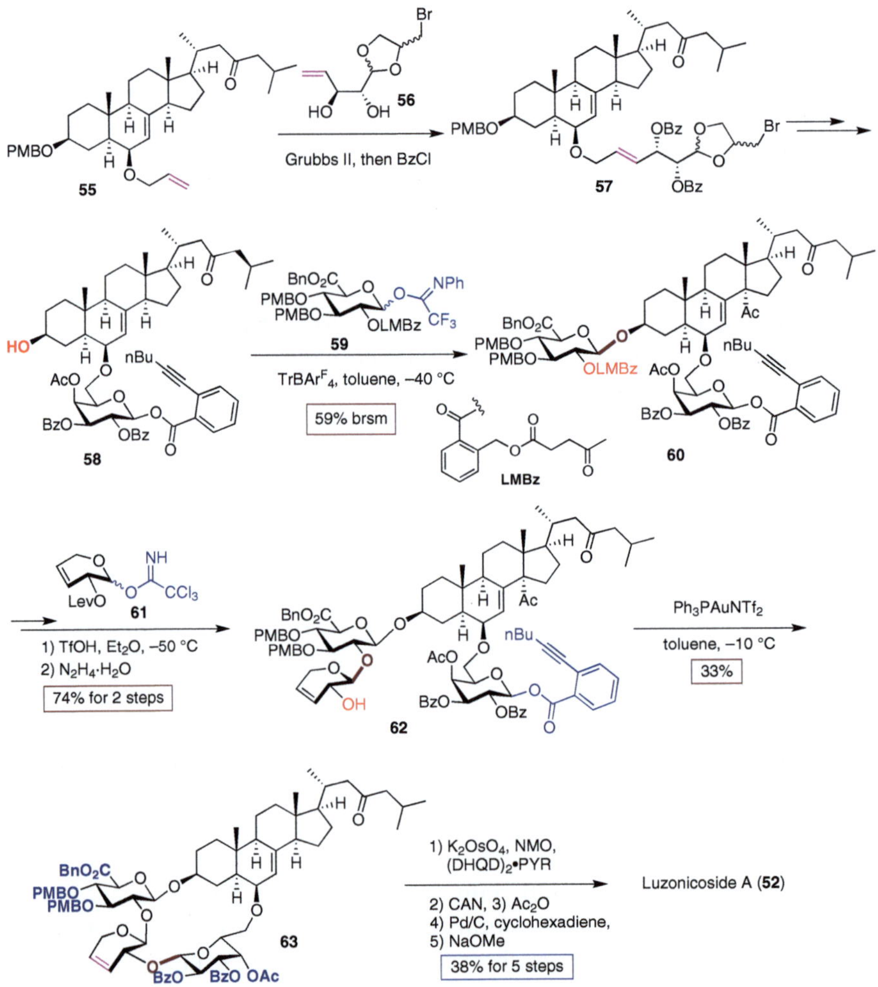


**Scheme 9.7** Total synthesis of luzonicoside A (**52**) by Yu and coworkers.

## Cardenolide Type Steroid Glycosides

### Digitoxin

Digitoxin (**64**, Figure 9.5), a cardiac glycoside derived from the leaves of *Digitalis purpurea* (purple foxglove), has long been used to decrease heart rate while increasing the contractility of heart muscle through the inhibition of $Na^+/K^+$-ATPase enzyme in myocardial cells [22]. In addition, digitoxin exhibits anticancer activity and holds promise as a potential neuroprotective agent. Structurally, digitoxin (**64**) consists of a steroid core substituted with an unsaturated lactone at C17 and a trisaccharide composed of 2,6-dideoxy sugars via β-linkages. The aglycone itself remains inactive without the attachment of the digoxose trisaccharide.

**Synthesis of digitoxin (64) by Wiesner and coworkers.** In 1985, Wiesner and coworkers reported the first synthesis of digitoxin (**64**) through a sequential glycosylation strategy, starting from the furyl precursor of digitoxigenin (Scheme 9.8) [23]. Glycosylation of the furyl-derived digitoxigenin **65** with hemiacetal **66** under the promotion of *p*-TsOH gave 3-*O*-glycoside **67** as an anomeric mixture (β/α, 3 : 1) in modest yield (48%). The C3′-*N*-methylurethane group was proposed to play a role in stereocontrol during this reaction. Subsequent replacement of the protecting group at the C3′-position, followed by benzyl-group deprotection, yielded 4′-OH acceptor **68**, which was then coupled with thioglycoside **69** under the promotion of $HgCl_2$ and $CdCO_3$. The resulting disaccharide **70** was obtained in 61% yield with excellent β selectivity, presumably due to a remote 1,3-participation effect of the ester group. After removal of the 4″-*O*-*p*-nitrobenzyl group, elongation of the sugar chain was realized using thioglycoside **71** under similar $HgCl_2/CdCO_3$-promoted conditions, which selectively afforded the protected β-linked trisaccharide **72** in moderate yield (58%). Finally, a three-step procedure involving (i) MBz deprotection with $LiAlH_4$, (ii) oxidation of the furyl residue with *m*-chloroperoxybenzoic acid, and (iii) reduction with $NaBH_4$ to generate the butenolide moiety, was employed to accomplish the synthesis of digitoxin (**64**) in 65% yield (for three steps).

**Synthesis of digitoxin (64) by McDonald and coworkers.** In 2001, McDonald and coworkers reported their synthesis of digitoxin (**64**), featuring a tungsten-catalyzed *endo*-selective alkynol cyclomerization methodology and acid-catalyzed glycosylations of disaccharide and trisaccharide glycals (Scheme 9.9) [24]. The acid ($Ph_3P$–HBr)-catalyzed glycosylation of glycal **74** with alkynyl alcohol

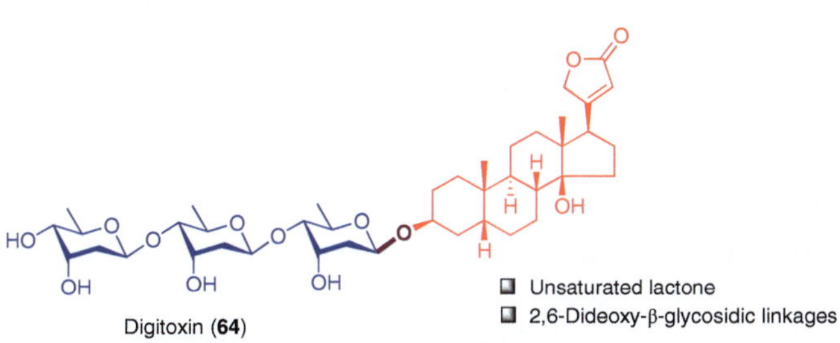


Digitoxin (**64**)
**Synthesis**: Wiesner (1985), McDonald (2001), O'Doherty (2006), Yu (2011), Zhu (2013)

**Figure 9.5** Digitoxin (**64**) and its structural features.

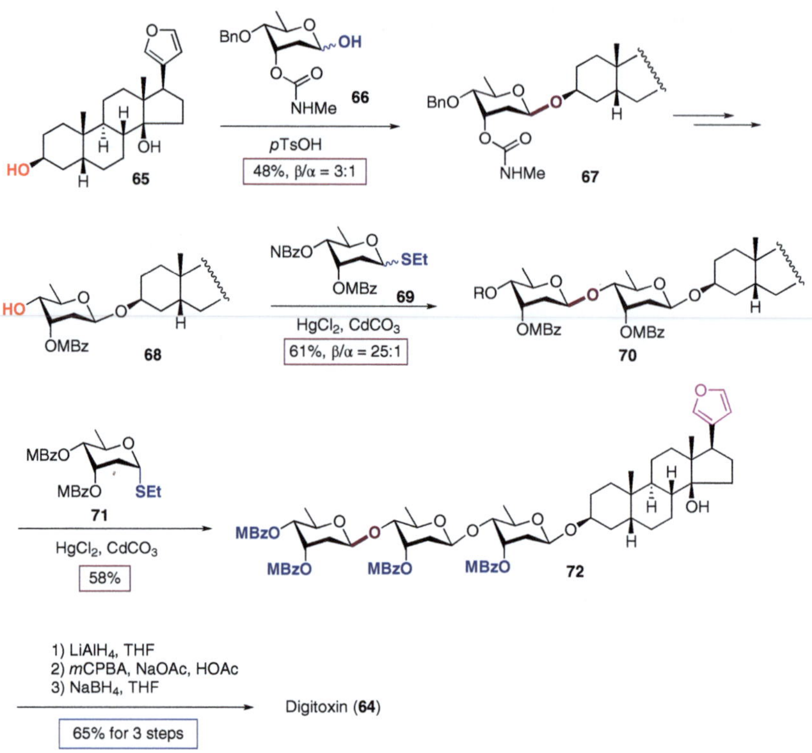


**Scheme 9.8** Synthesis of digitoxin (**64**) by Wiesner and coworkers.

**73** gave 2-deoxyglycoside **75** with high β-selectivity and high yield. Debenzoylation of **75** with DIBAL-H furnished the alkynol, which underwent subsequent tungsten-catalyzed *endo*-selective cycloisomerization to yield disaccharide glycal **76**. However, the iterative acid-catalyzed glycosylation of disaccharide **76** with alkynol alcohol **73** exhibited low yield and poor anomeric selectivity (β/α = 76 : 24, c. 30% combined yield). The authors observed that employing the acetate protective groups at the non-reducing sugar in the glycosylation of disaccharide **77** resulted in disaccharide **78** in excellent stereocontrol and good yield (β/α = 99 : 1, 71% yield). These results highlighted the sensitivity of the acid-catalyzed glycosylations to remote substituent effects. Upon removal of the ester groups, tungsten-catalyzed *endo*-selective cycloisomerization gave trisaccharide glycal **79**. The crucial acid-catalyzed coupling of **79** with digitoxigenin (**80**) afforded product **81** as an anomeric mixture in 82% yield, with the β-anomer as the major product (β/α = 3/2). Finally, global deprotection of the silyl and acetyl groups accomplished the synthesis of digitoxin (**64**) in 40% yield (for two steps).

**Synthesis of digitoxin (64) by O'Doherty and coworkers.** In 2006, O'Doherty and coworkers developed a de novo route to the synthesis of digitoxin, which featured the iterative application of the palladium-catalyzed glycosylation reaction, reductive 1,3-transposition, diastereoselective dihydroxylation, and regioselective protection (Scheme 9.10) [25]. Digitoxigenin (**80**) was first coupled with pyranone **82**

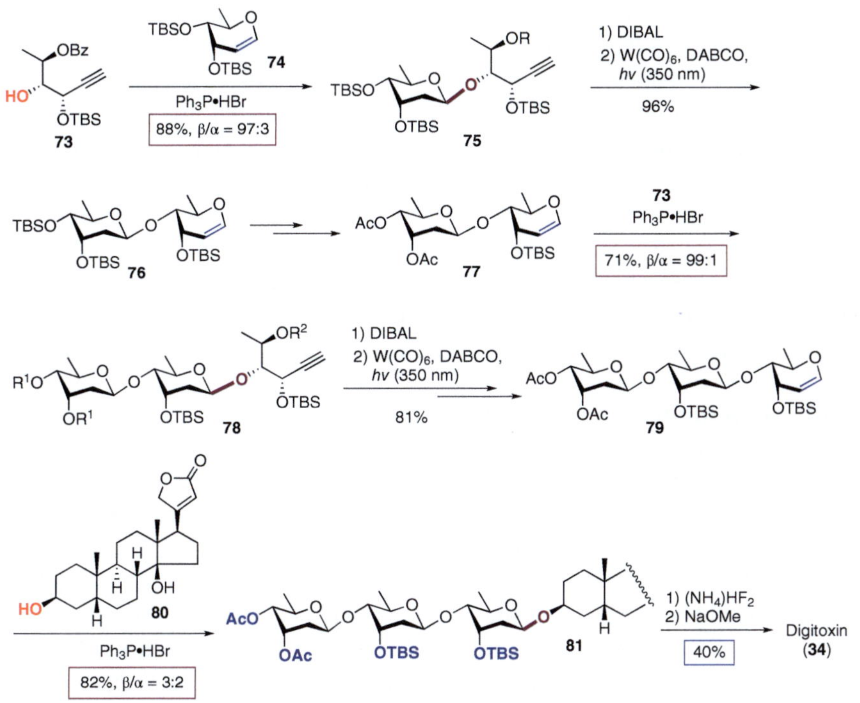


**Scheme 9.9** Synthesis of digitoxin (**64**) by McDonald and coworkers.

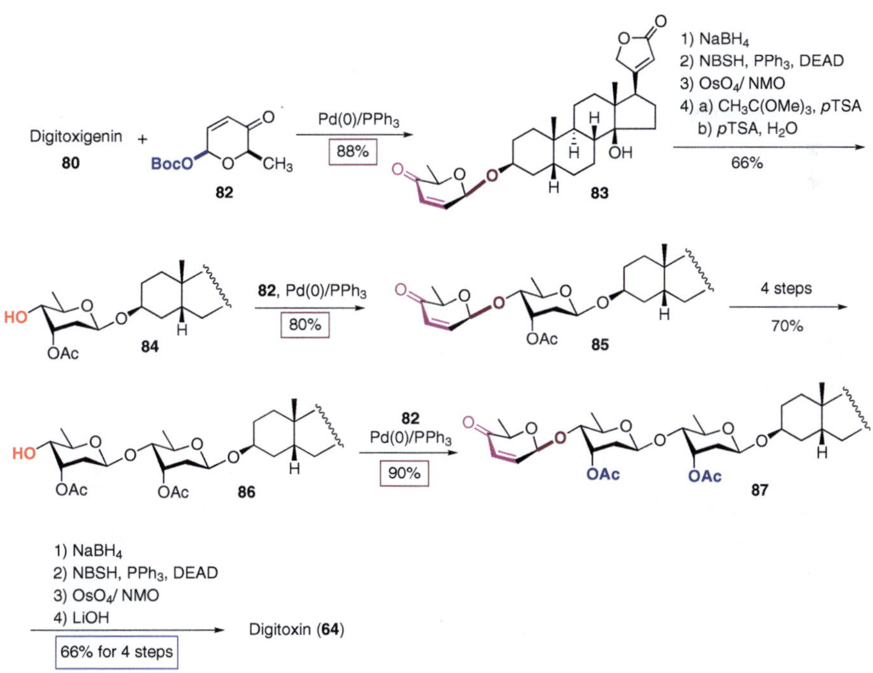


**Scheme 9.10** Synthesis of digitoxin (**64**) by O'Doherty and coworkers.

using palladium-catalyzed conditions, affording the corresponding 3-*O*-β-glycoside **83** stereoselectively in 86% yield. Through a four-step procedure including Luche reduction, Myers' reductive 1,3-allylic transposition, diastereoselective dihydroxylation, and a two-step regioselective protection, glycoside **83** was converted into the 4′-OH acceptor **84**. Iterative coupling of **84** with pyranone **82** under the same palladium-catalyzed conditions yielded the desired disaccharide **85** in 80% yield. The same four-steps manipulation of **85** were applied to provide disaccharide acceptor **86**, which was once again coupled with pyranone **82** using palladium catalysis to give trisaccharide **87** in 90% yield. Finally, after three steps involving the transformation of the enone group of **87**, the global deprotection of the *O*-acetyl groups under basic conditions completed the synthesis of digitoxin (**64**, 66% for four steps).

**Synthesis of digitoxin (64) by Yu and coworkers.** In 2011, Yu and coworkers reported their synthesis of digitoxin (**64**) through gold(I)-catalyzed glycosylation of glycosyl *o*-alkynylbenzoates (Scheme 9.11) [26]. Glycosylation of digitoxigenin (**80**) with 3,4-di-*O*-TBDPS-D-digitoxosyl *o*-cyclopropylethynylbenzoate **88** under the Au(I)-catalyzed conditions (0.1 equiv. $Ph_3PAuOTf$) led to the coupled glycoside **89** with nearly quantitative yield and high β/α selectivity (β/α = 6.6 : 1). The high β-selectivity could be attributed to the increased steric hindrance of the 3-*O* protecting group in the digitoxosyl donor. Removal of the TBDPS groups in **89**, followed by selective protection of the axial 3-OH with an acetyl group, gave 4′-OH acceptor **90**, which readily underwent the coupling with another *o*-alkynylbenzoate donor

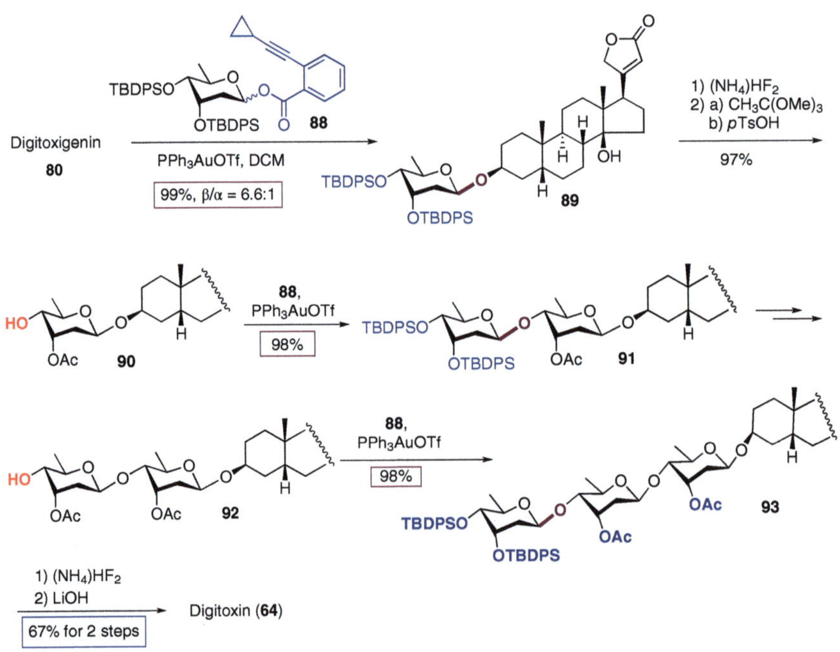


**Scheme 9.11** Synthesis of digitoxin (**64**) by Yu and coworkers.

**88** to give disaccharide **91**. The elongation of the saccharide chain was achieved by repetitive manipulation of protecting groups and Au(I)-catalyzed glycosylation with **88**, resulting in trisaccharide **93** in excellent yield. Finally, deprotection of the silyl groups with $(NH_4)_2F_2$ and the acetyl groups with LiOH accomplished the synthesis of digitoxin (**64**, 67% for two steps).

**Synthesis of digitoxin (64) by Zhu and coworkers.** In 2013, the Zhu group described their total synthesis of digitoxin (**64**) employing sequential rhenium(V)-catalyzed glycosylations with glycal donors to elaborate the sugar chain (Scheme 9.12) [27]. The Re(V)-catalyzed glycosylation of 6-deoxy D-allal glycal **94** with the C4-OH of thioglycoside acceptor **95** generated the desired β-disaccharide **96** in 64% yield and high diastereoselectivity (β/α = 8.3 : 1), the C4′-OH in **96** was left unmodified after removal of the TBS group. By repeating a similar two-step procedure between disaccharide **96** and glycal donor **94**, trisaccharide thioglycoside **97** was obtained in 77% yield with the β anomer as the major product (β/α = 7 : 1). Subsequently, reductive debenzylation and concurrent reductive lithiation–elimination with LiDBB, followed by TBS protection, afforded trisaccharide glycal **98** in 63% yield. The crucial glycosylation between glycal **98** and digitoxigenin (**80**) was carried out under the promotion of $Ph_3P$–HBr, generating the coupled digitoxin (**99**) in 57% yield with modest β-selectivity (β/α = 1.4 : 1). Interestingly, glycosylation with the corresponding per-*O*-TES protected trisaccharide glycal produced the undesired α-glycoside as the major product (β/α = 1 : 5) under the similar conditions. This result suggested that the larger size of the TBS ether, compared to the TES ether, could induce more severe 1,3-diaxial interaction, thus favoring the formation of the β anomer. Finally, global deprotection of the *O*-TBS groups completed the synthesis of digitoxin (**64**) in 85% yield.

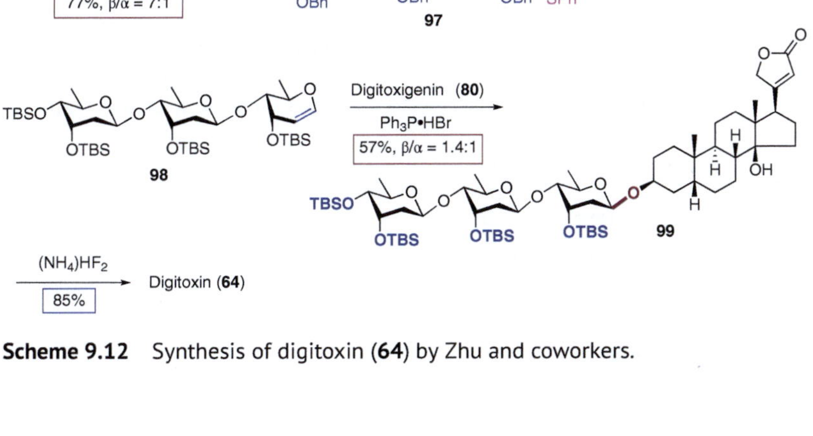


**Scheme 9.12** Synthesis of digitoxin (**64**) by Zhu and coworkers.

## Furostane Type Steroid Glycosides

### Furostan Saponin and Methyl Protodioscin

From the viewpoint of biosynthesis, furostan-type saponins are considered the precursors of the corresponding spirostan-type saponins in plants. Naturally occurring furostan saponins are typically bidesmosides, where a glucose monosaccharide is attached at the 26-OH group of the aglycone (with few exceptions) and an oligosaccharide chain commonly connected at 3-OH position. As representative congeners, furostan saponin **100** was isolated from the rhizomes of *Costus spicatus*, and methyl protodioscin (**101**, Figure 9.6) was first identified by Kawasaki and coworkers in 1974 from the fresh rhizomes of *Dioscorea gracillima MiQ* [28]. Notably, methyl protodioscin (**101**) demonstrated potent activity against the 60 NCI human cancer cell lines.

**Total synthesis of furostan saponin (100) by Yu, Hui, and coworkers.** In 2001, Yu, Hui, and coworkers reported the synthesis of **100**, representing the first synthetic route toward furostan saponins (Scheme 9.13) [29]. The synthesis commenced with diosgenin (**102**), where the C5,6 double bond was protected by dibromide formation, followed by the selective oxyfunctionalization of the ethereal C16 using dimethyldioxirane (DMDO) generated *in situ*, resulting in the formation of hemiketal **103**. Removal of the bromide by $Zn/NH_4Cl$ and subsequent manipulation of the protecting groups yielded aglycone **104**, which existed in equilibrium with its hemiketal form. The glycosylation of the aglycone mixture with glucosyl imidate **105** under the catalysis of TMSOTf afforded the C26-glycosylated product **106** in 86% yield. After removal of the C3-*O*-TBDPS group, the subsequent coupling with **105** under similar glycosylation conditions gave bisglycoside **107** in 96% yield. Finally, reduction of the ketone moiety, followed by global deprotection of the *O*-Bz groups, accomplished the synthesis of furostan saponin **100**.

**Synthesis of methyl protodioscin (101) by Yang and coworkers.** In 2003, Yang and coworkers reported the first synthesis of methyl protodioscin (**101**) starting from diosgenin in nine steps with an overall yield of 7.8% (Scheme 9.14) [30]. Coupling of phenyl 3,6-di-*O*-pivaloyl-1-thio-β-D-glucopyranoside (**108**) and α-L-rhamnopyranosyl trichloroacetimidate (**109**) under the promotion of $BF_3 \cdot Et_2O$ provided the desired trisaccharide **110** quantitatively. On the other hand, the C-26

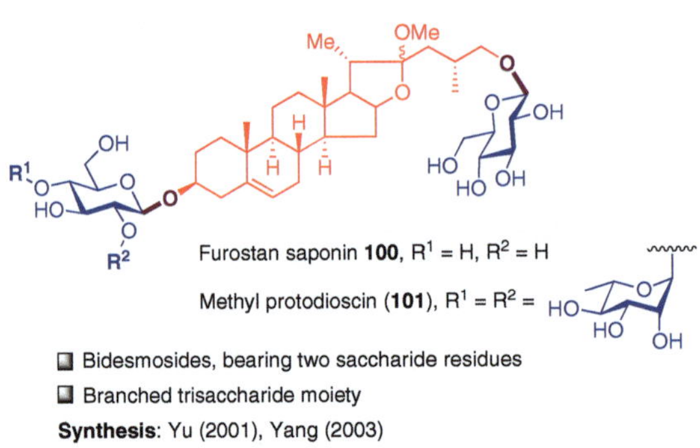


**Figure 9.6** Furostan saponin (**100**) and methyl protodioscin (**101**) and the structural features.

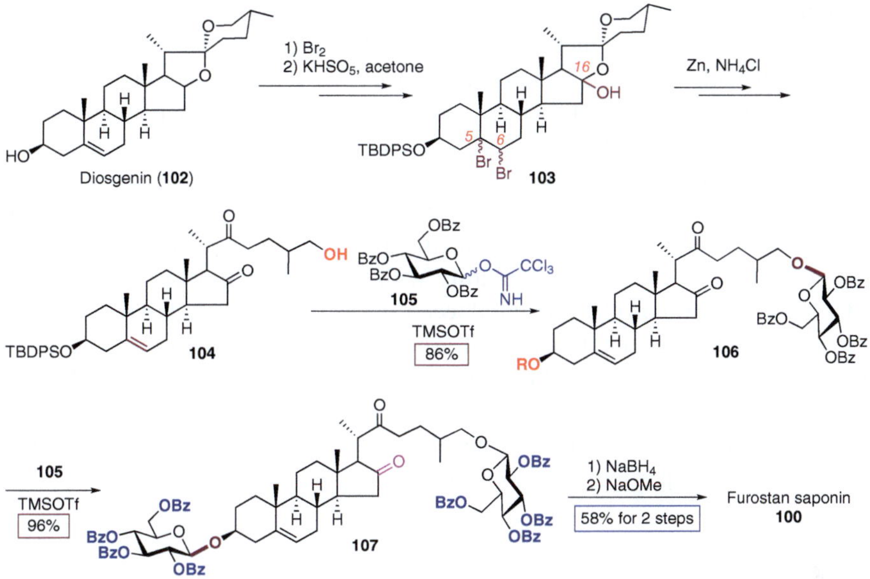

**Scheme 9.13** Synthesis of furostan saponin **100** by Yu, Hui, and coworkers.

glycosylated compound **106** was prepared through an optimized synthetic route from diosgenin (**102**) in four steps. The crucial coupling of **106** with trisaccharide thioglycoside **110**, under the promotion of *N*-iodosuccinimide (NIS) and TfOH, yielded the desired 3-*O*-β-glycoside **111** in 52%, in the absence of a C-2 participation group. Finally, selective reduction of the C16-ketone with $NaBH_4$, followed by the removal of all ester groups with sodium methoxide, successfully accomplished the synthesis of methyl protodioscin (**101**) in 58% yield (for two steps).

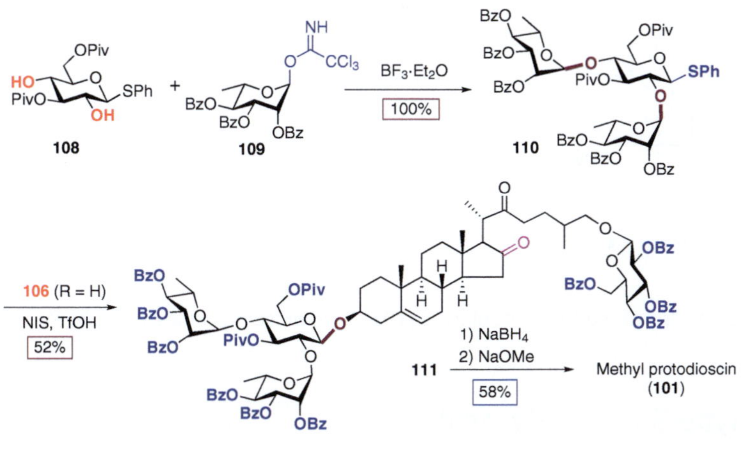

**Scheme 9.14** Total synthesis of methyl protodioscin (**101**) by Yang and coworkers.

**Synthesis of methyl protodioscin (101) by Yu and coworkers.** In 2006, Yu and coworkers reported a facile approach for converting spirostan saponins to furostan saponins, as exemplified by the transformation of the readily available dioscin pivalate (**112**) into methyl protodioscin (**101**) in six steps with 21% overall yield (Scheme 9.15) [31]. This approach involved the oxidative ring-opening of the spirostan E and F rings in dioscin pivalate (**112**) using *in situ* generated DMDO (Oxone, acetone), followed by the elimination of the 5,6-epoxide by *in situ* generated trimethylsilyl iodide (TMSI), resulting in the formation of the steroid 26-iodide **113** in high yield. Hydrolysis of the 26-iodide in **113** furnished 26-ol **114b** rapidly and quantitatively, which was equilibrated with the more favored hemiketal **114a** (1 : 4). Coupling of the resulting mixture with glucosyl trifluoroacetimidate **115**, under the promotion of TMSOTf, produced the desired glycoside **116** in 54% yield. Selective reduction of the 16-carbonyl group in **116** using $NaBH_4$ provided the hemiketal, which was subjected to global deprotection with LiOH to complete the synthesis of methyl protodioscin (**101**). The key advantage of this approach lies in the avoidance of stereocontrolled installation of the chacotriosyl moiety, which lacks a neighboring participatory group, onto the aglycone.

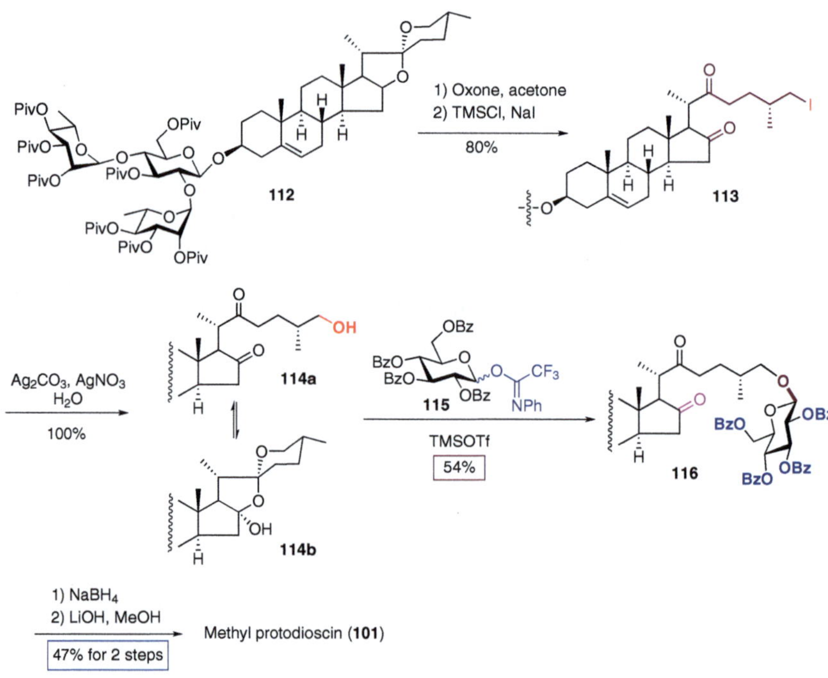


**Scheme 9.15** Synthesis of methyl protodioscin (**101**) from dioscin pivalate (**112**) by Yu and coworkers.

## Spirostan Type Steroid Glycosides

### Desgalactotigonin

Desgalactotigonin (**117**, Figure 9.7), isolated from *D. purpurea* L. [32], was characterized by the presence of a spiroketal linkage at C-21 in the aglycone and a

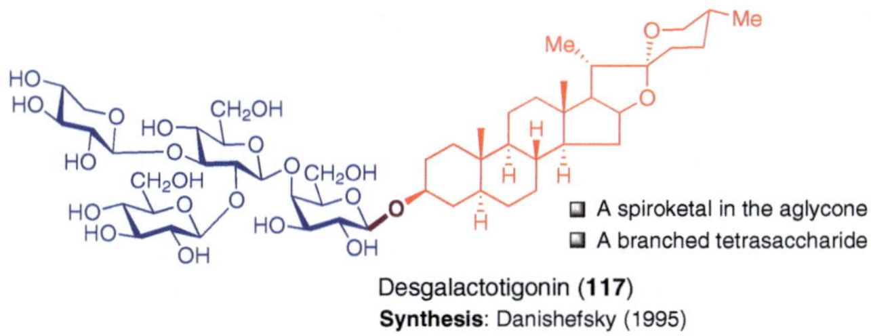


**Figure 9.7** Desgalactotigonin (**117**) and its structural features.

branched tetrasaccharide at C-3. Several digitalis saponins with similar structural features have also been identified, which have found use as biological detergents for the solubilization and isolation of membrane receptor. While exhibiting no cardiotonic activity, certain members of this class possess therapeutic potential as antiviral and antitumor agents.

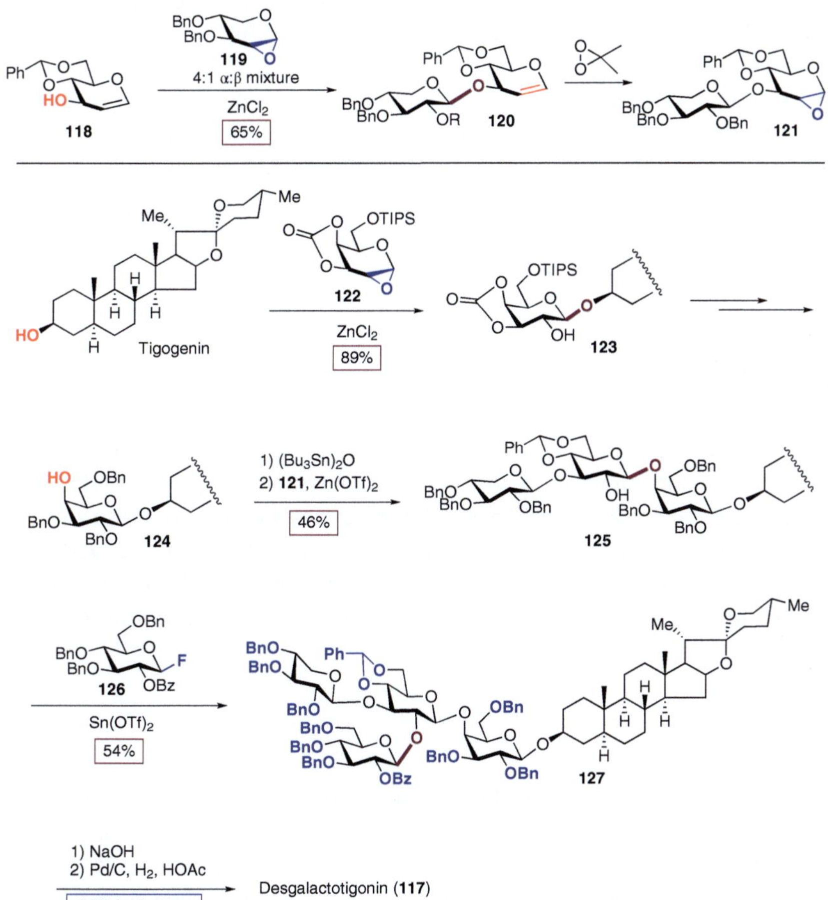


**Scheme 9.16** Synthesis of desgalactotigonin (**117**) by Danishefsky and coworkers.

**Synthesis of desgalactotigonin (117) by Danishefsky and coworkers.** In 1995, Danishefsky and coworkers reported the first synthesis of desgalactotigonin (**117**) utilizing glycosylation with 1,2-oxirane donors as key steps (Scheme 9.16) [33]. The coupling of an α/β mixture of xylal-derived epoxides **119** with D-glucal derivative **118** in the presence of $ZnCl_2$ provided a mixture containing the desired β-disaccharide **120** and the α-lyxo disaccharide originating from glycosylation with the β-epoxide. Following benzylation of the C2′-OH, the glycal **120** was subjected to epoxidation with 3,3-dimethyldioxirane (DMDO), yielding disaccharide epoxide **121**, which was further used for saccharide elongation. Simultaneously, glycosylation of tigogenin aglycone with D-galactal-derived 1,2-oxirane **122** proceeded under the promotion of $ZnCl_2$, affording the glycosylated product **123** in 89% yield. The high selectivities observed in both the epoxidation and galactosylation steps could be attributable to the conformational rigidity imposed by the cyclic carbonate. After several steps of protecting groups manipulations, treatment of **124** with bis(tributyltin) oxide followed by reaction with disaccharide epoxide **121** under the mediation of zinc triflate led to trisaccharide **125** in 46% yield. Interestingly, direct attempts to glycosylate **124** with **121** in the presence of $ZnCl_2$ failed to provide the desired trisaccharide. Further glycosylation of the C2″-OH of **125** via the glycal epoxide methodology with D-glucal proved unsuccessful, but this issue was solved by coupling of **125** with glucosyl fluoride donor **126** under the promotion of $Sn(OTf)_2$, resulting in the formation of tetrasaccharide **127** in 54% yield. Finally, removal of the *O*-Bz groups with NaOH, followed by catalytic hydrogenation to remove all *O*-Bn and benzylidene protection groups, accomplished the synthesis of desgalactotigonin (**117**) in 94% yield (for two steps).

## Forbeside E

Forbeside E3 (**128**, Figure 9.8) and E1 (**129**), two sulfated sterol monosaccharides, were isolated from the common starfish *Asterius forbesi* Desor in 1990 [34]. These compounds were known to be responsible for the general toxicity of these species. In 1993, Schmidt and coworkers reported the synthesis of forbeside E3 (**128**) and E1 (**129**) from 11α-hydroxyprogesterone and quinovopyranosyl trichloroacetimidate [35], which represents an early example in the synthesis of natural saponins.

## Osladin

Osladin (**130**, Figure 9.9), found in the rhizome of the European fern *Polypodium vulgare*, exhibits sweetness that is 500 times greater than sucrose. In 1995, Yamada and coworkers reported the synthesis and structure revision of osladin (**130**), which involved the sequential assembly of the disaccharide moiety at the C3-OH group, releasing of the lactol 26-OH, and subsequent glycosylation using glycosyl chloride as donors [36].

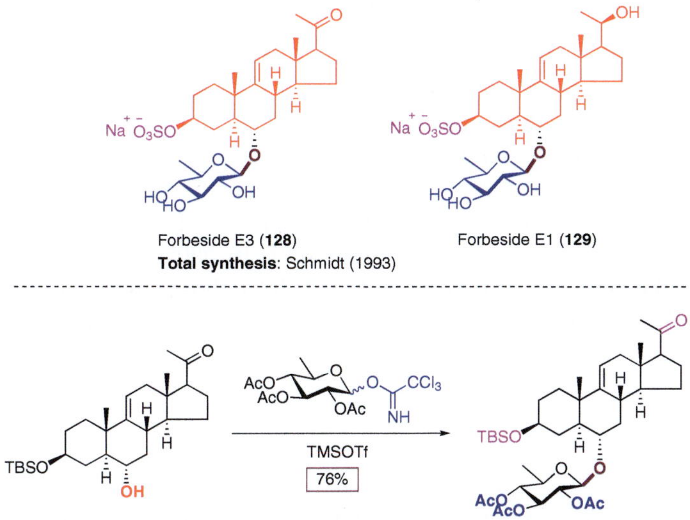


**Figure 9.8** Forbeside E3 (**128**) and E1 (**129**) and the glycosylation step in the synthesis.

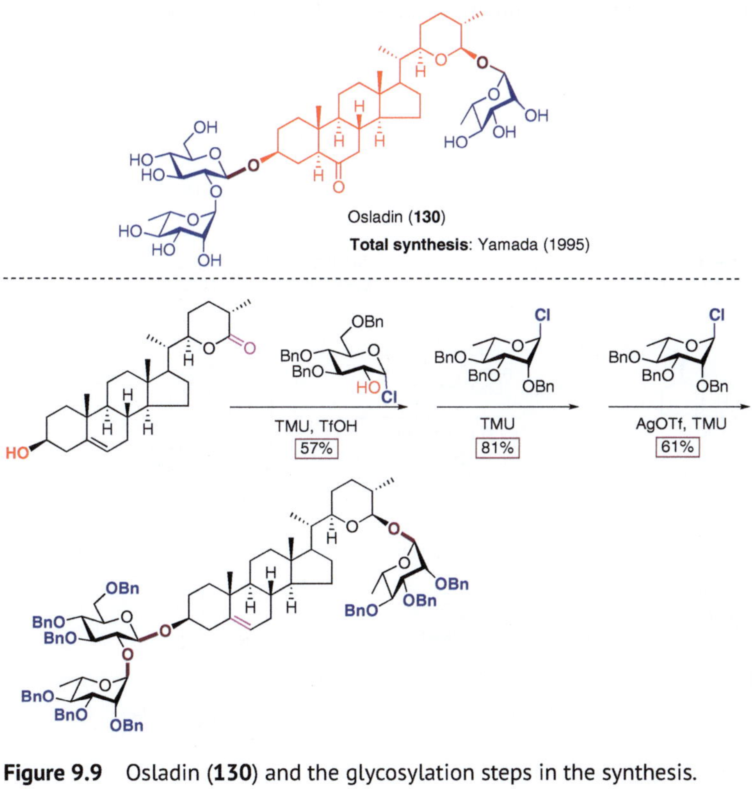


**Figure 9.9** Osladin (**130**) and the glycosylation steps in the synthesis.

## Pavonimin

Six steroidal *N*-acetylglucosaminides, including pavoninins-1 (**131**, Figure 9.10) and pavoninins-4 (**132**), were isolated in 1984 from a defense secretion of the sole *Pardachirus* spp., where they exhibited ichthyotoxic and shark repellent properties [37]. Structurally, these steroid glycosides are unique in their attachment of a β-glycoside at either C7 or C15 of their steroid aglycones. In 1997, Tachibana and coworkers reported the total synthesis of pavonimin-1 (**131**) employing glycosylation with a 2-azidoglycosyl sulfoxide donor [38]. In 2005, Williams and coworkers reported the synthesis of pavoninin-4 (**132**) from diosgenin in 12 steps with an overall yield of 21%, which featured the use of glycosyl fluoride as glycosyl donor [39].

## Dioscin

Dioscin (**133**, Figure 9.11), a representative spirostan-type saponin, has been isolated from various oriental vegetables and traditionally medicinal plants, making it one of the most prevalent saponins found in plants. The broad range of bioactivities exhibited by dioscin, including moderate to good antitumor, antivirus, antifungal, anti-inflammatory, and immunostimulant activities, has attracted

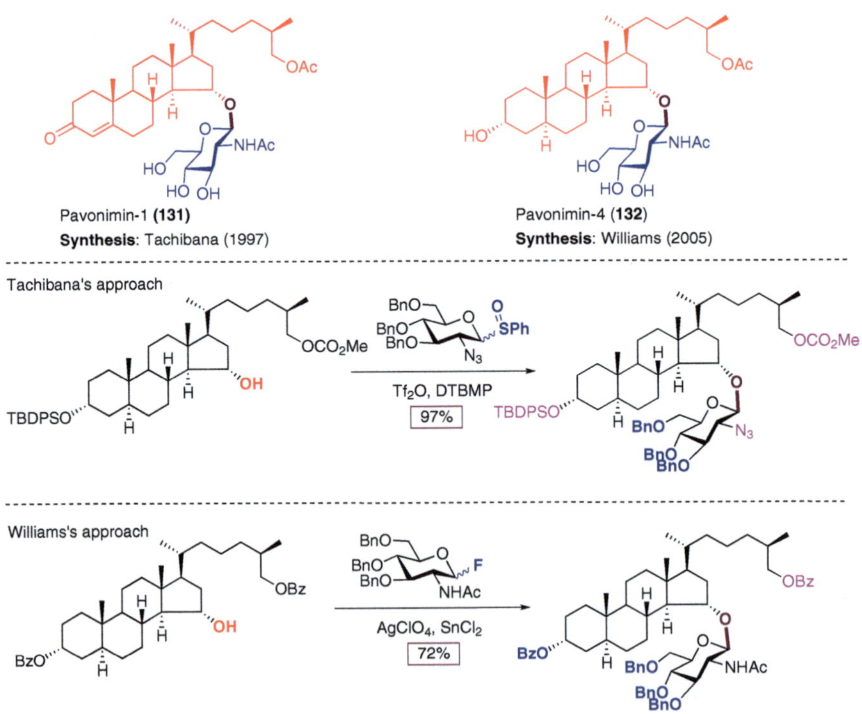


**Figure 9.10** Pavonimin-1 (**131**) and pavonimin-4 (**132**) and the glycosylation steps in the synthesis.

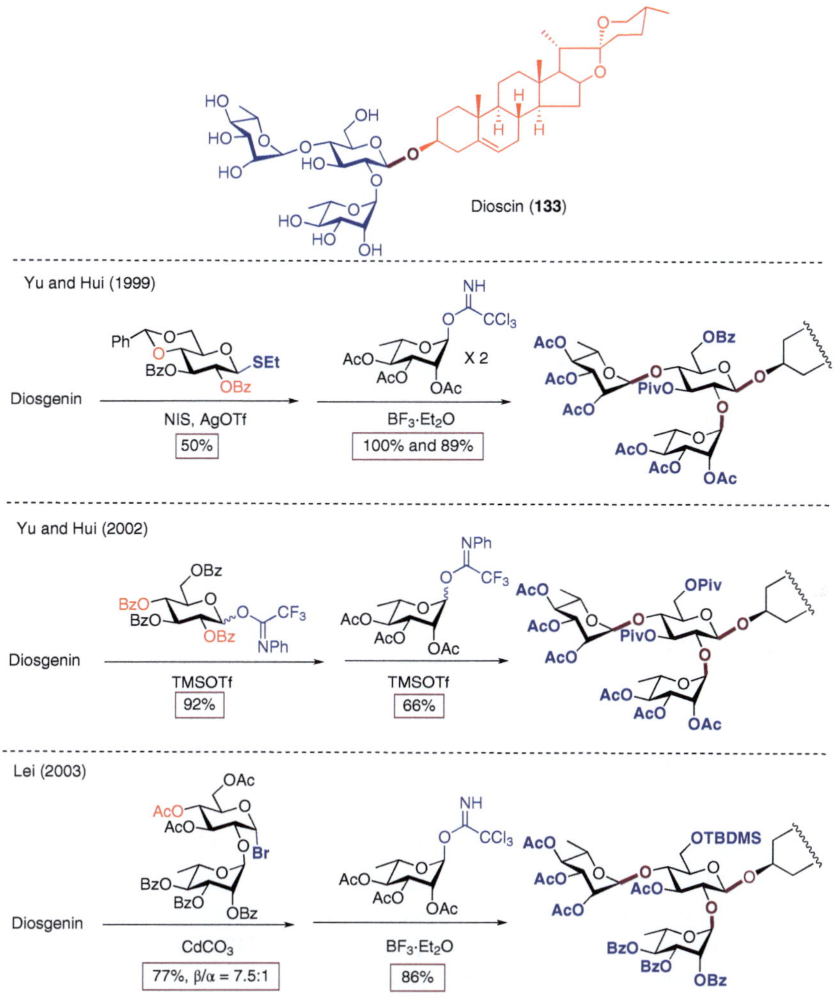


**Figure 9.11** Dioscin (**133**) and the glycosylation steps in its synthesis.

significant attention. In 1999, Yu and coworkers reported the first synthesis of dioscin (**133**) through a stepwise glycosylation strategy involving glycosyl sulfide and trichloroacetimidate donors [40]. In 2002, Yu and coworkers reported an improved synthesis by employing glycosyl *N*-phenyl trifluoroacetimidate donors [41]. In 2003, Lei and coworkers reported their synthesis of dioscin (**133**) using disaccharide bromide and glycosyl trichloroacetimidate donors [42].

## Polyphyllin D

Polyphyllin D (**134**, Figure 9.12) has been isolated from various *Paris* species, which have a historic use of medicine. This saponin displayed potent cytotoxic

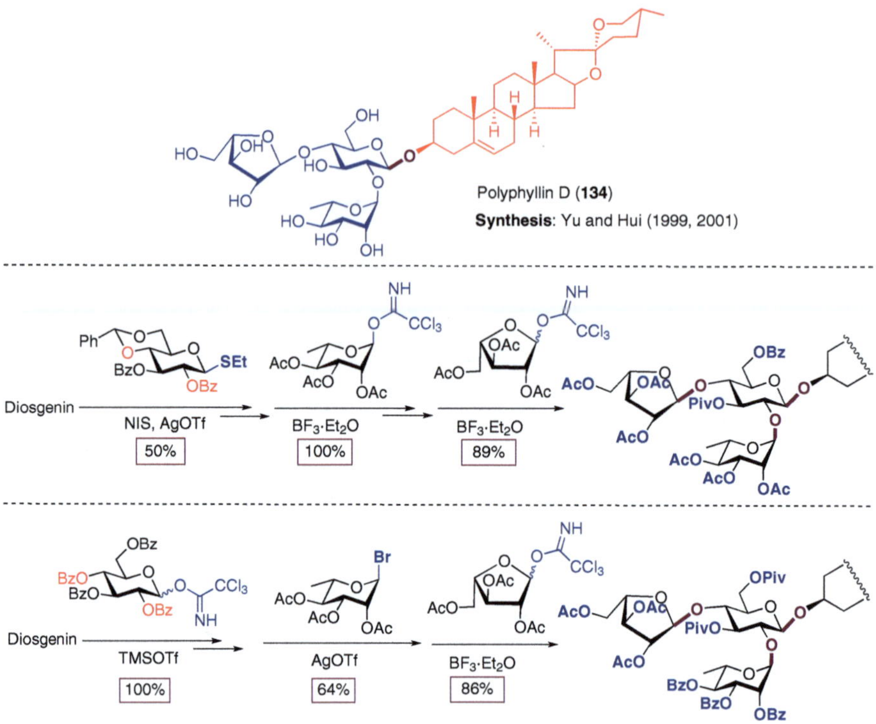


**Figure 9.12** Polyphyllin D (**134**) and the glycosylation steps in its synthesis.

effects against tumor cells (e.g. $ED_{50}$ = 0.94 μg/ml against P-388). Additionally, it showed stimulant effects on cell beating and calcium uptake by myocardial cells, as well as hemostatic and immunomodulating properties. In 1999, Yu and coworkers reported the synthesis of polyphyllin D (**134**) using a stepwise glycosylation strategy involving glycosyl sulfide and imidate donors [40]. In 2001, Yu, Hui, and coworkers presented an improved synthetic route toward the synthesis of polyphyllin D (**134**) in four steps and a 30% overall yield from diosgenyl-β-D-glucopyranoside [43].

## Maidong Saponin C

The tubers of certain *Ophiopogon* and *Liriope* plants, known as Maidong in traditional Chinese medicine, contain a diverse array of steroid saponins. Among them, Maidong saponin C (**135**, Figure 9.13) is a minor component, which demonstrated potent anti-inflammatory activity in a rat edema model. In 1998, Yu, Hui, and coworkers reported the synthesis of Maidong saponin C through a late-stage glycosylation strategy, which involved coupling with a trisaccharide trichloroacetimidate donor [44].

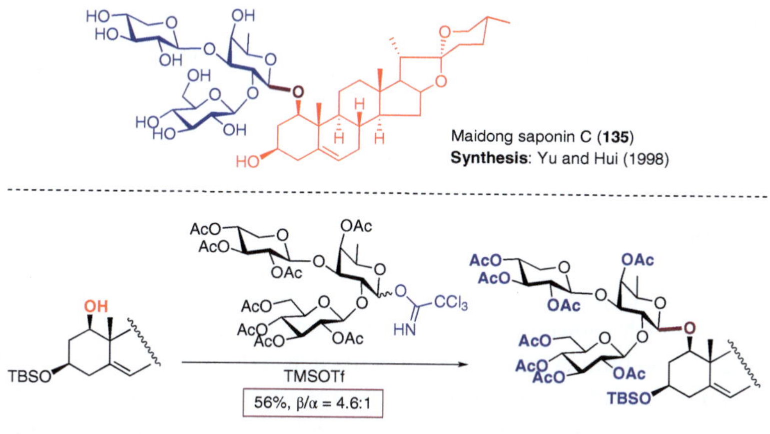


**Figure 9.13** Maidong saponin C (**135**) and the key glycosylation step in its synthesis.

# Xiebai Saponin I

Xiebai saponin I (**136**, Figure 9.14) was obtained from the bulbs of *Allium chinense*, which are commonly used as pickles and spices in China and Japan. Moreover, this compound also serves as the main source of a traditional Chinese medicine, "Xiebai," which is used for the treatment of chest pain, stenocardia, and heart asthma. Xiebai saponin I (**136**) demonstrated inhibitory activity against cyclic adenosine monophosphate (AMP) phosphodiesterase with an $IC_{50}$ value of

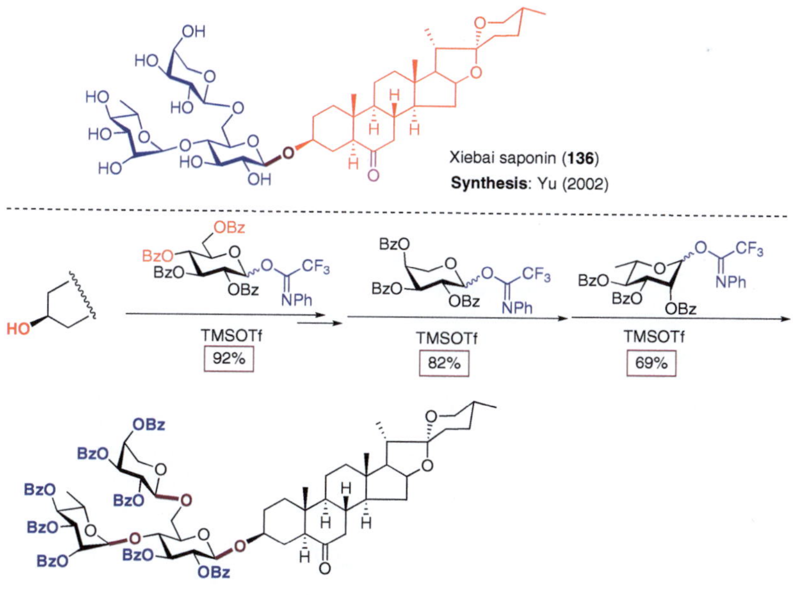


**Figure 9.14** Xiebai saponin (**136**) and the glycosylation steps in its synthesis.

$12.3 \times 10^{-5}$ M and strong cytotoxic activity toward Hela cells at 50 μg/mL dose. In 2002, Yu and coworkers reported the synthesis of Xiebai saponin I (**136**) in eight steps and 32% overall yield by utilizing glycosylation with glycosyl trifluoroacetimidate donors [41].

## Candicanoside A

Candicanoside A (**137**, Figure 9.15) was isolated from the bulbs of *Galtonia candicans* in 2000 [45], which showed potent antitumor activity, such as an $IC_{50}$ value of 32 nm against HL-60 human promyelocytic leukemia cells, being comparable to the clinically applied anticancer agents etoposide and methotrexate. Its structural distinctiveness arises from the unique fused-ring scaffold, formed by acetal between the aldehyde group at C23 and the diol at C16,18. In 2007, Yu and coworkers reported the total synthesis of candicanoside A (**137**) with the longest linear sequence of 27 steps and 1.0% overall yield, employing a late-stage and stepwise glycosylation strategy [46].

## Ouabain

Ouabain (**138**, Figure 9.16), a cardioactive glycoside, was originally isolated from the bark of the African *ouabio* tree (*Acokanthera ouabio*) by Arnaud in 1888. Its significance grew upon the discovery of a naturally occurring ouabain-like compound in

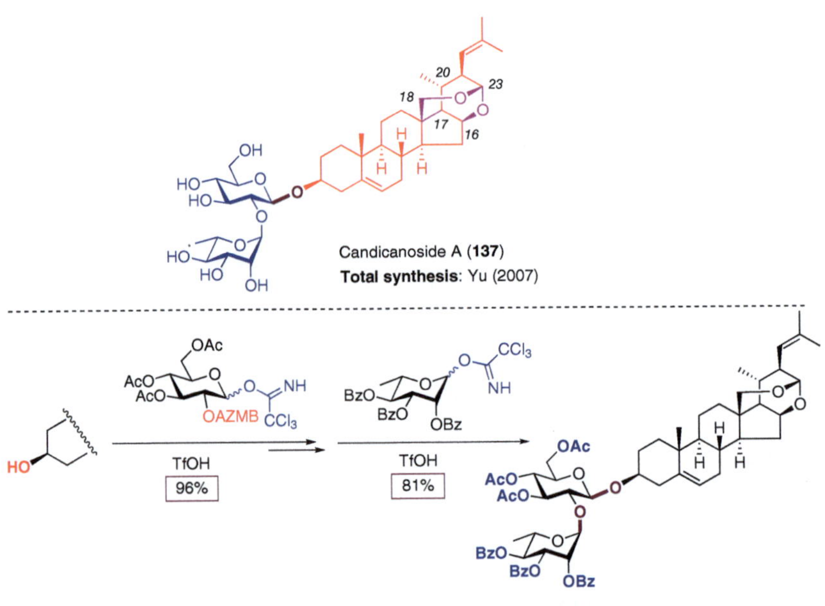


**Figure 9.15** Candicanoside A (**137**) and the glycosylation steps in its synthesis.

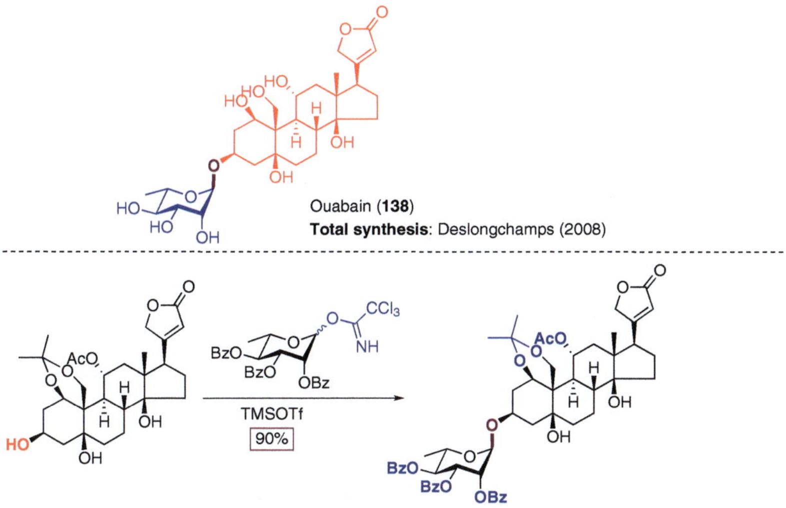


**Figure 9.16** Ouabain (**138**) and the glycosylation step in its synthesis.

mammals, which functions as an endogenous digitalis. Structurally, ouabain (**138**) is a highly oxygenated steroid with the A/B cis, B/C trans, and C/D cis ring junctions. In 2008, Deslongchamps and coworkers reported the first total synthesis of ouabain (**138**) through a polyanionic cyclization as the key step, along with a late-stage glycosylation utilizing a glycosyl trichloroimidate donor [47, 48]. In addition, the total synthesis of the aglycone of ouabain, namely ouabain, has been extensively studied by Baran [49, 50], Inoue [51], and Tang [52].

## Timosaponin BII

Timosaponin BII (**139**, Figure 9.17), also known as prototimosaponin AIII, was initially isolated from the rhizomes of *Anemarrhena asphodeloides Bunge* (*Liliaceae*) by Kawasaki and coworkers. This compound showed promising pharmaceutical properties, including effectiveness against dementia and stroke, as well as the ability to lower blood sugar levels, inhibit platelet aggregation, and scavenge free radicals. In 2009, Du and coworkers reported a synthesis of Timosaponin BII (**139**) in 10 steps and 18% overall yield, using a stepwise glycosylation strategy [53].

## Solamargine

Solamargine (**140**, Figure 9.18) is a major glycoalkaloid found in over 100 *Solanum* species. It could inhibit the growth of various cancer cell lines, including colon, prostate, breast, and human hepatoma cells. Structurally, solamargine is composed of two parts: an oxaza-spiro steroidal aglycone (solasodine) and a chacotriose

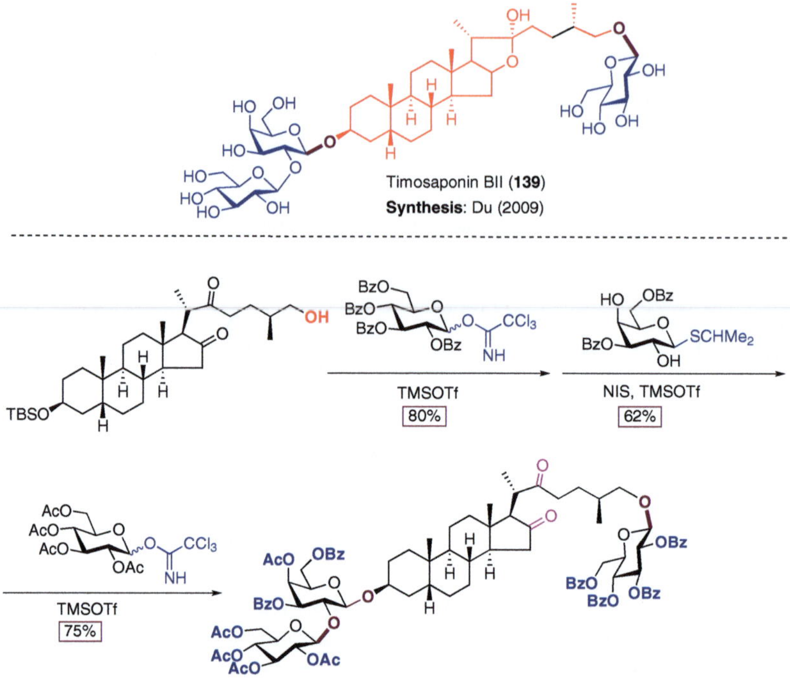


**Figure 9.17** Timosaponin BII (**139**) and the glycosylation steps in its synthesis.

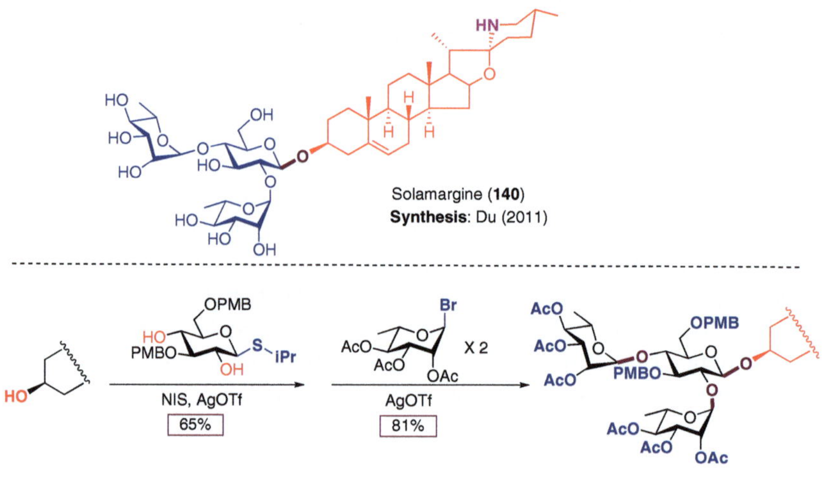


**Figure 9.18** Solamargine (**140**) and the glycosylation steps in its synthesis.

(2,4-bis-α-L-rhamnopyranosyl-β-D-glucopyranose) attached to the 3-hydroxy group of solasodine. In 2011, Du and coworkers reported the synthesis of solasodine (**140**) in 13 steps with 10.5% overall yield, employing glycosylation with glycosyl sulfide and bromide as donors [54].

## Rhodexins

Rhodexins were initially isolated from the leaves and roots of the Japanese evergreen *Rhodea japonica* in 1951, and late from several other plant species (Figure 9.19). Rhodexin A (**141**) displayed digitalis-like activity in the cat heart and showed efficacy against human leukemia K562 cells ($IC_{50}$ = 19 nM). Structurally, rhodexin A stands apart from typical steroids due to its cis fusion of the AB and CD rings. Moreover, it possesses a tertiary hydroxyl group at C14 and a β-butenolide substituent at C17. In 2011, the Jung group reported the first total synthesis of rhodexin A (**141**) by employing the inverse-electron-demand Diels–Alder reaction as a key reaction, along with a late-stage glycosylation with glycosyl imidate donor [55]. On the other hand, rhodexin B (**142**) features a β16-acetoxy-substituted aglycone. In 2021, the Nagorny group reported a total synthesis of rhodexin B (**142**) using a Pd(II)-catalyzed glycosylation with a L-rhamnopyranosyl trichloroacetimidate as donor [56].

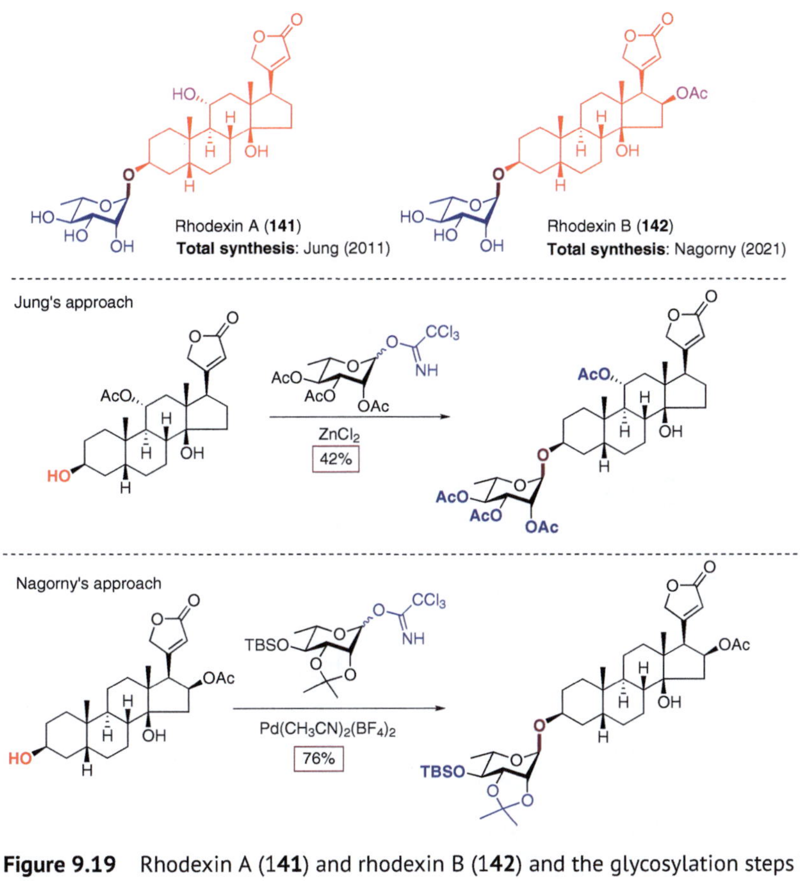


**Figure 9.19** Rhodexin A (**141**) and rhodexin B (**142**) and the glycosylation steps in the synthesis.

## Goniopectenoside B

Goniopectenoside B (**143**, Figure 9.20) was isolated from *Goniopecten demonstrans*, a large pinkish starfish inhabiting deep waters of the Gulf of Mexico, in 2000 [57]. Structurally, goniopectenoside B (**143**) consists of a $\Delta^{9(11)}$-3β,6α-dihydroxysteroidal nucleus with a sulfate residue at C3 and a branched pentasaccharide at C6. In 2013, Yu and coworkers reported the total synthesis of goniopectenoside B (**143**) with a linear sequence of 21 steps and a 4.3% overall yield, by employing late-stage glycosylation with a pentasaccharide *ortho*-alkynylbenzoate donor under gold(I) catalysis [58].

## Astrosterioside A

Astrosterioside A (**144**, Figure 9.21) was isolated from *Astropecten monacanthus,* an edible starfish in Vietnam, in 2013. This compound could inhibit the

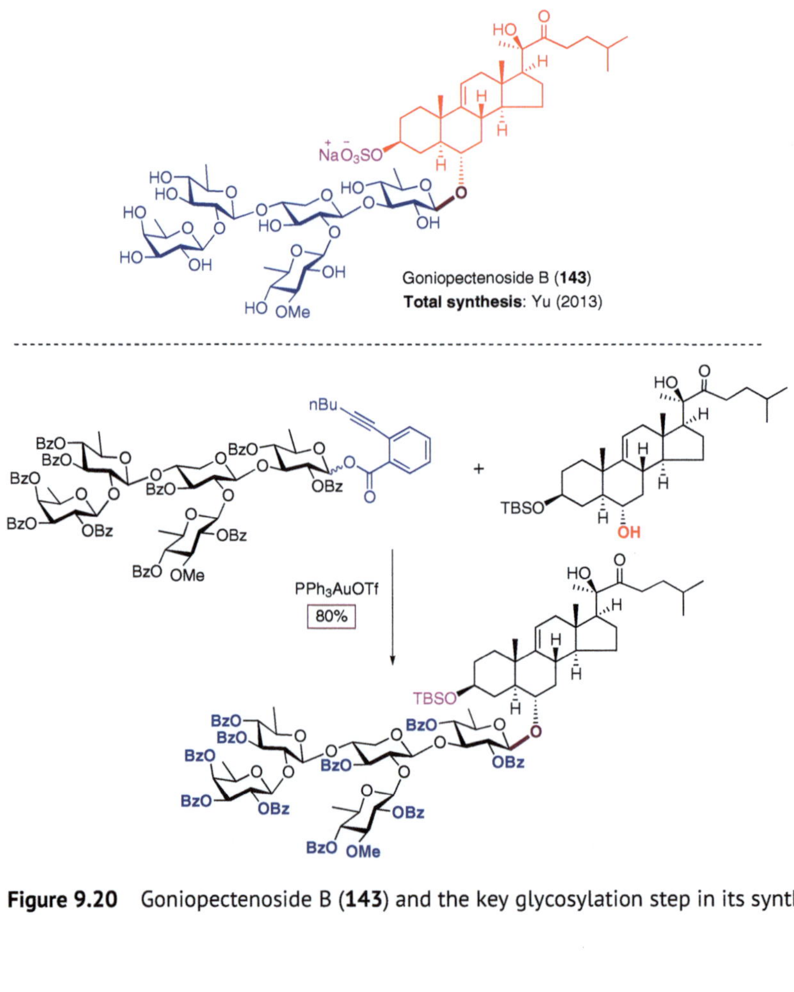


**Figure 9.20** Goniopectenoside B (**143**) and the key glycosylation step in its synthesis.

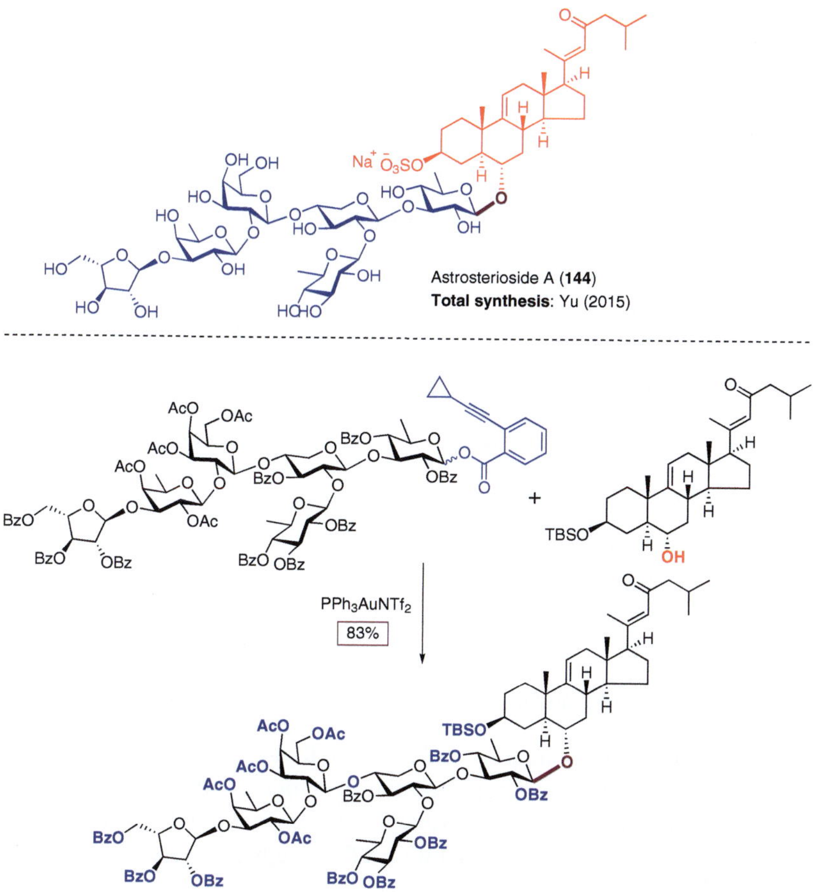


**Figure 9.21** Astrosterioside A (**144**) and the key glycosylation step in its synthesis.

pro-inflammatory cytokine IL-6 production in LPS-stimulated bone marrow-derived dendritic cells ($IC_{50}$ = 3.17 mM). Structurally, astrosterioside A (**144**) possesses a 20(22)*E*-ene-23-one motif in the steroidal side chain and a terminal α-L-arabinofuranose in its hexasaccharide component. In 2015, Yu and coworkers reported the total synthesis of astrosterioside A (**144**) in a convergent linear sequence of 24 steps and in 6.8% overall yield, featuring late-stage glycosylation with a hexasaccharide *ortho*-alkynylbenzoate donor under gold(I) catalysis [59].

## Linckoside

Linckosides A (**145**, Figure 9.22) and B (**146**) were discovered in 2002 from *Linckia laevigata,* an Okinawan starfish, during a screening for neuritogenic natural products with potential as preventive and therapeutic agents for neurodegenerative diseases [60]. Structurally, linckosides consists of a shared cholesta

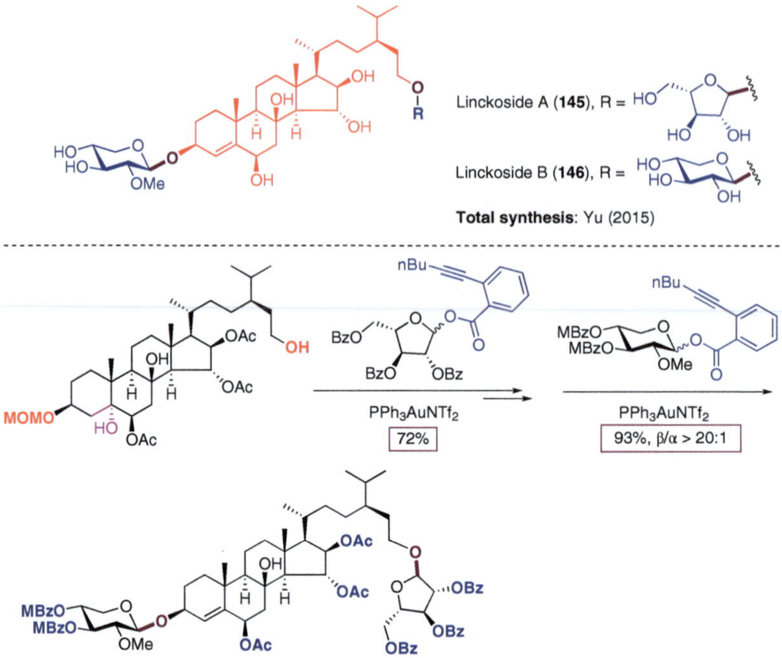


**Figure 9.22** Linckoside A (**145**) and B (**146**) and the glycosylation steps in the synthesis.

3β,6β,8β,15α,16β-pentahydroxy nucleus and a 2-*O*-methyl-β-D-xylopyranose residue at C3. In 2015, Yu and coworkers reported the first total synthesis of linckoside A (**145**) and B (**146**) utilizing readily available starting materials in a modular sequence, involving a longest linear operation of 32 steps (44 steps in total) and resulting in an overall yield of 0.5% [61].

## Trewianin

Trewianin (**147**, Figure 9.23), a member of cardenolide family, was isolated from the stem bark of *Trewia nudiflora* in 2010 [62]. Structurally, trewianin consists of a 6/6/6/5-membered ABCD-ring system, a β-oriented C17-butenolide, C11,14,19-hydroxy groups, and a C3-*O*-L-rhamnoside moiety. In 2018, Inoue and coworkers reported the total synthesis of trewianin (**147**) using a late-stage glycosylation strategy without the need to protect groups on the tetraol aglycone [63].

## P57

The discovery of the appetite suppressant property of *Hoodia* plants in the South African desert has garnered significant attention, resulting in numerous commercial preparations. Within the *Hoodia* species, a series of pregnane glycosides have been identified, with P57 (**148**, Figure 9.24) being the only reported compound exhibiting

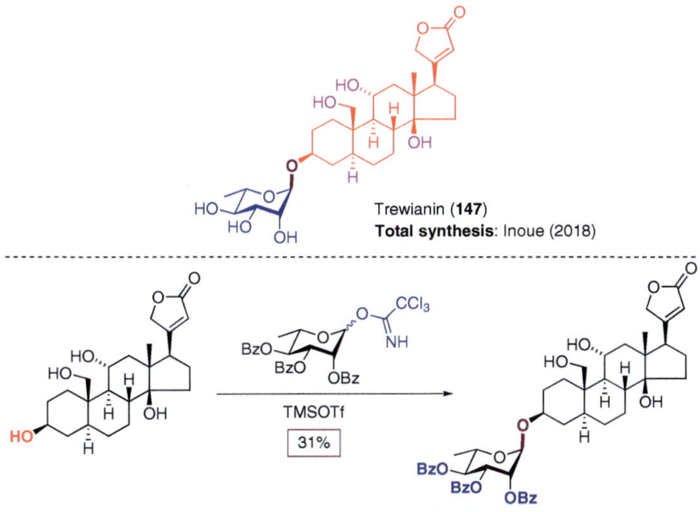

**Figure 9.23** Trewianin (**147**) and the glycosylation step in its synthesis.

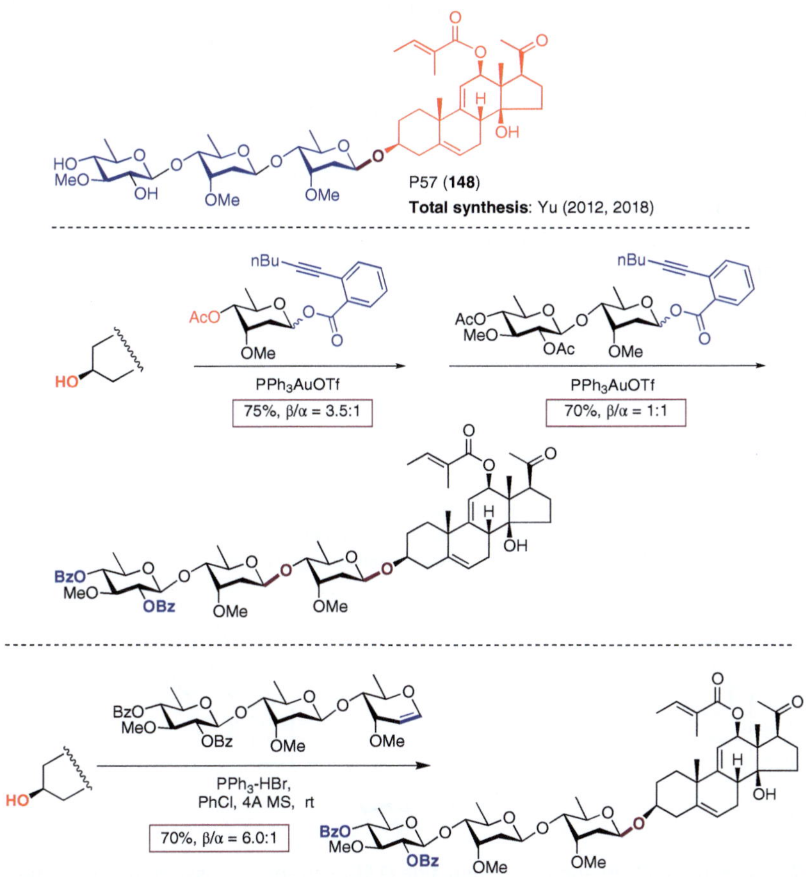

**Figure 9.24** Saponin P57 (**148**) and the glycosylation steps in its synthesis.

the desired activity [64]. In 2012, Yu and coworkers reported the synthesis of P57 (**148**) through a linear 20 steps and 2.4% yield from digoxin, employing the gold(I)-catalyzed glycosylation method with *ortho*-alkynylbenzoate donors for the assembly of the glycan [65]. In 2018, Yu and coworkers reported a convergent synthesis of P57 (**148**), which featured the direct β-selective glycosylation (β/α = 6.0 : 1) of the aglycone (hoodigogenin A) with trisaccharide glycal donor in the presence of triphenylphosphine hydrobromide (TPHB) [66].

## Oleandrin

Oleandrin (**149**, Figure 9.25) is the major constituent derived from the extracts of *Nerium oleander*, a Mediterranean plant known for its medicinal properties [67]. The extracts containing oleandrin have been investigated for their potential in treating cancer and viral infection [68]. Moreover, oleandrin showed promise for the treatment of neurodegenerative disorders due to its unique ability to cross the blood–brain barrier and reduce cellular prion protein (PrPC) in the brain [69]. In addition, the α-oleandrose-containing motifs are commonly observed in other cardiotonic steroids, such as beaumontoside (**150**). From synthetic standpoint, the incorporation of the α-oleandrose moiety represents a significant challenge, as direct methods for α-glycosylation with oleandrose donors would result in low yields and poor α/β-selectivity. In 2023, Nagorny and coworkers reported the total synthesis of cardiotonic steroids oleandrin (**149**) and beaumontoside (**150**), featuring an efficient and stereoselective α-glycosylation of the aglycone with L-oleandral-based glycal donor enabled by HBr-$PPh_3$ as the promoter [70].

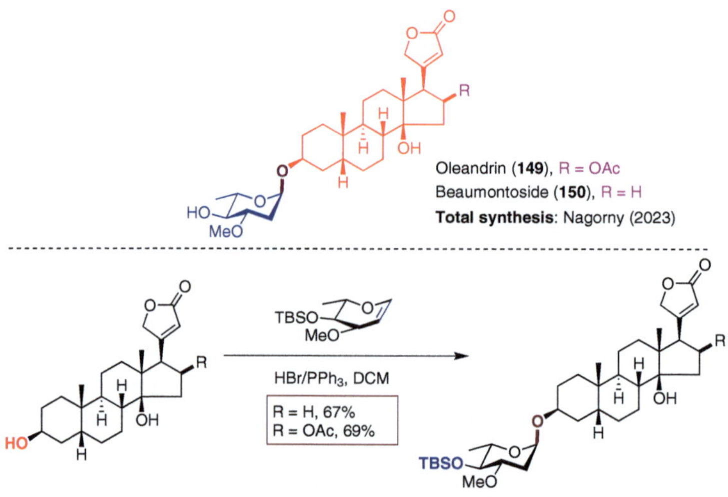


**Figure 9.25** Oleandrin (**149**) and beaumontoside (**150**) and the glycosylation step in the synthesis.

## References


1 Pellissier, H. (2004). *Tetrahedron* 60: 5123–5162.
2 Yu, B., Zhang, Y., and Tang, P. (2007). *Eur. J. Org. Chem.* 2007: 5145–5161.
3 Yang, Y., Laval, S., and Yu, B. (2014). *Adv. Carbohydr. Chem. Biochem.* 71: 137–226.
4 Zu, L., Zhao, Y., and Gu, G. (2014). *J. Carbohydr. Chem.* 33: 269–297.
5 Kubo, S., Mimaki, Y., Terao, M. et al. (1992). *Phytochemistry* 31: 3969–3973.
6 Mimaki, Y., Kuroda, M., Kameyama, A. et al. (1997). *Bioorg. Med. Chem. Lett.* 7: 633–636.
7 Tang, Y., Li, N., Duan, J.-a., and Tao, W. (2013). *Chem. Rev.* 113: 5480–5514.
8 Deng, S., Yu, B., Lou, Y., and Hui, Y. (1999). *J. Org. Chem.* 64: 202–208.
9 Shi, B., Tang, P., Hu, X. et al. (2005). *J. Org. Chem.* 70: 10354–10367.
10 Yu, W. and Jin, Z. (2001). *J. Am. Chem. Soc.* 123: 3369–3370.
11 Yu, W. and Jin, Z. (2002). *J. Am. Chem. Soc.* 124: 6576–6583.
12 Xue, J., Liu, P., Pan, Y., and Guo, Z. (2008). *J. Org. Chem.* 73: 157–161.
13 Hikino, H., Oshima, Y., and Hirota, T. (1987). *Heterocycles* 26: 2093–2098.
14 Zhang, X., Zhou, Y., Zuo, J., and Yu, B. (2015). *Nat. Commun.* 6: 5879.
15 De Simone, F., Dini, A., Finamore, E. et al. (1981). *J. Chem. Soc., Perkin Trans. 1* 1855–1862.
16 Kicha, A.A., Kalinovsky, A.I., Malyarenko, T.V. et al. (2015). *J. Nat. Prod.* 78: 1397–1405.
17 Malyarenko, O.S., Dyshlovoy, S.A., Kicha, A.A. et al. (2017). *Mar. Drugs* 15: 227.
18 Chen, Y., Xiao, G., Zhu, D., and Yu, B. (2023). *Chin. J. Chem.* 41: 897–902.
19 Shi, B., Chen, Y., Geng, M. et al. (2023). *Chin. J. Chem.* 41: 891–896.
20 Zhu, D., Geng, M., Shen, D., and Yu, B. (2023). *Chin. J. Chem.* 41: 903–908.
21 Zhu, D., Geng, M., and Yu, B. (2022). *Angew. Chem. Int. Ed.* 61: e202203239.
22 Schmiedeberg, J.E.O. (1875). *Arch. Exp. Pathol. Pharmakol.* 3: 16.
23 Wiesner, K., Tsai, T.Y.R., and Jin, H. (1985). *Helv. Chim. Acta* 68: 300–314.
24 McDonald, F.E. and Reddy, K.S. (2001). *Angew. Chem. Int. Ed.* 40: 3653–3655.
25 Zhou, M. and O'Doherty, G.A. (2006). *Org. Lett.* 8: 4339–4342.
26 Ma, Y., Li, Z., Shi, H. et al. (2011). *J. Org. Chem.* 76: 9748–9756.
27 Baryal, K.N., Adhikari, S., and Zhu, J. (2013). *J. Org. Chem.* 78: 12469–12476.
28 Kawasaki, T., Komori, T., Miyahara, K. et al. (1974). *Chem. Pharm. Bull.* 22: 2164–2175.
29 Yu, B., Liao, J., Zhang, J., and Hui, Y. (2001). *Tetrahedron Lett.* 42: 77–79.
30 Cheng, M.S., Wang, Q.L., Tian, Q. et al. (2003). *J. Org. Chem.* 68: 3658–3662.
31 Li, M. and Yu, B. (2006). *Org. Lett.* 8: 2679–2682.
32 Kawasaki, T. and Nishioka, I. (1964). *Chem. Pharm. Bull.* 12: 1311–1315.
33 Randolph, J.T. and Danishefsky, S.J. (1995). *J. Am. Chem. Soc.* 117: 5693–5700.
34 Findlay, J.A. and He, Z.-Q. (1990). *J. Nat. Prod.* 53: 710–712.
35 Jiang, Z.-H., Han, X.-B., and Schmidt, R.R. (1993). *Liebigs Ann. Chem.* 1993: 1179–1184.
36 Yamada, H. and Nishizawa, M. (1995). *J. Org. Chem.* 60: 386–397.
37 Tachibana, K., Sakaitani, M., and Nakanishi, K. (1984). *Science* 226: 703–705.

**38** Ohnishi, Y. and Tachibana, K. (1997). *Bioorg. Med. Chem.* 5: 2251–2265.
**39** Williams, J.R., Gong, H., Hoff, N., and Olubodun, O.I. (2005). *J. Org. Chem.* 70: 10732–10736.
**40** Deng, S., Yu, B., Hui, Y. et al. (1999). *Carbohydr. Res.* 317: 53–62.
**41** Yu, B. and Tao, H. (2002). *J. Org. Chem.* 67: 9099–9102.
**42** Zou, C.-C., Hou, S.-J., Shi, Y. et al. (2003). *Carbohydr. Res.* 338: 721–727.
**43** Li, B., Yu, B., Hui, Y. et al. (2001). *Carbohydr. Res.* 331: 1–7.
**44** Liu, M., Yu, B., and Hui, Y. (1998). *Tetrahedron Lett.* 39: 415–418.
**45** Mimaki, Y., Kuroda, M., Sashida, Y. et al. (2000). *Helv. Chim. Acta* 83: 2698.
**46** Tang, P. and Yu, B. (2007). *Angew. Chem. Int. Ed.* 46: 2527–2530.
**47** Zhang, H., Reddy, M.S., Phoenix, S., and Deslongchamps, P. (2008). *Angew. Chem. Int. Ed.* 47: 1272–1275.
**48** Reddy, M.S., Zhang, H., Phoenix, S., and Deslongchamps, P. (2009). *Chem. Asian J.* 4: 725–741.
**49** Renata, H., Zhou, Q., Dünstl, G. et al. (2015). *J. Am. Chem. Soc.* 137: 1330–1340.
**50** Renata, H., Zhou, Q., and Baran, P.S. (2013). *Science* 339: 59–63.
**51** Mukai, K., Kasuya, S., Nakagawa, Y. et al. (2015). *Chem. Sci.* 6: 3383–3387.
**52** Sun, J., Chen, Y., Ragab, S.S. et al. (2023). *Angew. Chem. Int. Ed.* 62: e202303639.
**53** Cheng, S., Du, Y., Ma, B., and Tan, D. (2009). *Org. Biomol. Chem.* 7: 3112–3118.
**54** Hau, C.-K., Chui, S.S.-Y., Lu, W. et al. (2011). *Chem. Sci.* 2: 1068–1075.
**55** Jung, M.E. and Yoo, D. (2011). *Org. Lett.* 13: 2698–2701.
**56** Fejedelem, Z., Carney, N., and Nagorny, P. (2021). *J. Org. Chem.* 86: 10249–10262.
**57** De Marino, S., Iorizzi, M., Zollo, F. et al. (2000). *Eur. J. Org. Chem.* 2000: 4093–4098.
**58** Xiao, G. and Yu, B. (2013). *Chem. Eur. J.* 19: 7708–7712.
**59** Dai, Y. and Yu, B. (2015). *Chem. Commun.* 51: 13826–13829.
**60** Qi, J., Ojika, M., and Sakagami, Y. (2002). *Bioorg. Med. Chem.* 10: 1961–1966.
**61** Zhu, D. and Yu, B. (2015). *J. Am. Chem. Soc.* 137: 15098–15101.
**62** Shilpi, J.A., Gray, A.I., and Seidel, V. (2010). *Fitoterapia* 81: 536–539.
**63** Urabe, D., Nakagawa, Y., Mukai, K. et al. (2018). *J. Org. Chem.* 83: 13888–13910.
**64** van Heerden, F.R., Marthinus Horak, R., Maharaj, V.J. et al. (2007). *Phytochemistry* 68: 2545–2553.
**65** Zhang, J., Shi, H., Ma, Y., and Yu, B. (2012). *Chem. Commun.* 48: 8679–8681.
**66** Liu, C., Ma, Y., Pei, C. et al. (2018). *Chin. J. Chem.* 36: 1007–1010.
**67** Newman, R., Sastry, K., Arav-Boger, R. et al. (2020). *J. Exp. Pharmacol.* 12: 503–515.
**68** Plante, K.S., Dwivedi, V., Plante, J.A. et al. (2021). *Biomed. Pharmacother.* 138: 111457.
**69** Mehrabian, M., Wang, X., Eid, S. et al. (2022). *PLoS One* 17: e0270915.
**70** Carney, N., Perry, N., Garabedian, J., and Nagorny, P. (2023). *Org. Lett.* 25: 966–971.

# 10

# Triterpenoid Glycosides

Triterpene glycosides, also named triterpene saponins, constitute the most diverse class of natural glycosides. The triterpene skeletons can be classified into four major types: pentacyclic oleanane, ursane, lupane, and tetracyclic dammarane types (Figure 10.1). Among these, oleanane-type saponins are the most prevalent, with oleanolic acid being the most commonly observed triterpenoid aglycone. Based on the number of the glycosylation site, these glycosides can be further categorized as (i) monodesmosides, which feature a single saccharide residue, predominantly attached at the 3-OH position of the triterpene aglycones, (ii) bidesmosides, which possess two saccharide residues, with the second one often linked through an ester bond at the C-28 carboxylic acid of the aglycones, and (iii) tridesmosides, which possess three saccharide residues although they are relative uncommon. A large number of the crude extracts of plant triterpene glycosides have long been employed as medicinal remedies for various human ailments, despite limited understanding of their structure–activity relationships and the mechanisms of action. Chemical synthesis of triterpenoid glycosides provides a practical approach for accessing homogenous triterpenoid saponins [1–5]. Over the past two decades, some remarkable achievements in the total synthesis of intricate triterpenoid glycosides have been recorded [6–10].

## Ciwujianoside C3

Ciwujianoside C3 (**1**, Figure 10.2) was initially isolated from the leaves of the commonly utilized Chinese medicinal herb *Acanthopanax*, which displayed anti-inflammatory effects [11]. Structurally, ciwujianoside C3 (**1**) represents a typical bidesmosidic triterpene saponin, featuring one glycan tethered via an acetal linkage at C-3 and another through an ester linkage at C-28.

**Total synthesis of ciwujianoside C3 (1) by the Yu and Hui group.** In 1999, Yu and Hui et al. reported the first chemical synthesis of ciwujianoside C3 (**1**) through the implementation of four consecutive glycosylation steps. This synthetic approach demonstrated a conceptual advance by employing a two-flask system and combining the one-pot glycosylation with two-directional glycosylation strategies (Scheme 10.1) [12]. Thus, the assembly of the acyl-protected saponin **10** involved the

*Carbohydrate Chemistry in the Total Synthesis of Naturally Occurring Glycosides*, First Edition. Biao Yu and Xiaoyu Yang.
© 2024 WILEY-VCH GmbH. Published 2024 by WILEY-VCH GmbH.

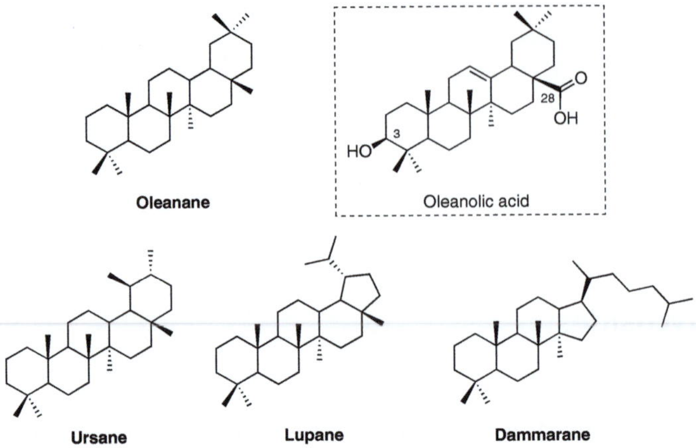


**Figure 10.1** Four representative scaffolds of triterpenoid aglycones.

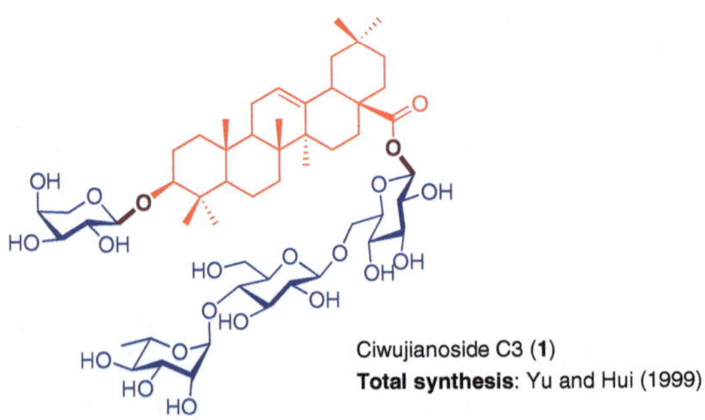


**Figure 10.2** Ciwujianoside C3 (**1**).

following four glycosylation steps: (i) the coupling of oleanolic ester **2** (1.0 equiv.) with trichloroacetimidate **3** was completed within 20 minutes at low temperature (−60 °C) in the presence of a catalytic amount of TMSOTf, providing the desired glycoside **4** after removal of the Tr group by warming the reaction mixture to room temperature for 20 minutes; (ii) addition of a DCM solution of trichloroacetimidate **5** into the above mixture at 0 °C to provide **6**; (iii) in a separate flask, phenyl thio-disaccharide **9** was prepared through the glycosylation of thioglycoside **7** with L-rhamnopyranosyl trichloroacetimidate **8** under the promotion of TMSOTf; (iv) subsequent addition of the second flask's mixture into the first flask, along with TMSOTf and NIS at −15 °C, led to the desired protected saponin **10** in 62% yield (based on oleanolic ester **2**). Finally, selective removal of the acetate and benzoate protecting groups in the presence of the C-28 ester glycosidic linkage (NaOMe/MeOH/DCM) afforded ciwujianoside C3 (**1**) in 73% yield.

## QS-21

QS-21 was isolated from the bark of *Quillaja saponaria* in South America as the 21st fraction of a RP-HPLC separation [13]. QS-21 stands as one of the

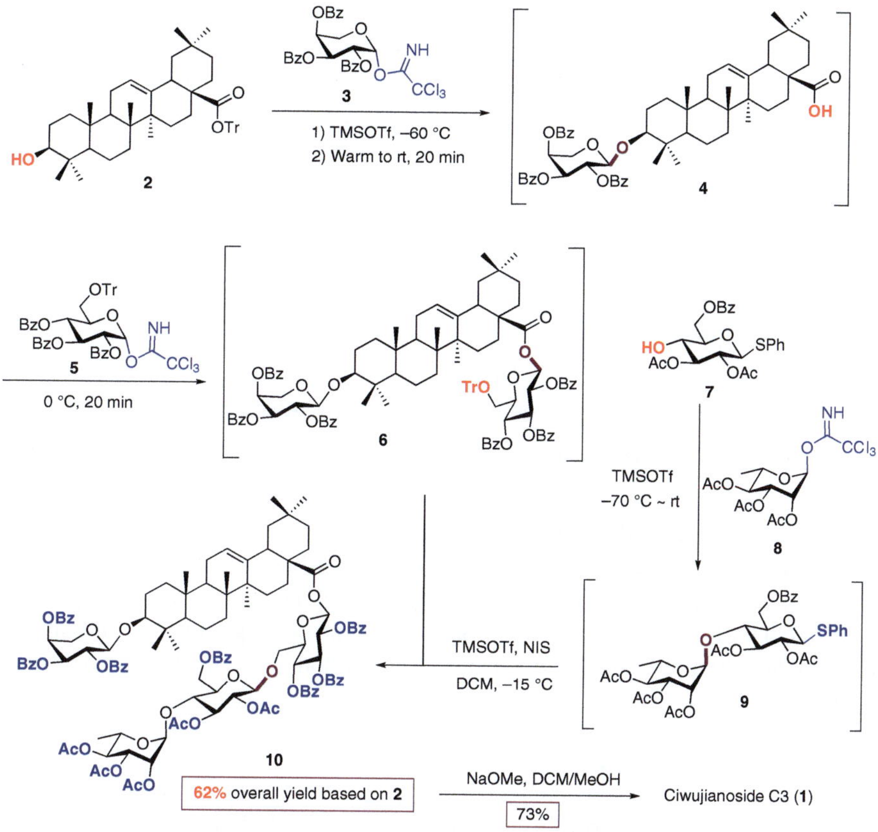


**Scheme 10.1** Total synthesis of ciwujianoside C3 (**1**) by Yu, Hui, and coworkers.

most promising adjuvants for immune response potentiation and reducing vaccine dosage due to its exceptional potency and favorable toxicity profile. However, QS-21 is a heterogeneous substance consisting of at least two principal constituents: QS-21-Api (**11**) and QS-21-Xyl (**12**), which differ in the terminal sugar residue of the linear tetrasaccharide segment (Figure 10.3). Structurally, QS-21 is composed of a quillaic acid aglycone bearing a trisaccharide with glucuronic acid at the 3-OH position and a linear tetrasaccharide elongated by an L-arabinofuranose-terminated fatty acyl chain at the 28-$CO_2H$ group.

**Total synthesis of QS-21A$_{api}$ (11) by the Gin group.** In 2006, Gin and coworkers reported the total synthesis of QS-21A$_{api}$ (**11**) by employing a modular assembly strategy [14, 15]. The synthesis of the branched trisaccharide and its conjugation to quillaic acid was elucidated in Scheme 10.2. Galactose hemiacetal **14** was activated with $Ph_2SO$ and $Tf_2O$ [16], followed by the addition of glucuronate acceptor **13**, resulting in the exclusive formation of β-disaccharide **15** in 80% yield. After removal of the TBS group in **15**, the subsequent coupling with 2,3,4-tri-*O*-benzyl-D-xylose (**16**) in the presence of $Ph_2SO$ and $Tf_2O$ afforded trisaccharide **17** in 92% yield with an anomeric

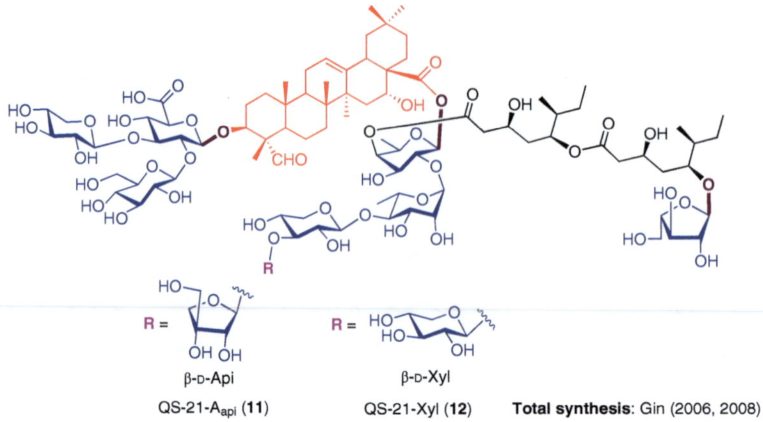

**Figure 10.3** QS-21-$A_{api}$ (**11**) and QS-21-Xyl (**12**).

ratio of 1 : 2 (α/β). After the removal of the anomeric PMB group, trichloroacetimidate **18** was prepared. Subsequently, glycosylation of quillaic acid C28-*O*-allyl derivative **19** with **18** under the catalysis of $(C_6F_5)_3B$ (3 mol%) led to the desired glycoside in 59% yield with a 7 : 1 β/α selectivity. Subsequent protecting group manipulations provided **20**, which was ready for further glycosylation at the C-28 carboxylic acid residue.

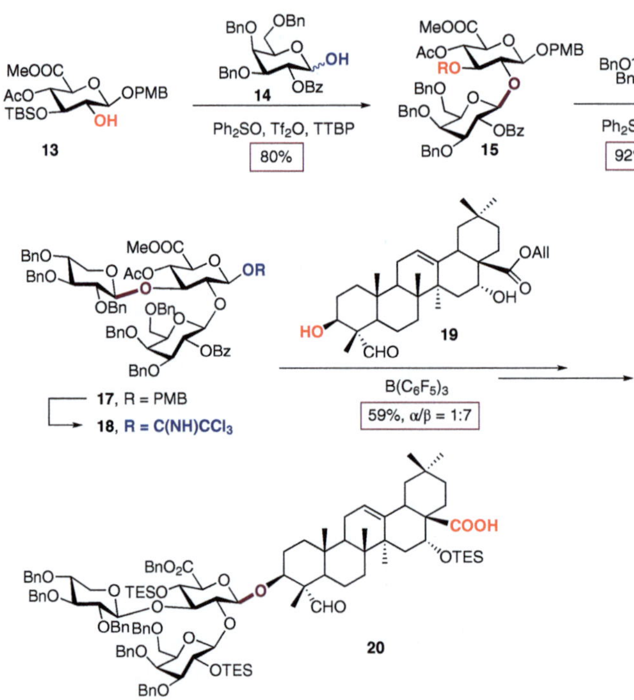

**Scheme 10.2** Synthesis of triterpenoid 3-*O*-trisaccharide **20**.

The completion of the total synthesis of QS-21$_{api}$ (**11**) was described in Scheme 10.3. 2,4-di-*O*-benzyl-D-xylopyranose (**22**) was activated with excess $Ph_2SO$ and $Tf_2O$, and the subsequent addition of L-rhamnopyranose (**21**) resulted in the 1→4-β-linked disaccharide **23** in 55% yield as a single constitutional and stereoisomer. Glycosylation of acceptor **23** with acetyl furanose **24** in the presence of TESOTf provided the β-apiose anomer of the fully protected trisaccharide **25** in 51% yield. Considering the instability of QS-21 toward base hydrolytic, the

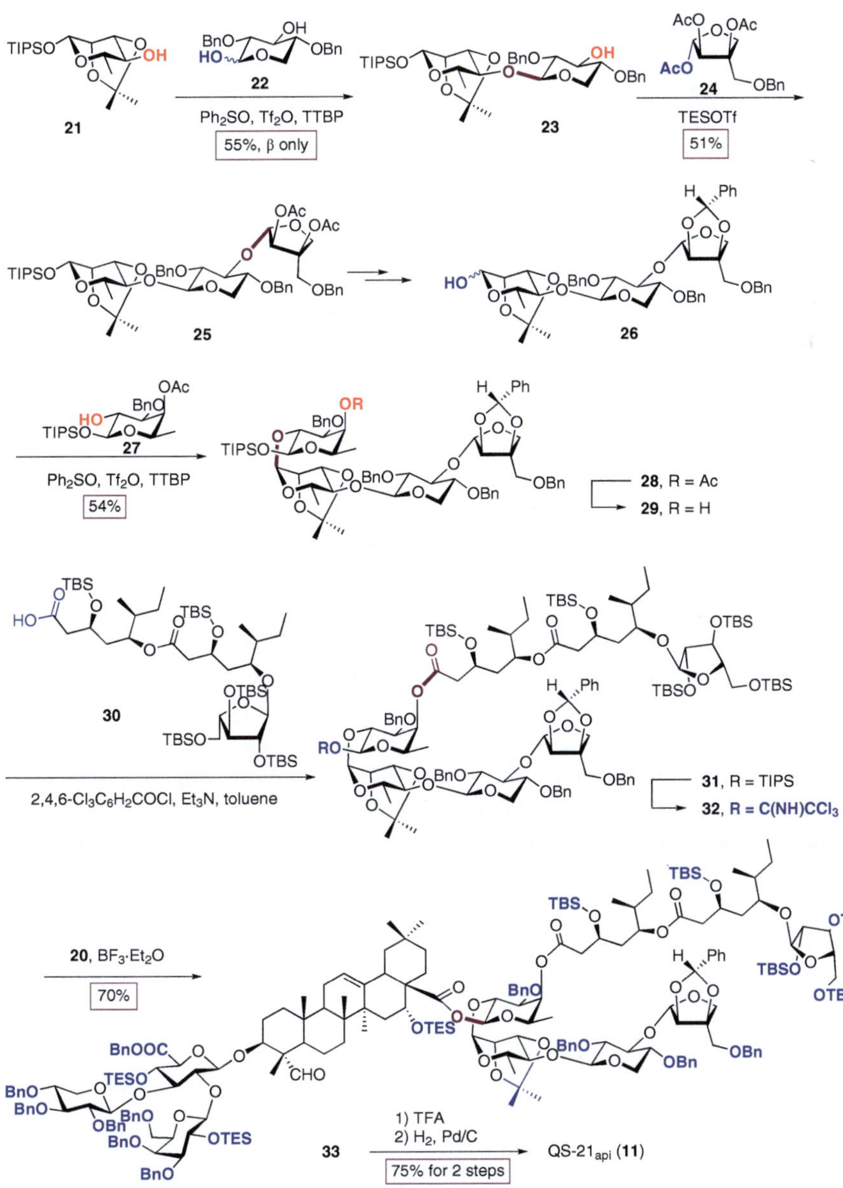

**Scheme 10.3** Completion of the total synthesis of QS-21$_{api}$ (**11**).

acetate-protecting groups, which controlled the anomeric selectivity in the glycosylation, were replaced with benzylidene acetal. Subsequently, the anomeric TIPS group on the rhamnose residue was removed, resulting in trisaccharide hemiacetal **26** after three steps. Dehydrative glycosylation ($Ph_2SO$, $Tf_2O$) of fucopyranoside **27** with trisaccharide hemiacetal **26** provided the fully protected tetrasaccharide fragment **28** in 54% yield, whose fucose C4-acetyl group was removed under basic conditions to afford tetrasaccharide **29**. Esterification of the fucose C4-hydroxyl in **29** with the mixed 2,4,6-trichlorobenzoyl anhydride of **30** (*in situ* generated by treatment with 2,4,6-trichlorobenzoyl chloride) yielded pentasaccharide **31** in high yield. After the conversion of the anomeric TIPS ether in **31** to the corresponding trichloroacetimidate **32**, the coupling of trisaccharide-triterpene **20** with **32** under the catalysis of $BF_3{\cdot}OEt_2$ afforded the fully protected QS-21$A_{api}$ **33** in a good 70% yield. Finally, through mild acidic hydrolysis of the rhamnose isopropylidene ketal and all the silyl ethers (TBS and TES) with trifluoroacetic acid (TFA) and water, followed by hydrogenolysis of all benzylic protecting groups with $H_2$ and Pd/C, the total synthesis of QS-21-$A_{api}$ (**11**) was accomplished in 75% yield over two steps.

In 2008, Gin and coworkers reported the total synthesis of QS-21-Xyl (**15**), employing a similar synthetic approach [17]. The successful synthesis of the active ingredient of QS-21 enabled the creation of novel synthetic analogs of QS-21, with enhanced immune responsiveness and reduced toxicity [18–21].

## Lobatoside E

Lobatoside E (**34**, Figure 10.4) belongs to a small group of triterpene saponins called cyclic bisdesmosides, which consist of two oligosaccharides attached to a pentacyclic triterpene and bridged with 3-hydroxy-3-methyl glutarate. Only 10 such compounds have been disclosed from two Chinese medicinal plants, namely *Bolbostemma paniculatum* and *Actinostemma lobatum* (Cucurbitaceae) [22, 23]. Cyclic bisdesmosides exhibit cytotoxic properties and have the potential to induce apoptosis in tumor cells. Removal of the glutarate bridge eliminated their activity, highlighting the essential role of the cyclic scaffold in their antitumor action. Among its counterparts, lobatoside E (**34**) demonstrated the highest potency against tumor cell growth and

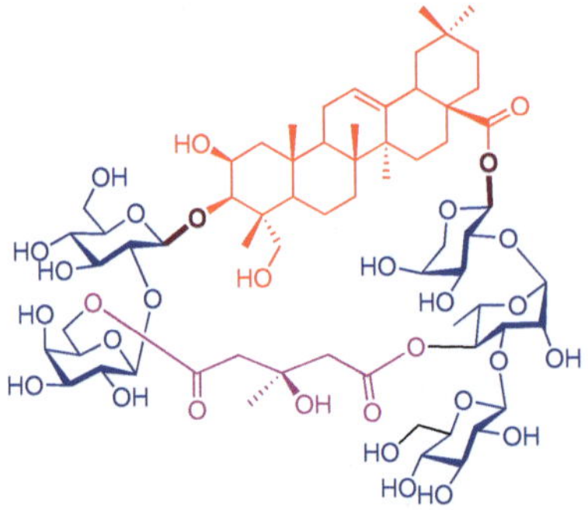

Lobatoside E (**34**)
**Total synthesis**: Yu (2008)

**Figure 10.4** Lobatoside E (**34**).

exhibited remarkable sensitivity toward lung cancer cell A549, colon cancer cell SW-620, and melanoma cell SK-MEL-5 [24].

**Total synthesis of lobatoside E (34) by the Yu group.** In 2008, Yu and coworkers reported the first total synthesis of lobatoside E (**34**) using a modular synthetic route from cheap starting materials, with the longest linear sequence of 31 steps and 1.2% overall yield (Scheme 10.4) [25]. The bayogenin derivative **35** was synthesized from oleanolic acid in 17 steps by selective introduction of the 23-OH and 2-α-OH groups. Regioselective coupling of the 3-ol-28-carboxylic acid **35** with α-arabinosyl bromide **36** under phase transfer conditions gave the desired ester α-arabinoside in 80% yield. Subsequent glycosylation of the remaining 3-OH with glucosyl imidate **37** under the catalysis of TMSOTf provided β-glucoside **38** in 96% yield. Following the selective removal of the CA group (vs. the Ac group), glycosylation with galactosyl imidate **39** (with TMSOTf) gave β-galactoside **40** in 65% yield. Selective removal of the Ac group (vs. the Bz group) was followed by glycosylation with thiodisaccharide **41** in the presence of NIS/TMSOTf, resulting in stereoselective formation of the α-rhamnosyl-linked pentasaccharide **42** in 81% yield. Upon steps of protecting group manipulations, the condensation and subsequent macrocyclization of pentasaccharide **42** with the racemic diacid **43** under the Yamaguchi conditions provided macrocycle **44** in good yield. At this stage, the two C3′ epimers (**44a** and **44b**), formed without a preference, could be separated by chromatography on silica gel. Finally, removal of all Bn groups via hydrogenolysis over $Pd(OH)_2$ furnished lobatoside E (**34**) and its C-3′ epimer in ~80% yield.

A simplified cyclic triterpenoid saponin **52** was efficiently synthesized by Yu and coworkers in 2010, testifying the newly developed gold(I)-catalyzed glycosylation methodology (Scheme 10.5) [26]. The assembly commenced with a successful chemoselective glycosylation of the 28-carboxylic acid of oleanolic acid with arabinosyl *ortho*-hexynylbenzoate **45**. This reaction was catalyzed by $Ph_3PAuOTf$ (0.1 equiv.) with the assistance of $BF_3{\cdot}OEt_2$ and DBU, leading to ester α-L-arabinoside **46** predominantly in a satisfactory yield of 72%. Subsequently, a gold-catalyzed one-pot sequential glycosylation with 1,2-anhydroglucose **47** and galactosyl *ortho*-hexynylbenzoate **48** was performed, leading to the assembly of the left-handed disaccharide and formation of bisglycoside **49** in 60% yield. Selective removal of the 2-*O*-acetyl group on the arabinose unit in **49** was followed by glycosylation with rhamnosyl *ortho*-hexynylbenzoate **50** under the standard gold-catalyzed conditions, which provided tetrasaccharide **51** in nearly quantitative yield. After several steps, involving protection group manipulation and macrocyclization with glutaric acid, the cyclic triterpene saponin **52** was assembled, requiring only 11 steps and in ~11% overall yield from six readily available building blocks.

## Ginsenosides

Ginsenosides, a major type of tetracyclic dammarane-type saponins, are found abundantly in plants of the *Panax genus*, which have been widely used in traditional Chinese medicine as potent tonics for over 2000 years (Figure 10.5) [27]. Ginsenosides

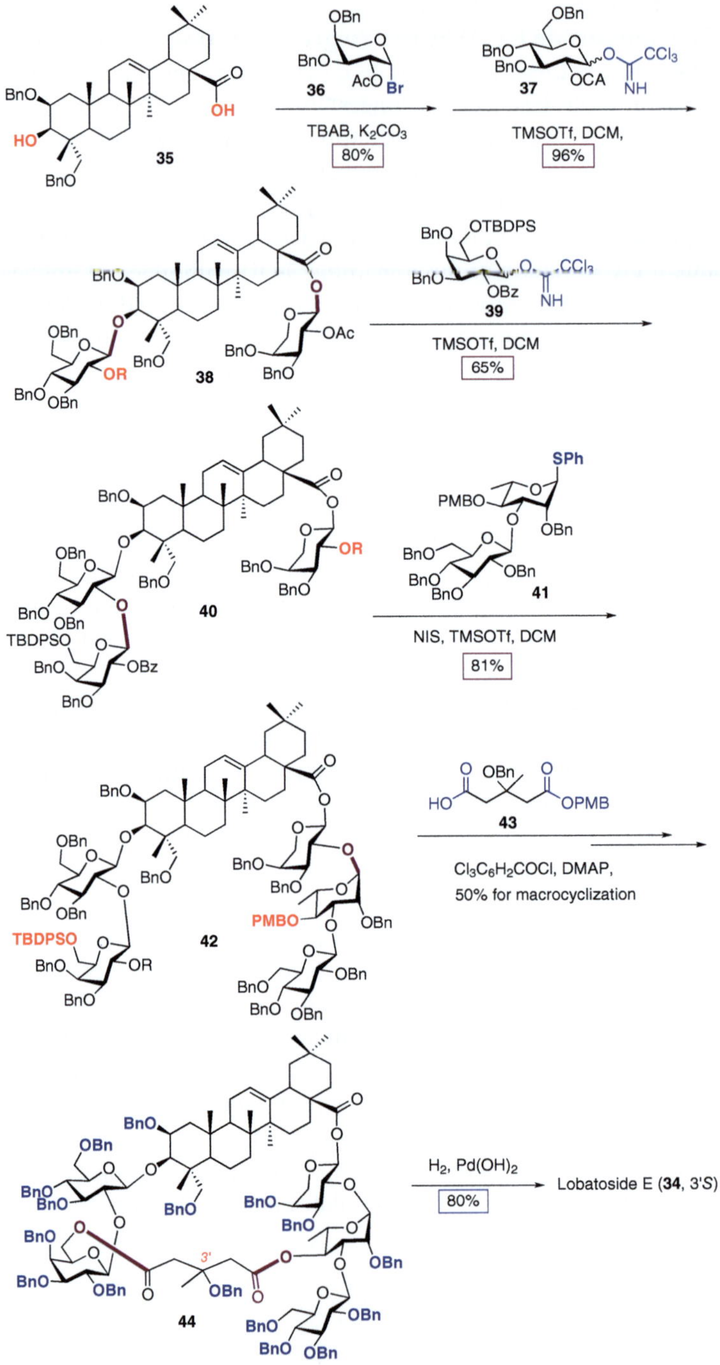


**Scheme 10.4** Total synthesis of lobatoside E (**34**) by Yu and coworkers.

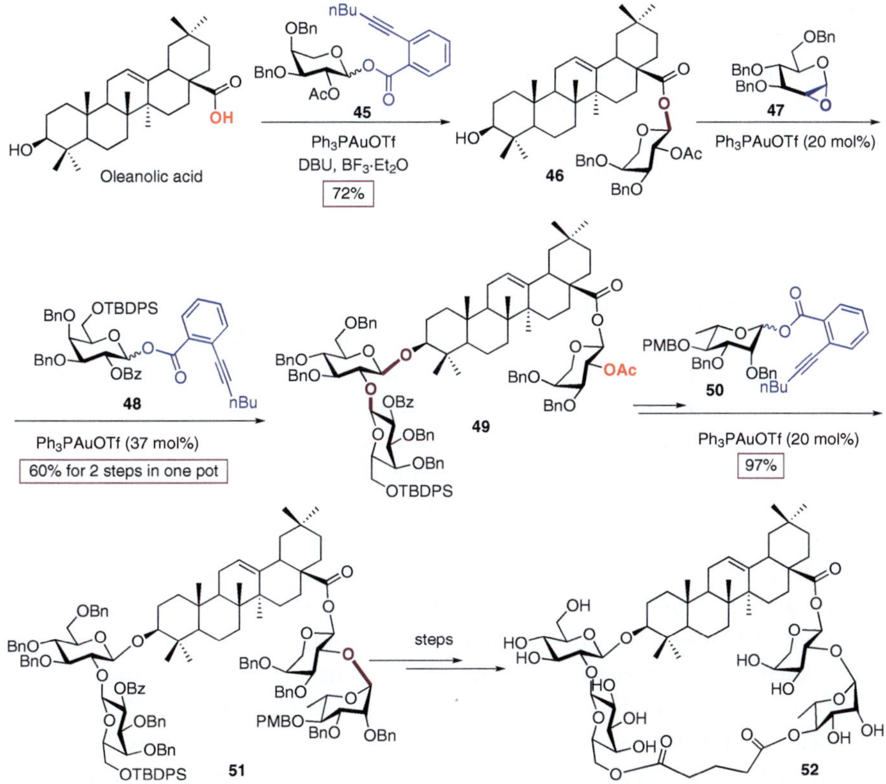


**Scheme 10.5** Synthesis of cyclic triterpene saponin **52** employing the gold(I)-catalyzed glycosylations by Yu and coworkers.

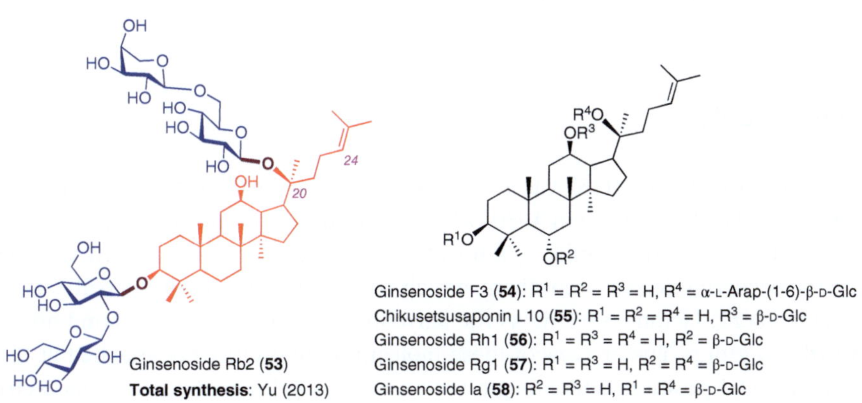


**Figure 10.5** Representative ginsenosides.

exhibited a wide range of pharmacological activities, including immunostimulatory, antiatherosclerotic, antihypertensive, antidiabetic, anticarcinogenic, and CNS protective effects [28, 29].

**Synthesis of diverse ginsenosides by Yu and coworkers.** In 2013, Yu and coworkers reported the divergent synthesis of all possible types of protopanaxatriol and protopanaxadiol glycosides, via selective protection of the aglycone and sequential glycosylation using the gold-catalyzed glycosylation methodology (Scheme 10.6) [30]. The synthesis of ginsenoside Rb2 (**53**) commenced with 3-*O*-chloroacetyl-12-*O*-allyl-protopanaxadiol (**59**), which was glycosylated with glucosyl *ortho*-alkynylbenzoate **60** under the catalysis of $PPh_3AuNTf_2$ (15 mol%), yielding 20-*O*-glucoside **61** in a good 75% yield. Interestingly, the stability of the dammarane 20-*O*-glycosidic linkage was found to be highly dependent on the protecting group at the proximal 12-OH position. The 12-*O*-allyl-20-*O*-glycosides were prone to elimination in the presence of trace acid, whereas the 12-*O*-acetyl counterparts remained intact even in the presence of 0.5 equiv. of TMSOTf. Therefore, the 12-*O*-allyl group in **62** was replaced with an acetyl group, and subsequent coupling of the C6′-OH with arabinose alkynylbenzoate **62** provided disaccharide **63** in nearly quantitative yield. After selective removal of the 3-*O*-chloroacetyl group on **63**, the following coupling with glucosyl alkynylbenzoate **64** under the action of gold(I) catalyst furnished trisaccharide **65** in 80%. The neighboring-participating group (AZMB) was then cleaved with $n\text{Bu}_3\text{P}$ to release the 2″-OH, which was glycosylated with perbenzoyl glucosyl alkynylbenzoate **66** under the standard gold(I)-catalyzed conditions, yielding the desired tetrasaccharide **67** in a good 84% yield. Finally, through successive removal of one allyl group with $PdCl_2$, one benzylidene acetal with *p*-TsOH, one acetyl and 10 benzoyl groups upon treatment of KOH, the synthesis of ginsenoside Rb2 (**53**) was achieved. Similarly, ginsenosides **54–58** and their analogs have been synthesized. Later, the Yu group also completed the synthesis of ginsenoside Rh2 [31] and several numbers of the ocotillol-type ginsenosides [32] by employing the gold(I)-catalyzed glycosylation method with glycosyl *ortho*-alkynylbenzoate donors.

## Echinoside A

Echinoside A (**68**, Figure 10.6) was initially isolated from the Okinawa sea cucumber *Actinopyga echinites* (Jaeger) in 1980, which showed potent antifungal and antitumor activities [33]. Structurally, echinoside A (**68**) represents a typical triterpene glycoside occurring in sea cucumbers, which consists of a holostane, specifically a lanostane-18(20)-lactone, serving as the aglycone, with a $\Delta^{9,11}$ double bond and a 12α hydroxy group. The glycans of these molecules are exclusively attached at the lanostane 3-OH position, which begin with D-xylose (Xyl) and extends at its 2-OH position with D-quinovose (Qui) and D-glucose (Glc). Additionally, common structural features include a sulfonyl residue at the 4-OH position of the Xyl unit and a methyl substituent at the 3-OH position of the terminal Glc unit.

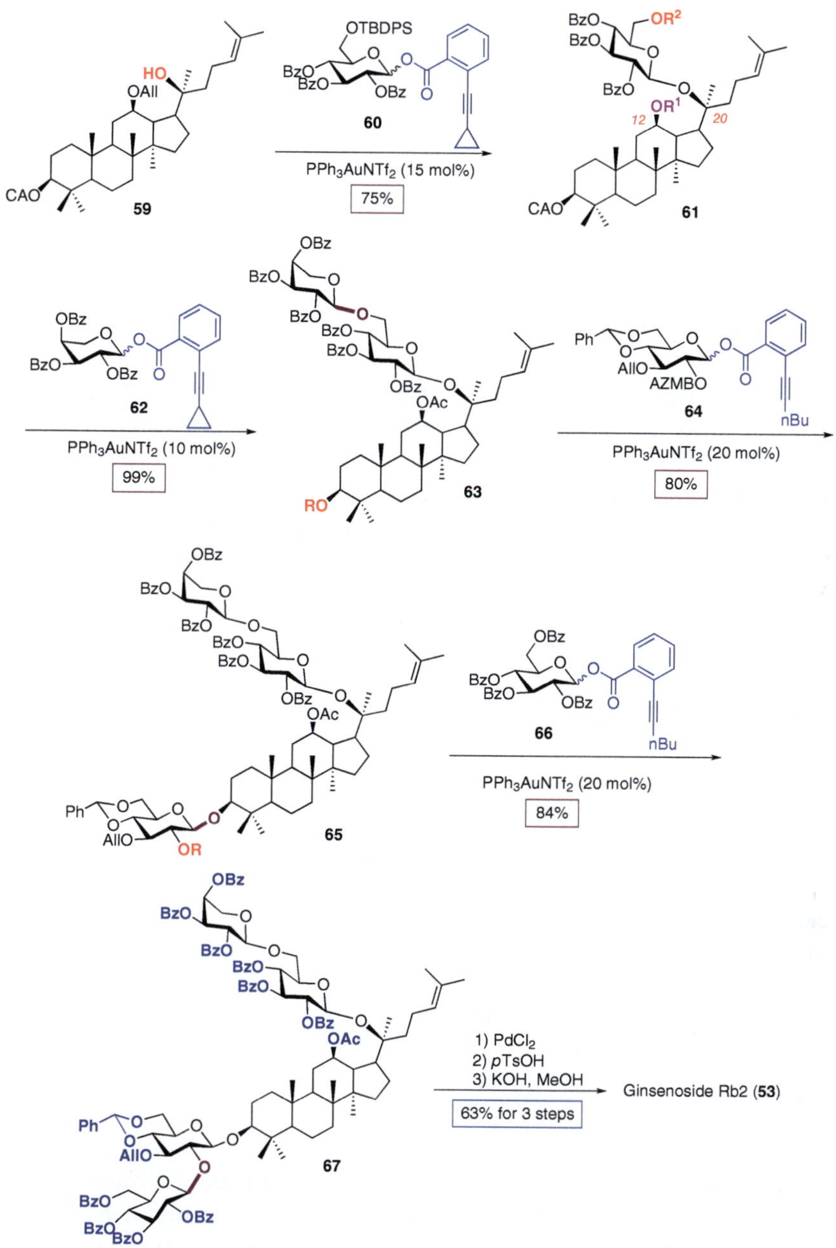


**Scheme 10.6** Synthesis of ginsenoside Rb2 (**53**) by Yu and coworkers.

**Total synthesis of echinoside A (68) by the Yu group.** In 2017, Yu and coworkers reported the first total synthesis of echinoside A (**68**) in a longest linear sequence of 35 steps with an overall yield of 0.6% [34]. The syntheses of the trisaccharide unit began with the coupling of quinovoside acceptor **69** with thioglucoside **70** under the action of NIS/TMSOTf, which was followed by the removal of the TES protecting

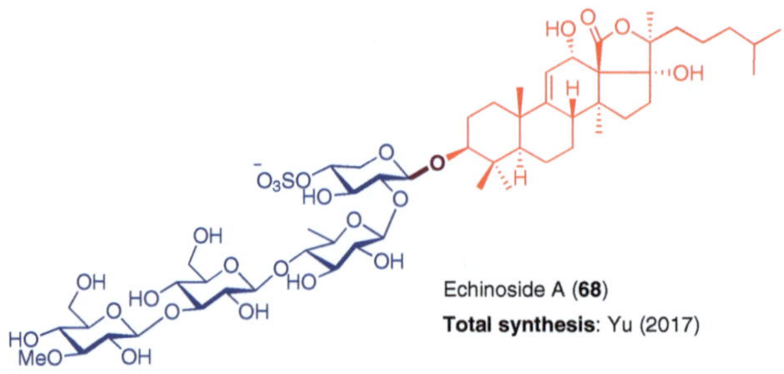


**Figure 10.6** Echinoside A (**68**).

group to yield the desired β-disaccharide **71** in 67% yield (Scheme 10.7). The coupling of disaccharide **71** with thioglycoside **72** under similar conditions provided β-trisaccharide **73** in 83% yield. Switching the 4,6-*O*-benzylidene groups to benzoyl groups was followed by the cleavage of the anomeric MP group and subsequent condensation with 2-(cyclopropylethynyl)benzoic acid, providing the desired trisaccharide donor **74**.

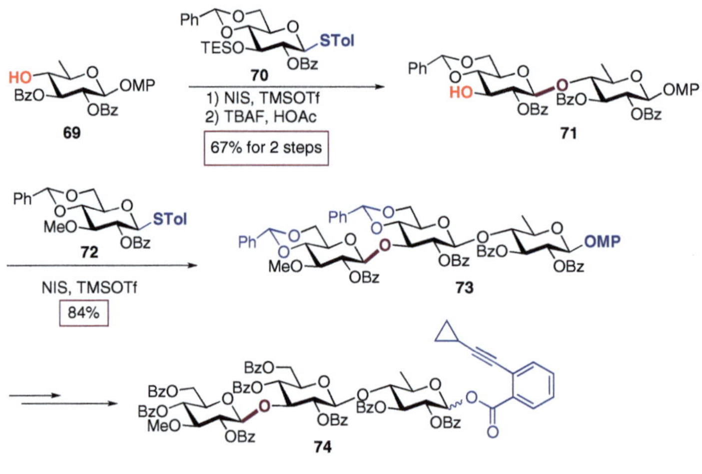

**Scheme 10.7** Synthesis of trisaccharide donor **74**.

The syntheses of the echinoside A aglycone began with the utilization of the known lanosterol derivative **75**, which upon irradiation of a tungsten lamp in the presence of iodine and $Pb(OAc)_4$ led to the desired 18-iodo-20-oxo derivative **76** (Scheme 10.8). The installation of the side chain was realized through the α-alkylation of the methyl ketone moiety with isopentyl iodide in the presence of LDA, producing compound **77**. After conversion of the $\Delta^{8(9)}$ double bond to $\Delta^{9(11)}$, the allylic C12 methylene group was oxidized with $CrO_3$·DMP to afford the desired enone **78**. The glycosylation of 3-hydroxyholostane **78** was performed using xylosyl *o*-hexynylbenzoate **79** in the presence of $PPh_3AuNTf_2$ (0.1 equiv.), which provided β-xyloside **80** in 73% yield after the selective removal of the 2′-*O*-Azmb group with $PMe_3$. The trisaccharide fragment was successfully introduced using the Au(I)-catalyzed glycosylation of **80** with previously prepared trisaccharide donor **74**, leading to β-tetrasaccharide **81** in a high 85% yield. Following the stereoselective reduction of the C12-ketone group and removal of the 4′-*O*-Nap group in the Xyl unit, the resultant compound underwent treatment with $SO_3$· pyridine and global deprotection of all the acyl groups with KOH, thus accomplishing the total synthesis of echinoside A (**68**) in an overall yield of 85% for the final two steps.

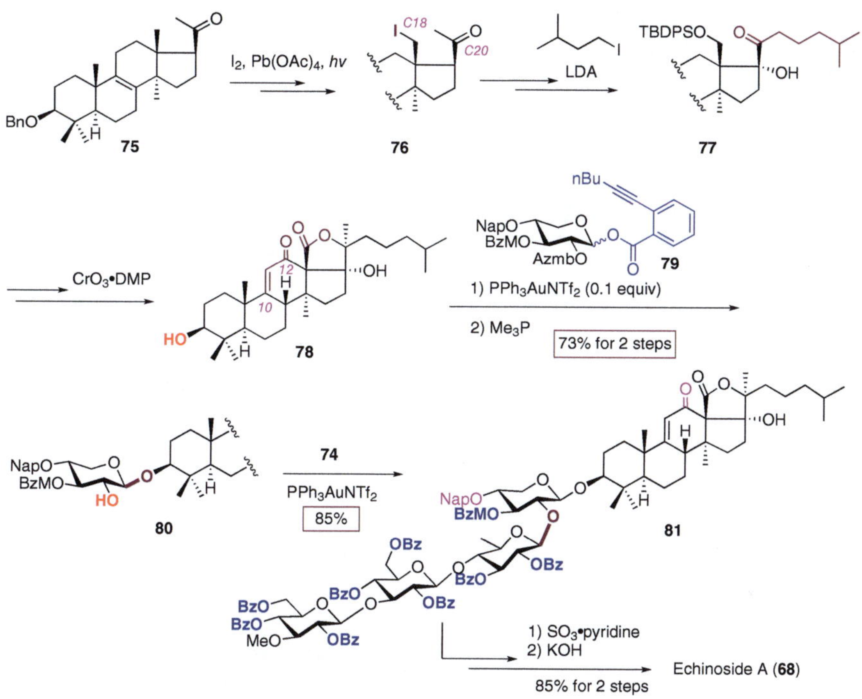


**Scheme 10.8** Total synthesis of echinoside A (**68**) by the Yu group.

## Anemoclemoside B

Anemoclemoside B (**82**, Figure 10.7) was originally isolated from the roots of *Anemoclema glaucifolium*, a traditional medicinal plant found in the Yangtse River valley region of China, by Li et al. in 1995 [35]. Structurally, anemoclemoside B (**82**) possesses a distinctive open-chain cyclic acetal linkage connecting the 2-*O*-α-L-rhamnopyranosyl-(1 → 2)-L-arabinose moiety to the 3,23-diol of hederagenin. In 2005, Yu and coworkers developed an effective approach to the synthesis of the open-chain cyclic acetal linkage, employing TMSOTf as a catalyst for the condensation between a linear sugar aldehyde and the aglycone's 1,3-diols. This method enabled the synthesis of anemoclemoside B (**82**) [36].

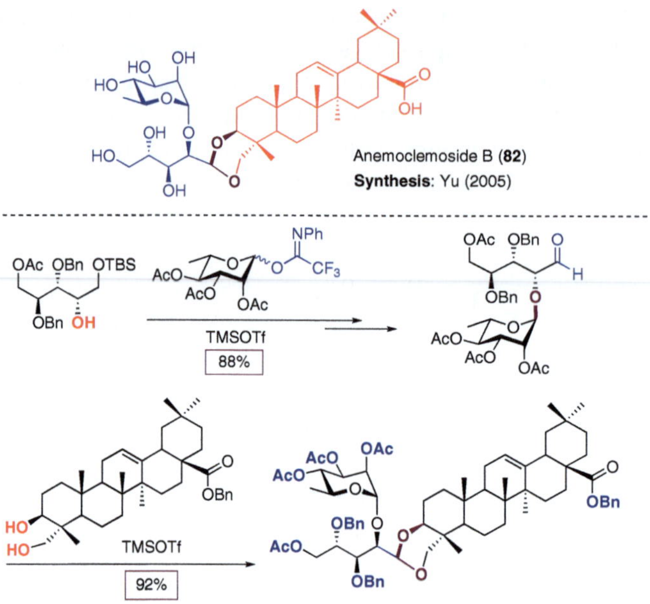


**Figure 10.7** Anemoclemoside B (**82**) and the glycosylation step in the synthesis.

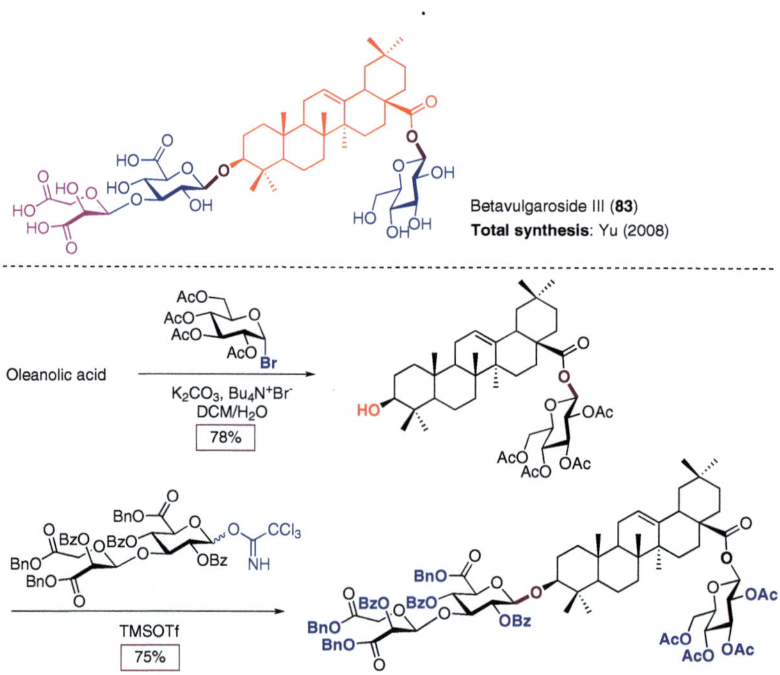


**Figure 10.8** Betavugaroside III (**83**) and the glycosylation steps in the syntheses.

## Betavugaroside III

Betavulgaroside III (**83**, Figure 10.8), classified as a triterpene secoglycoside, was isolated from *Beta Vulgaris* and *Achyranthes faurie* in 1994 [37]. It exhibits a unique structural characteristic, characterized by the oxidative fragmentation of a terminal monosaccharide unit. These triterpene secoglycosides were recognized as mimics of the sialyl Lewis X (sLe$^x$) structure, a saccharide epitope found on cell surfaces that plays a vital role in a variety of biological functions. In 2008, Yu and colleagues accomplished the total synthesis of betavulgaroside III (**83**) with a longest linear sequence of 23 steps and an overall yield of 0.9%, starting from L-arabinose, D-glucose, and oleanolic acid [38]. This synthesis notably focused on the elaboration of the seco-saccharide unit before the assembly of the triterpene 3,28-bisglycoside.

## $\Delta^{20}$-Ginsenosides

$\Delta^{20}$-Ginsenosides possess a carbon–carbon double bond at the position C-20 within the side chain of the dammarane-type sapogenins. The double bond may be located

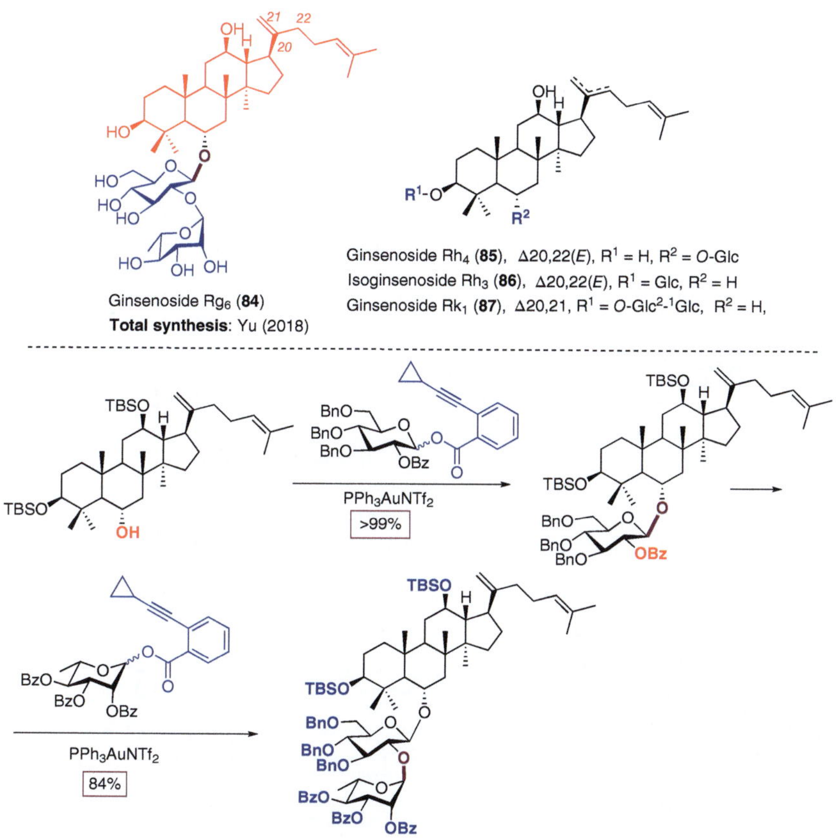


**Figure 10.9** Ginsenoside Rg$_6$ (**84**), Rh$_4$ (**85**), Rh$_3$ (**86**), and Rk$_1$ (**87**) and the glycosylation steps in the syntheses.

between the C20–C21 or C20–C22, exhibiting *E*- or *Z*-configuration, as exemplified by ginsenosides $Rg_6$ (**84**, Figure 10.9), $Rh_4$ (**85**), (20*E*)-$Rh_3$ (**86**), and $Rk_1$ (**87**) [39]. These compounds showed potent antitumor activities in various cancer cell lines. In 2018, Yu and coworkers reported the chemical syntheses of $\Delta^{20}$-ginsenosides ginsenosides **84–87** [40]. The syntheses were achieved through direct glycosylation of the appropriate $\Delta^{20}$-sapogenin acceptor and a glycosyl *ortho*-alkynylbenzoate donor under gold(I)-catalysis. The use of neutral conditions in this methodology was imperative to preserve the acid-labile nature of the C20–C21 double bond, ensuring its integrity throughout the synthetic process.

## Flaccidoside II

Flaccidoside II (**88**, Figure 10.10) was isolated from the alcohol extracts of *Anemone flaccida* Fr. Schmidt, commonly known as DiWu in Chinese folk medicine, which is used for its detoxifying properties. It is characterized by the presence of a 2′-OH

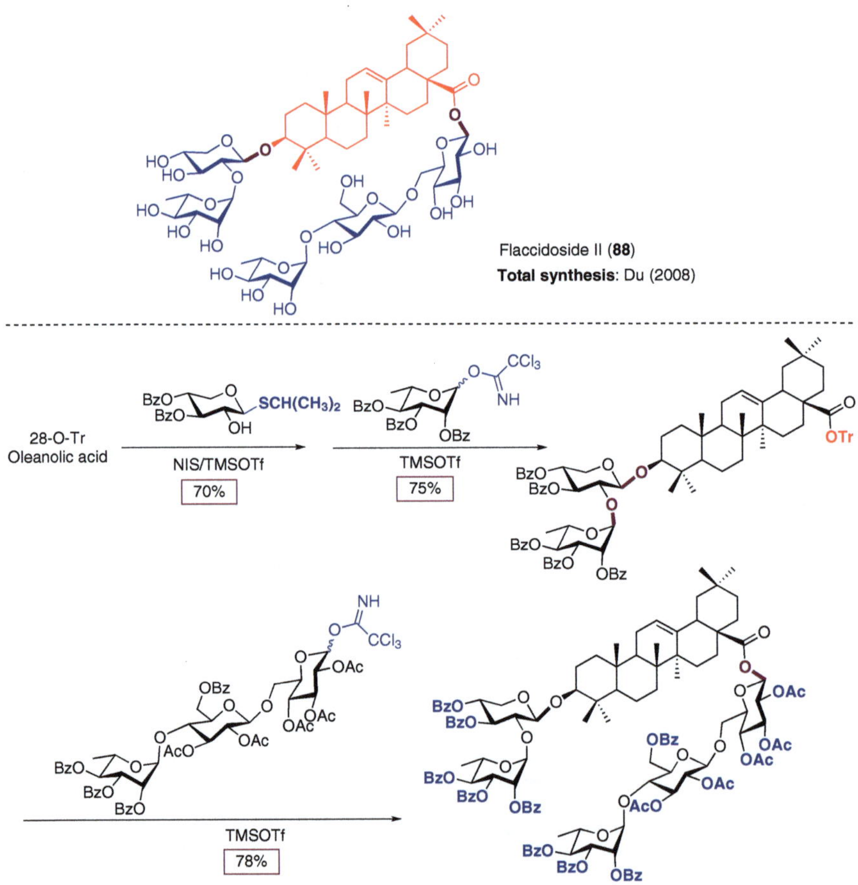


**Figure 10.10** Flaccidoside II (**88**) and the glycosylation steps in the synthesis.

branched sugar chain attached to the C-3 position of oleanolic acid. In 2008, Du and coworkers reported the synthesis of flaccidoside II (**88**), featuring the glycosylation of oleanolic acid 28-ester with a thio-xylopyranoside bearing an unprotected 2-OH group [41].

## Asiaticoside

Asiaticoside (**89**, Figure 10.11) is an ursane-type triterpene glycoside which has been used as a biomarker of the traditional medicinal plant *Centella asiatica* [42]. It could enhance collagen synthesis, potentially by activating the Smad pathway or inhibiting glycogen phosphorylase. Structurally, asiaticoside (**89**) consists of an aglycone known as 2α,3β,23-trihydroxy-urs-12-en-28-oic acid (also referred to asiatic acid), with a trisaccharide glycan moiety attached through ester linkage at position C28. In 2017, Yu and coworkers reported the synthesis of asiaticoside (**89**), featuring the condensation of a triterpenoid acid with a trisaccharide *ortho*-hexynylbenzoate under gold(I) catalysis [43].

## Pulsatilla Saponin D

Pulsatilla saponin D (**90**, Figure 10.12), an oleanane-type saponin, was initially isolated from *Pulsatilla japonica* in 1978 [44]. It demonstrated remarkable *in vivo*

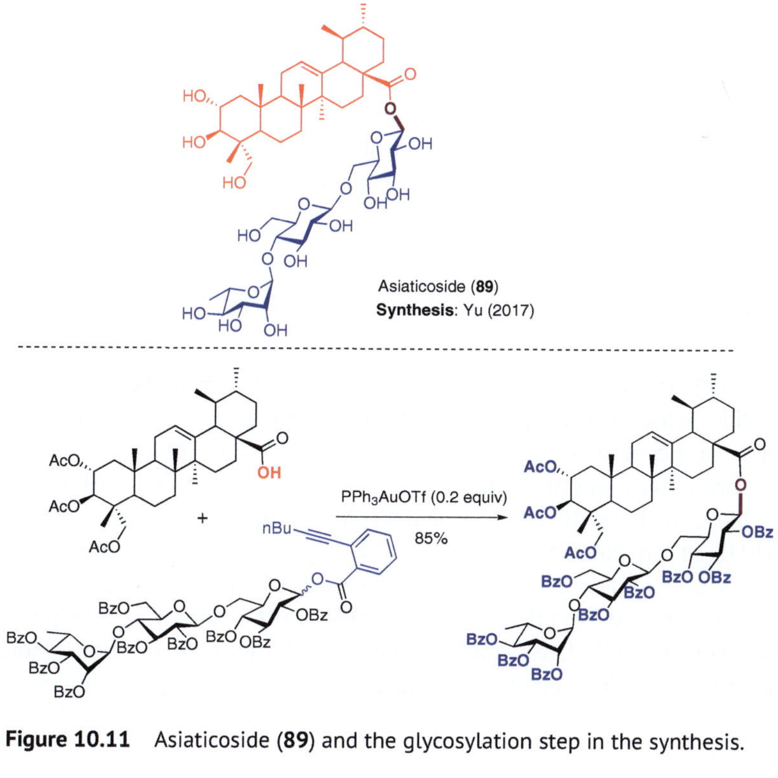


**Figure 10.11** Asiaticoside (**89**) and the glycosylation step in the synthesis.

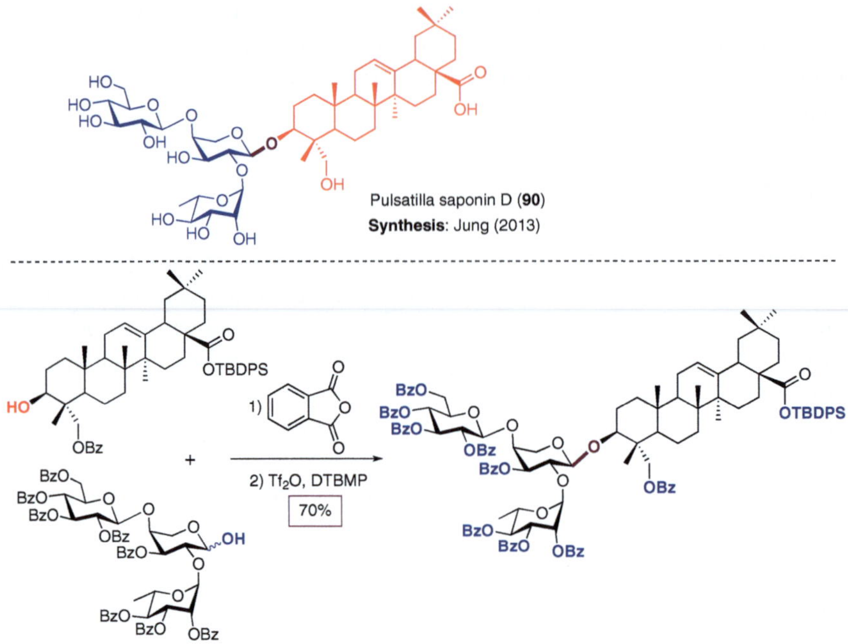


**Figure 10.12** Pulsatilla saponin D (**90**) and the glycosylation step in the synthesis.

antitumor activity, surpassing that of taxol and doxorubicin, in mice bearing Lewis lung carcinoma. Structurally, pulsatilla saponin D (**90**) consists of hederagenin as aglycone and a branched trisaccharide at C3. In 2013, Jung and coworkers reported the synthesis of pulsatilla saponin D (**92**), featuring a β-selective glycosylation mediated by phthalic anhydride and $Tf_2O$ between a trisaccharide hemiacetal and aglycone derivative [45].

## Lotoidoside D

Lotoidoside D (**91**, Figure 10.13) is a typical *N*-acetylglucosamine-bearing triterpenoid glycoside, which was isolated from the roots of *Glinus lotoides*, a medical plant growing in Egyptian desert, by Hamed et al. in 2005 [46]. Lotoidoside D exhibited prominent antiproliferative activity against various murine and human cell lines, with an $IC_{50}$ value of 0.36 μM against murine fibrosarcoma cell line WEHI164. In 2003, Yu and coworkers reported the first synthesis of such type of *N*-acetylglucosamine-bearing oleanolic acid [47]. In 2006, Cheng and coworkers reported the synthesis of lotoidoside D (**91**), employing a stepwise glycosylation strategy [48]. Later, more congeners of this type of triterpenoid glycosides have been synthesized by the Liang group [49, 50] and the Guo and Song group [51].

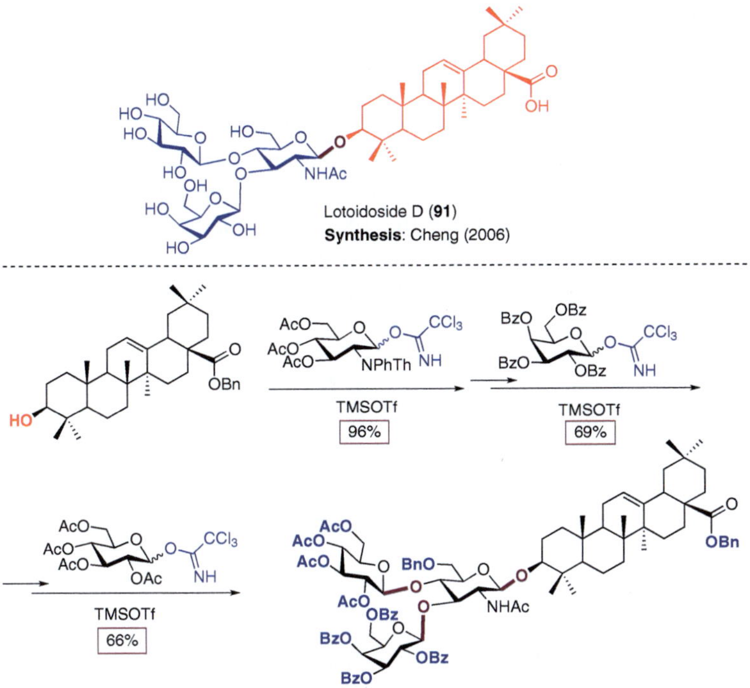


**Figure 10.13** Lotoidoside D (**91**) and the glycosylation steps in its total synthesis.

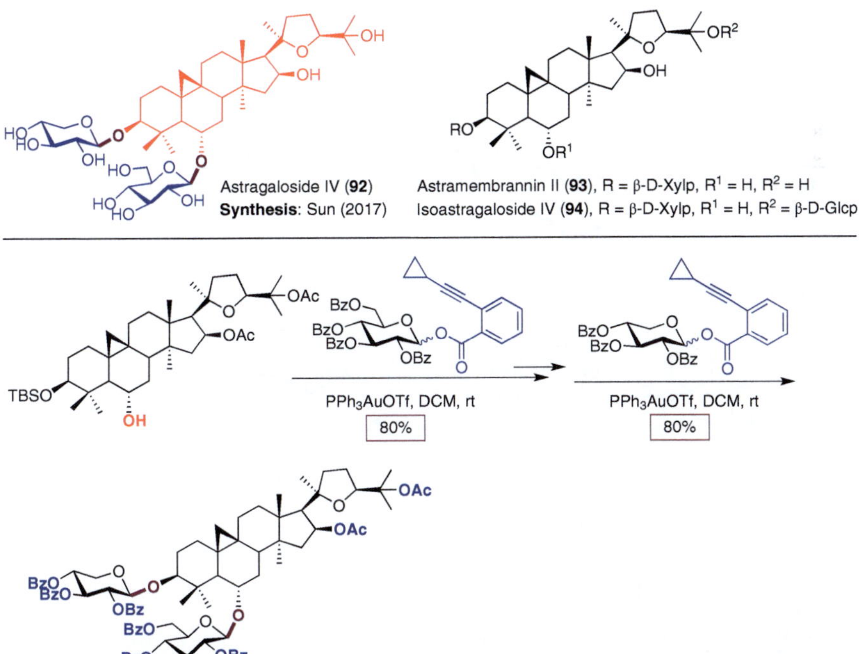


**Figure 10.14** Astragaloside IV (**92**), astramembrannin II (**93**), and isoastragaloside IV (**94**), and the glycosylation steps in the synthesis.

## Astragalosides

Astragaloside IV (**92**, Figure 10.14), astramembrannin II (**93**), and isoastragaloside IV (**94**) belong to the sub-family of cycloastragenol 3-*O*-glycosides, cycloastragenol 3,6-*O*-bisglycosides, and cycloastragenol 3,25-*O*-bisglycosides respectively, which are the major cycloartane glycosides found in *Astragalus membranaceus* [52]. Astragali Radix (Huangqi), derived from the root of *A. membranaceus*, has a long history of use as a vital tonic herb in the traditional Chinese medicine, displaying diverse biological functions, such as hepatoprotection, neuroprotection,

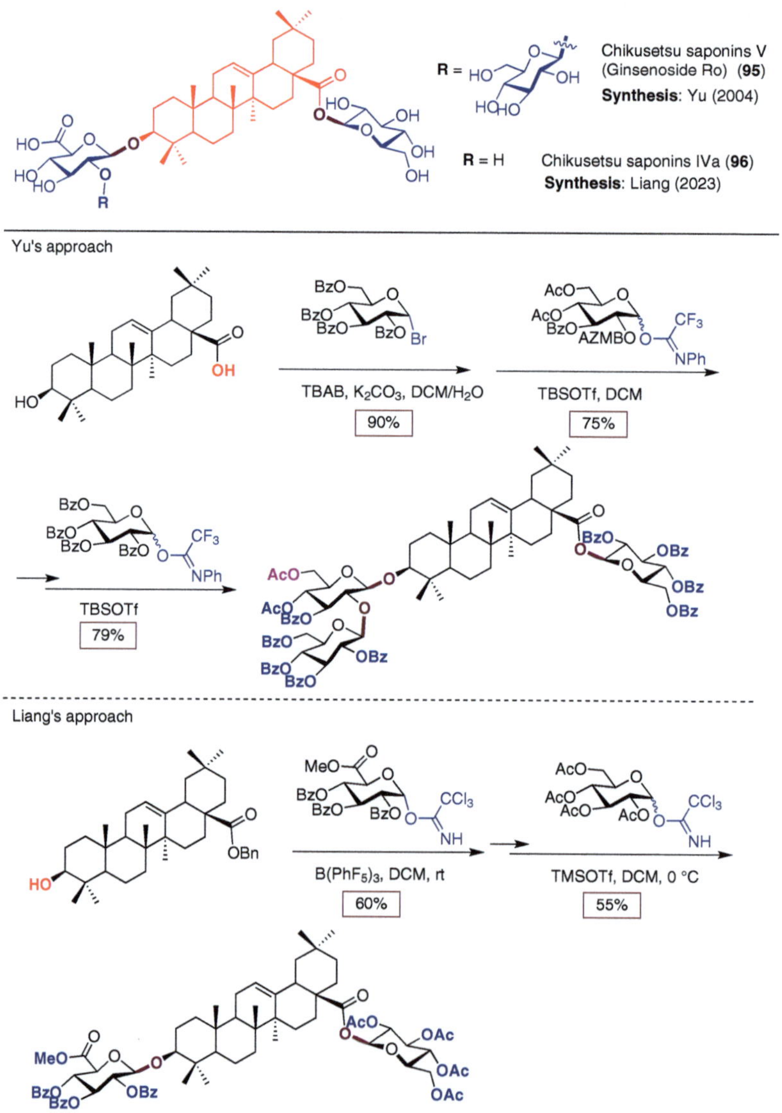


**Figure 10.15** Chikusetsu saponins Iva (**95**) and the glycosylation steps in the synthesis.

immunomodulatory activity, and antitumoral effects. In 2017, Sun and coworkers reported a collective synthesis of these cycloastragenol glycosides, featuring glycosylation with the Au(I)-catalyzed glycosylation with glycosyl *ortho*-alkynylbenzoates as donors [53].

## Chikusetsu Saponins

Chikusetsu saponins V (**95**, also named as ginsenoside Ro) and Iva (**96**, Figure 10.15), simple triterpene 3-*O*-β-glucuronides, were isolated from *Panax* [54]. In fact, triterpene 3-*O*-β-glucuronides have gained recognition in herbal medicine and pharmaceutical applications, which demonstrated anti-cancer, anti-inflammatory, and anti-hyaluronidase effects. In 2004, Yu and coworkers reported the synthesis of chikusetsu saponins V (**95**), as well as its congener, by employing a combination of glycosyl bromide and glycosyl imidate donors, in which the C6′-carboxylic acid moiety was introduced at the later stage of the synthesis [55, 56]. In 2023, Liang et al. report the synthesis of chikusetsu saponins Iva (**96**) and a range of derivatives, employing glycosylation with glucuronyl imidate donors under the catalysis of $B(PhF_5)_3$ [57].

## References


**1** Zu, L., Zhao, Y., and Gu, G. (2014). *J. Carbohydr. Chem.* 33: 269–297.
**2** Yang, Y., Laval, S., and Yu, B. (2014). *Adv. Carbohydr. Chem. Biochem.* 71: 137–226.
**3** Yu, B., Zhang, Y., and Tang, P. (2007). *Eur. J. Org. Chem.* 2007: 5145–5161.
**4** Pellissier, H. (2004). *Tetrahedron* 60: 5123–5162.
**5** Xu, P. and Yu, B. (2021). *Advances in Carbohydrate Chemistry and Biochemistry*, vol. 79 (ed. D.C. Baker), 1–62. Academic Press.
**6** Yu, B. and Sun, J. (2009). *Chem. Asian J.* 4: 642–654.
**7** Charles, G., Jean, L., and Andre, P. (2009). *Mini-Rev. Org. Chem.* 6: 321–344.
**8** Xiao, G., Shao, X., Zhu, D., and Yu, B. (2019). *Nat. Prod. Rep.* 36: 769–787.
**9** Zhu, D. and Yu, B. (2018). *Chin. J. Chem.* 36: 681–691.
**10** Liao, J.-X., Hu, Z.-N., Liu, H., and Sun, J.-S. (2021). *Synthesis* 53: 4389–4408.
**11** Shao, C.-J., Kasai, R., Xu, J.-D., and Tanaka, O. (1988). *Chem. Pharm. Bull.* 36: 601–608.
**12** Yu, B., Xie, J., Deng, S., and Hui, Y. (1999). *J. Am. Chem. Soc.* 121: 12196–12197.
**13** Kensil, C.R., Patel, U., Lennick, M., and Marciani, D. (1991). *J. Immunol.* 148: 431–437.
**14** Kim, Y.-J., Wang, P., Navarro-Villalobos, M. et al. (2006). *J. Am. Chem. Soc.* 128: 11906–11915.
**15** Deng, K., Adams, M.M., and Gin, D.Y. (2008). *J. Am. Chem. Soc.* 130: 5860–5861.
**16** Garcia, B.A., Poole, J.L., and Gin, D.Y. (1997). *J. Am. Chem. Soc.* 119: 7597–7598.
**17** Deng, K., Adams, M.M., Damani, P. et al. (2008). *Angew. Chem. Int. Ed.* 47: 6395–6398.

**18** Fernández-Tejada, A., Tan, D.S., and Gin, D.Y. (2016). *Acc. Chem. Res.* 49: 1741–1756.

**19** Fernández-Tejada, A., Chea, E.K., George, C. et al. (2014). *Bioorg. Med. Chem.* 22: 5917–5923.

**20** Chea, E.K., Fernández-Tejada, A., Damani, P. et al. (2012). *J. Am. Chem. Soc.* 134: 13448–13457.

**21** Ragupathi, G., Gardner, J.R., Livingston, P.O., and Gin, D.Y. (2011). *Expert Rev. Vaccines* 10: 463–470.

**22** Fujioka, T., Iwamoto, M., Iwase, Y. et al. (1989). *Chem. Pharm. Bull.* 37: 2355–2360.

**23** Fujioka, T., Iwamoto, M., Iwase, Y. et al. (1989). *Chem. Pharm. Bull.* 37: 1770–1775.

**24** Fujioka, T., Kashiwada, Y., Okabe, H. et al. (1996). *Bioorg. Med. Chem. Lett.* 6: 2807–2810.

**25** Zhu, C., Tang, P., and Yu, B. (2008). *J. Am. Chem. Soc.* 130: 5872–5873.

**26** Li, Y., Yang, X., Liu, Y. et al. (2010). *Chem. Eur. J.* 16: 1871–1882.

**27** Liu, Z.-Q. (2012). *Chem. Rev.* 112: 3329–3355.

**28** Qi, L.-W., Wang, C.-Z., and Yuan, C.-S. (2011). *Nat. Prod. Rep.* 28: 467–495.

**29** Park, J.D., Rhee, D.K., and Lee, Y.H. (2005). *Phytochem. Rev.* 4: 159–175.

**30** Yu, J., Sun, J., Niu, Y. et al. (2013). *Chem. Sci.* 4: 3899–3905.

**31** Liao, J., Sun, J., Niu, Y., and Yu, B. (2011). *Tetrahedron Lett.* 52: 3075–3078.

**32** Shen, R., Cao, X., Laval, S. et al. (2016). *J. Org. Chem.* 81: 10279–10294.

**33** Kitagawa, I., Inamoto, T., Fuchida, M. et al. (1980). *Chem. Pharm. Bull.* 28: 1651–1653.

**34** Chen, X., Shao, X., Li, W. et al. (2017). *Angew. Chem. Int. Ed.* 56: 7648–7652.

**35** Li, X.-C., Yang, C.-R., Liu, Y.-Q. et al. (1995). *Phytochemistry* 39: 1175–1179.

**36** Sun, J., Han, X., and Yu, B. (2005). *Org. Lett.* 7: 1935–1938.

**37** Ida, Y., Satoh, Y., Katsumata, M. et al. (1995). *Chem. Pharm. Bull.* 43: 896–898.

**38** Zhu, S., Li, Y., and Yu, B. (2008). *J. Org. Chem.* 73: 4978–4985.

**39** Lee, S.M., Shon, H.J., Choi, C.-S. et al. (2009). *Chem. Pharm. Bull.* 57: 92–94.

**40** Shen, R., Laval, S., Cao, X., and Yu, B. (2018). *J. Org. Chem.* 83: 2601–2610.

**41** Cheng, S., Du, Y., Bing, F., and Zhang, G. (2008). *Carbohydr. Res.* 343: 462–469.

**42** Bylka, W., Znajdek-Awiżeń, P., Studzińska-Sroka, E. et al. (2014). *Phytother. Res.* 28: 1117–1124.

**43** Shao, W., Cao, X., Shen, L. et al. (2017). *Asian J. Org. Chem.* 6: 1270–1276.

**44** Shimizu, M., Shingyouchi, K., Morita, N. et al. (1978). *Chem. Pharm. Bull.* 26: 1666–1671.

**45** Kim, M., Lim, E., and Jung, M. (2013). *Tetrahedron* 69: 5481–5486.

**46** Hamed, A.I., Piacente, S., Autore, G. et al. (2005). *Planta Med.* 71: 554–560.

**47** Sun, J., Han, X., and Yu, B. (2003). *Carbohydr. Res.* 338: 827–833.

**48** Yan, M.-C., Liu, Y., Chen, H. et al. (2006). *Bioorg. Med. Chem. Lett.* 16: 4200–4204.

**49** Lin, Y.-Y., Chan, S.-H., Juang, Y.-P. et al. (2018). *Eur. J. Med. Chem.* 143: 1942–1958.

**50** Zeng, Y.-B., Hsiao, H.-M., Chan, S.-H. et al. (2014). *Mol. Diversity* 18: 13–23.

51 Guo, T., Song, Y., Lu, Y. et al. (2019). *J. Chem. Res.* 44: 42–49.
52 He, Z.-Q. and Findlay, J.A. (1991). *J. Nat. Prod.* 54: 810–815.
53 Liu, T., Liao, J.-X., Hu, Y. et al. (2017). *J. Org. Chem.* 82: 4170–4178.
54 Kondo, N., Marumoto, Y., and Shoji, I. (1971). *Chem. Pharm. Bull.* 19: 1103.
55 Peng, W., Sun, J., Lin, F. et al. (2004). *Synlett* 2004: 0259–0262.
56 Peng, W., Han, X., and Yu, B. (2004). *Synthesis* 1641–1647.
57 Lai, Y.-H., Chou, Y.-T., Lin, Y.-Y. et al. (2023). *J. Org. Chem.* 88: 9946–9958.

# 11

# Miscellaneous Glycosides

## Allosamidin

Allosamidin (**1**, Figure 11.1) was isolated from the mycelial extracts of *Streptomyces* sp. 1713 in 1987 by Sakuda et al. [1]. *In vitro* experiments revealed strong inhibition of allosamidin against chitinases from the silkworm *Bombyx mori*, whereas a 500-fold decrease against the bacterial chitinases from *Streptomyces griseus* and *Serratia marcescens*. Structurally, allosamindn (**1**) consists of a β-linked 3,3′-epi-chitobiose unit attached to a novel aglycone known as "allosamizoline," which could mimic the transition state of the hydrolyzing chitin.

**Total synthesis of allosamidin by Danishefsky and coworkers.** In 1991, Danishefsky and coworkers reported the first total synthesis of allosamidin (**1**, Scheme 11.1) [2, 3]. The aglycone was prepared starting from *meso* diacetate **2**, which, upon treatment with electric eel acetylcholinesterase, provided chiral alcohol **3** in 95% yield with >95% ee. Subsequent steps involving carbamate group installation, epoxidation, and selective protection led to the suitable protected aglycone **4**. Meanwhile, treatment of glycal **5** with *N,N*-dibromobenzenesulfonamide provided glycosyl bromosulfonamide **6**, which was then coupled with glycal acceptor **7** in the presence of KHMDS, furnishing the desired β-disaccharide **8** in 81% yield, with the *N*-sulfonamide group migrating to the C2′-position [4]. After converting **8** into glycosyl bromosulfonamide **9**, the azaglycosylation with aglycone **4** under basic conditions afforded the coupled product **10** in 42% yield (76% based on recovered starting material). Finally, the SEM and benzylidene protecting groups were removed with HCl/MeOH, followed by deprotection of the sulfonamides and benzyl ethers with Na/$NH_3$, peractylization, and removal of the *O*-Ac groups with $NH_3$/MeOH, ultimately achieving the total synthesis of allosamidin (**1**).

## Staurosporine

Staurosporine (**11**, Figure 11.2), a prototypical indolocarbazole glycoside, was isolated from *Streptomyces staurosporeus* by Omura et al. in 1977 during their exploration for new alkaloids derived from actinomycetes [5]. Notably, staurosporine (**11**) stands out as one of the most potent inhibitors of protein kinase C (PKC) with

*Carbohydrate Chemistry in the Total Synthesis of Naturally Occurring Glycosides*,
First Edition. Biao Yu and Xiaoyu Yang.
© 2024 WILEY-VCH GmbH. Published 2024 by WILEY-VCH GmbH.

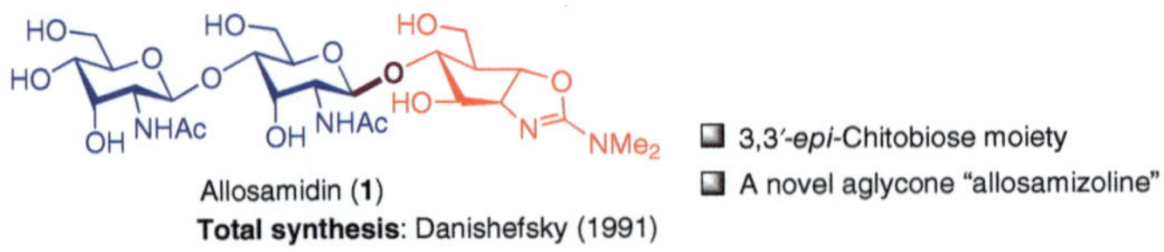


**Figure 11.1** Allosamidin (**1**) and its structural features.

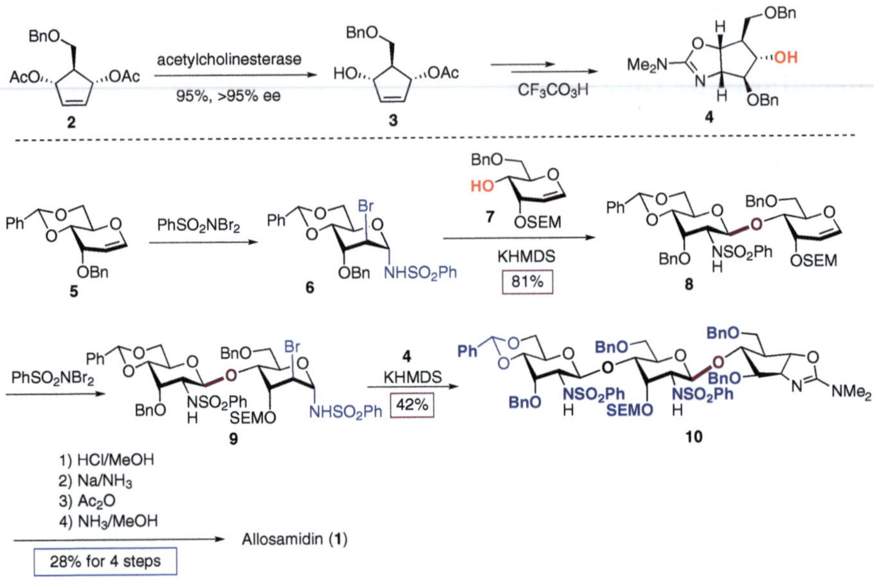


**Scheme 11.1** Total synthesis of allosamidin (**1**) by Danishefsky and coworkers.

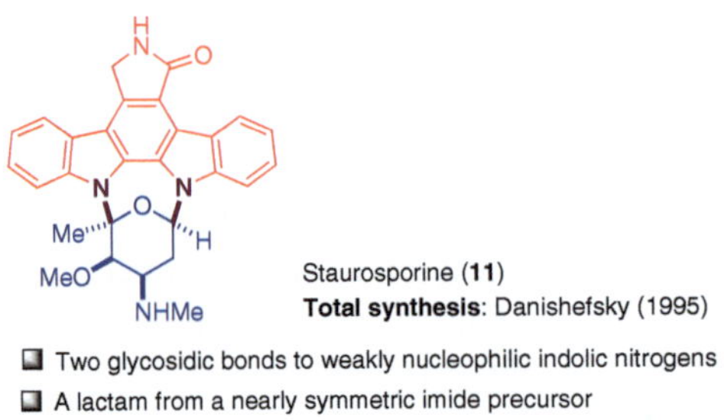


**Figure 11.2** Staurosporine (**11**) and its structural features.

an $IC_{50}$ value of 2.7 nM, which could interact with the ATP binding site within the catalytic domain of PKC. The total synthesis of staurosporine posed two major challenges: construction of two glycosidic bonds with weakly nucleophilic indolic nitrogens and the regioselectivity issue in the formation of a lactam from a nearly symmetric imide moiety.

**Total synthesis of staurosporine by Danishefsky and coworkers.** In 1995, Danishefsky and coworkers reported the first total synthesis of staurosporine (**11**) and *ent*-staurosporine (Scheme 11.2) [6, 7]. The aglycone **13** was efficiently synthesized from the addition of benzyloxymethyl (BOM) dibromomaleimide (**12**) and indole Grignard reagents. Glycosyl epoxide **14** was obtained from the corresponding glucal by treatment with 2,2-dimethyldioxirane, which was coupled with aglycone **13** under basic conditions to afford the desired *N*-glycosylated product **15** in 47% yield. Subsequent removal of the C2′ hydroxy group and elimination of the C6′-OH functionality gave exo-glycal **16**, which upon treatment of KO*t*Bu and $I_2$ afforded the cyclized product **17** in 65% yield. After removal of the C6′-iodide and introduction of the *O*-methyl and *N*-methyl groups, the resultant **18** was subjected to removal of the *N*-BOM group with $H_2/Pd(OH)_2$ and *N*-Boc group with TFA, providing compound **19** with an overall yield of 82% for the two steps. Finally, the combined use of $NaBH_4$ and PhSeH led to monoreduction of the imide moiety in **19**, thus completing the synthesis of staurosporine (**11**, in 39% yield) and isostaurosporine (**20**, in 36% yield). Ever since, many indolocarbazole glycosides have been synthesized [8].

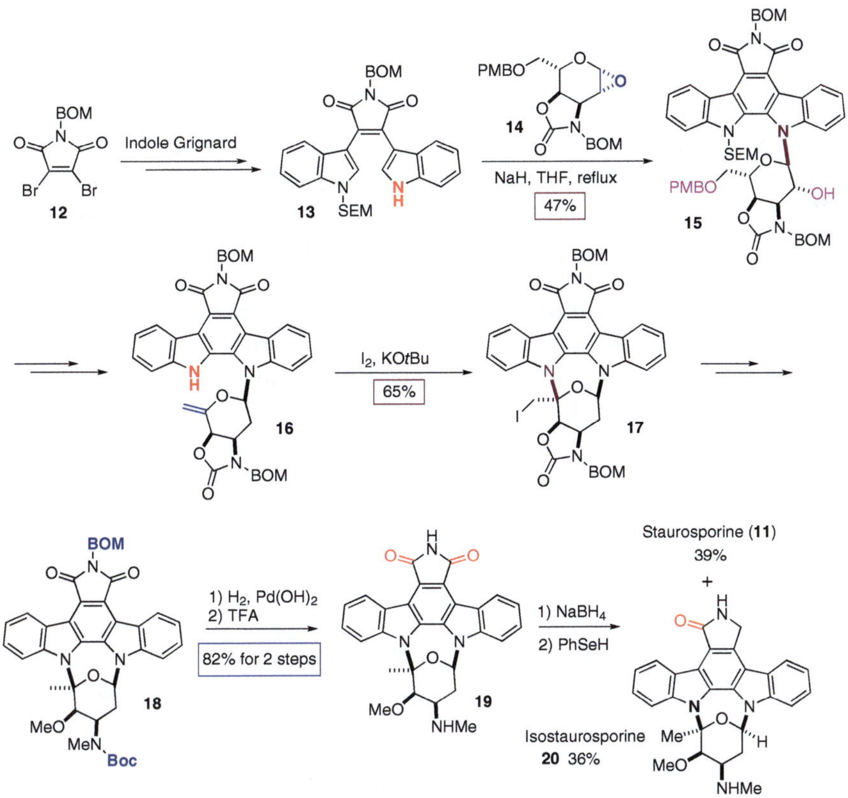


**Scheme 11.2** Total synthesis of staurosporine (**11**) by Danishefsky and coworkers.

## Everninomicin 13,384-1

Everninomicin 13,384-1 (**21**, Figure 11.3), belonging to the orthosomicin class of antibiotics [9], was first isolated from *Micromonospora carbonacea* var. *africana* (collected from the banks of the Nyiro river in Kenya) in 1989 [10]. This antibiotic exhibited promising antibacterial properties against drug-resistant pathogens,

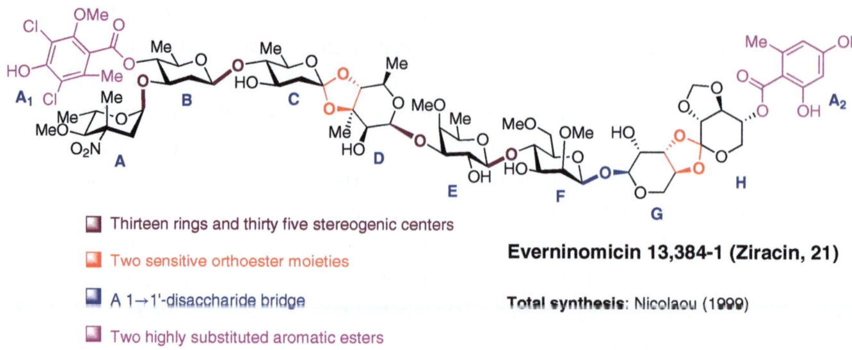


**Figure 11.3** Eveminomicin 13,384-1 (**21**) and its structural features.

including methicillin-resistant *Staphylococci* as well as vancomycin-resistant *Streptococci* and *Enterococci*, and has undergone extensive clinical trials to evaluate its efficacy against both susceptible and resistant strains. Structurally, everninomicin 13,384-1 (**21**) possesses a distinctive oligosaccharide structure featuring two orthoester moieties and terminating with two highly substituted aromatic esters. In addition, everninomicin contains a $1 \rightarrow 1'$-disaccharide bridge, a nitrosugar (evernitrose), 13 rings, and a remarkable 35 stereogenic centers integrated within its molecular architecture.

**Total synthesis of everninomicin 13,384-1 by the Nicolaou group.** In 1999, Nicolaou and coworkers reported the total synthesis of everninomicin 13,384-1 (**21**) [11–17]. The synthesis of the A1B(A)C fragment of everninomicin 13,384-1 began with the coupling of C4-OH glycoside **22** with C2-SPh-substituted glycosyl fluoride **23** in the presence of $SnCl_2$, which provided the desired β-disaccharide **24** in 71% yield as a single anomer (Scheme 11.3). Following subsequent steps of protecting

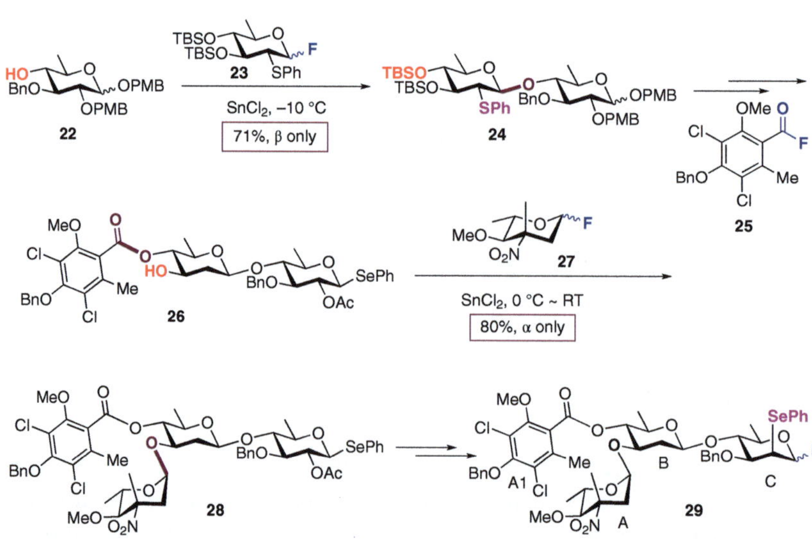

**Scheme 11.3** Synthesis of the A1B(A)C fragment **29** of everninomicin 13,384-1.

group manipulations and esterification using substituted benzoyl fluoride **25**, disaccharide acceptor **26** was obtained. Glycosylation of **26** with evernitrosyl fluoride **27** under the promotion of $SnCl_2$ led to trisaccharide **28** in 80% yield, with the newly formed glycoside bond (A → B, α-anomer) being controlled by the anomeric effect. Lastly, the acetate was removed and subsequent treatment with DAST resulted in glycosyl fluoride donor **29**, with a notable stereochemical inversion at C2 of ring C, which facilitated the subsequent elongation of the saccharide chain.

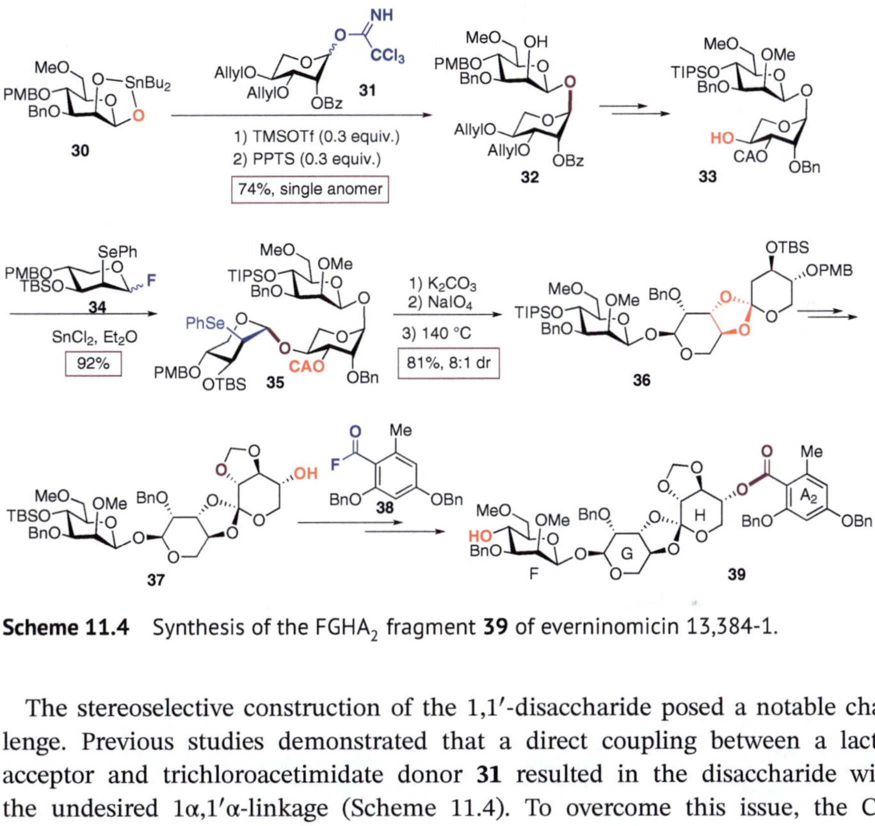

**Scheme 11.4** Synthesis of the FGHA$_2$ fragment **39** of everninomicin 13,384-1.

The stereoselective construction of the 1,1′-disaccharide posed a notable challenge. Previous studies demonstrated that a direct coupling between a lactol acceptor and trichloroacetimidate donor **31** resulted in the disaccharide with the undesired 1α,1′α-linkage (Scheme 11.4). To overcome this issue, the C-2 hydroxyl group in the acceptor was leveraged to fix the anomeric oxygen in the desired β-configuration using a five-membered stannane ring strategy [11]. Thus, coupling of tin acetal **30** with imidate **31** in the presence of TMSOTf, followed by PPTS-induced cleavage of the TMS ether, yielded the desired 1β,1′α-linked disaccharide **32** in 74% yield with complete stereocontrol. Upon protecting group manipulations, $SnCl_2$-mediated coupling of the disaccharide acceptor **33** with glycosyl fluoride **34** resulted in the stereoselective formation of trisaccharide **35** in 92% yield. After removal of the acetyl group, oxidation of the phenylseleno group with $NaIO_4$, followed by heating the crude selenoxide at 140 °C, led to sequential *syn*-elimination at the anomeric position and ring closure to furnish orthoester **36** in 81%, with diastereoselectivity of approximately 8 : 1. Extensive model studies demonstrated that the correct orthoester stereochemistry could only be obtained by first attaching the H ring to the C4 oxygen atom of ring G and then constructing the second oxygen bridge (C3), as the reverse order would yield the opposite orthoester stereochemistry. Following subsequent protecting group manipulations and introduction of α-hydroxyl group at the C2 position of ring H, esterification of alcohol **37** with substituted benzoyl fluoride **38** furnished the desired $FGHA_2$ fragment **39** after removal of the TBS group.

The final stage of the synthesis involved the $S_N2$-type glycosylation of D ring acceptor **40** with glycosyl sulfoxide **41** under the action of $Tf_2O$ and DTBMP, resulting in the β-mannoside-type disaccharide **42** in 71% yield (Scheme 11.5). After constructing the C-3 tertiary alcohol moiety of the D unit and performing necessary protecting group manipulations, the resultant disaccharide imidate donor **43** was coupled

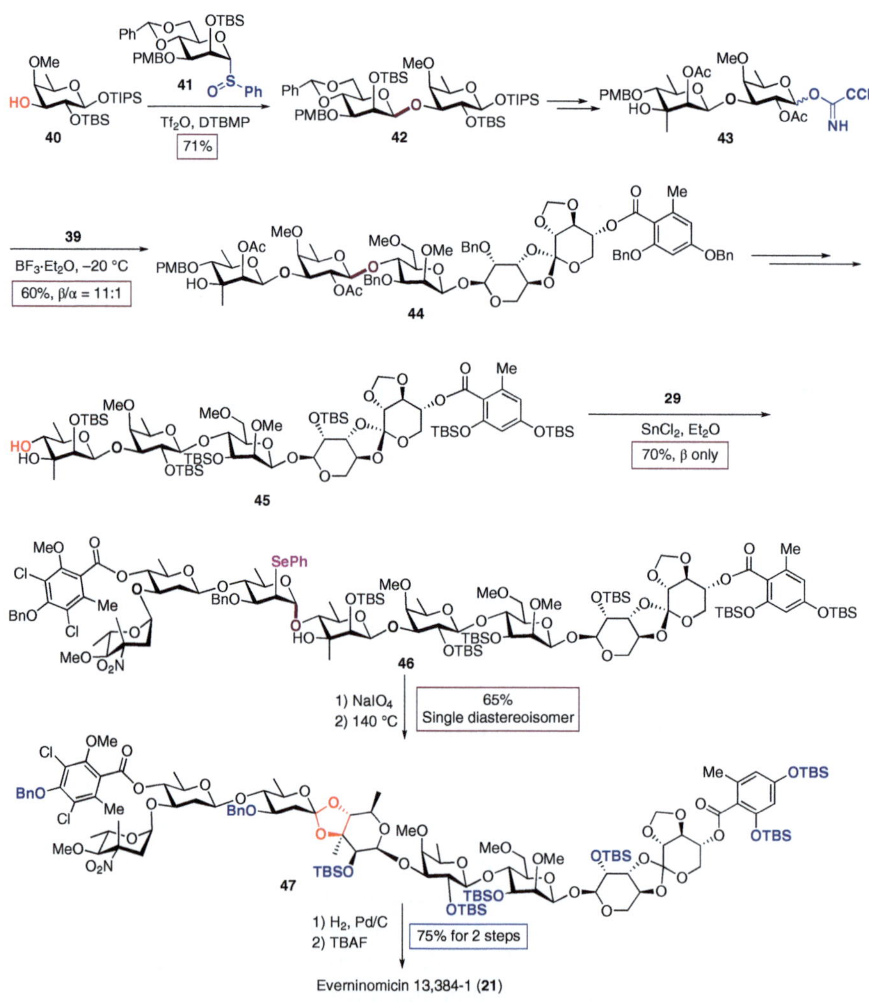

**Scheme 11.5** Synthesis of the DE fragment and completion of the total synthesis of everninomicin 13,384-1 (**21**) by Nicolaou and coworkers.

with FGH trisaccharide acceptor **39** under the promotion of $BF_3 \cdot Et_2O$ at −20 °C, affording the desired β-pentasaccharide **44** in 55% yield, along with a minor 5% of the α-anomer. Further protecting group manipulations led to the coupling of pentasaccharide **45** with the ABC trisaccharide fluoride donor **29** in the presence of $SnCl_2$, resulting in the desired oligosaccharide **46** with complete stereocontrol facilitated by the C2-SePh group. The remaining orthoester motif was efficiently formed using the Sinaÿ conditions ($NaIO_4$ oxidation followed by heating at 140 °C) [18], yielding the fully protected everninomicin derivative **47** as a single stereoisomer in 65% yield. Finally, hydrogenation of the Bn groups, followed by removal of the TBS groups, accomplished the total synthesis of everninomicin 12,384-1 (**21**).

## Brasilicardins

Brasilicardins A–D (**48–51**, Figure 11.4) were isolated by Kobayashi and coworkers in 1998 from the cultured broth of the pathogenic actinomycete *Nocardia brasiliensis* [19]. Among these congeners, brasilicardin A (**48**) demonstrated the most potent immunosuppressive activity in a mouse mixed lymphocyte reaction assay ($IC_{50}$ = 0.057 μg/ml). Furthermore, it exhibited cytotoxic activity against adriamycin-resistant murine lymphoma cells ($IC_{50}$ = 0.078 μg/ml). Structurally, brasilicardins represent a unique class of diterpenoid-saccharide-amino acid hybrid metabolites, featuring a highly strained *trans/syn/trans*-perhydrophenanthrene skeleton, a mono- or disaccharide attachment, and an appended amino acid side chain.

**Total synthesis of brasilicardins A and C by Anada and coworkers.** In 2017, Anada et al. reported the first total synthesis of brasilicardins A (**48**) and C (**51**) in 29 steps, achieving overall yields of 2.1% and 1.4%, respectively (Scheme 11.6) [20]. The Diels–Alder reaction between α-cyano-α,β-enone **52** with silyloxydiene **53** in a sealed tube at 140 °C proceeded smoothly to give tricyclic product **54** with a high yield of 98%. The crucial reductive methylation step on **54** involved treating it sequentially with lithium naphthalenide and MeI, resulting in the C8-methylated product **55** in 74% yield, which possessed the desired *trans/syn/trans*

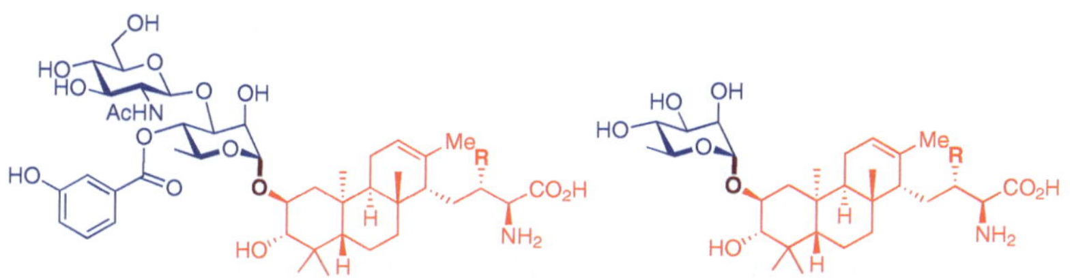

Brasilicardin A, R = OMe (**48**)
Brasilicardin B, R = H (**49**)
Brasilicardin C, R = OMe (**50**)
Brasilicardin D, R = H (**51**)

**Total synthesis**: Anada (2017), Yoshimura & Tanino (2018)

- A highly strained *anti-syn-anti*-fused perhydrophenanthrene diterpenoid scaffold
- Two angular methyl groups with the B-ring boat conformer

**Figure 11.4** Brasilicardins and their structural features.

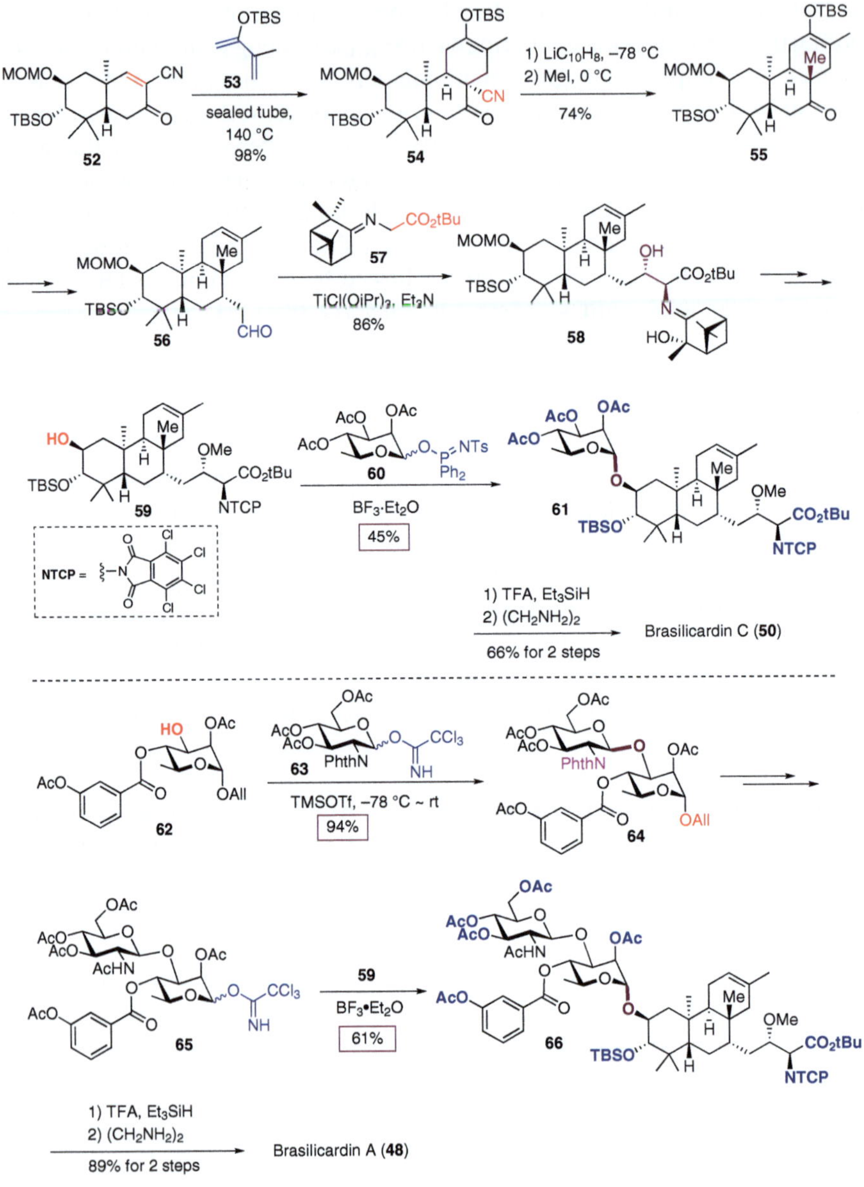


**Scheme 11.6** Total synthesis of brasilicardins A and C by Anada and coworkers.

stereochemistry. Following several transformations, an aldol coupling between aldehyde **56** and the titanium enolate generated from **57** afforded the *anti*-aldol product **58** as a single product in 86% yield. After removal of the chiral auxiliary group, glycosylation of **59** with *P,P*-diphenyl-*N*-(*p*-toluenesulfonyl)-phosphinimidate **60** in the presence of $BF_3{\cdot}Et_2O$, gave the desired coupling product **61** in 45% yield. In contrast, glycosylation with the corresponding trichloroacetimidate donor led to the coupled product in only 23% yield. Finally, simultaneous removal of the *tert*-butyl and TBS groups using TFA/$Et_3SiH$, followed by treatment with 1,2-ethylenediamine to remove the TCP group and the five acetyl groups, accomplished the synthesis of brasilicardin C (**50**).

The disaccharide unit was prepared following Jung's approach [21]. Coupling of glycosyl acceptor **62** with 2-phthalimido glucosamine imidate donor **63** under the catalysis of TMSOTf afforded the desired β-disaccharide **64** in 94% yield. After

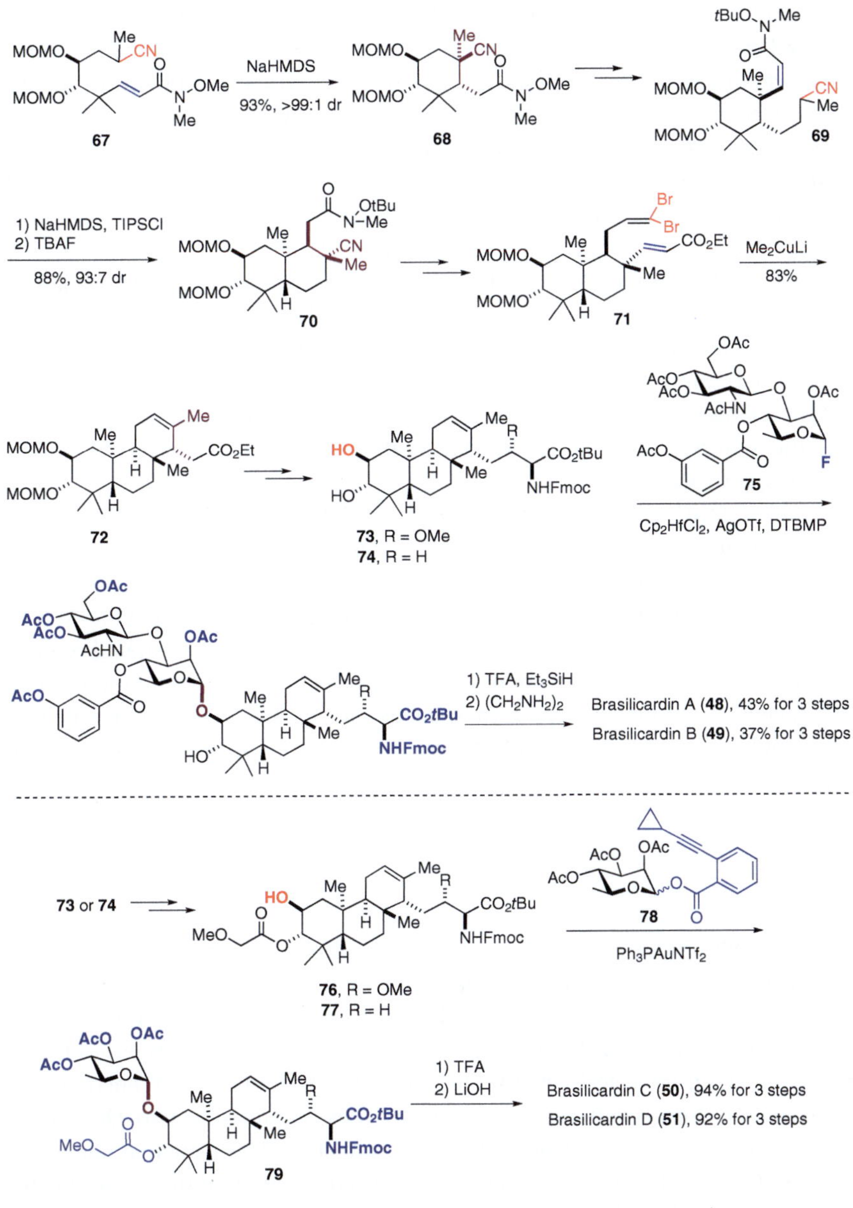

**Scheme 11.7** Total synthesis of brasilicardins A, B, C and D by Yoshimura, Tanino, and coworkers.

conversion of the NPhth group to NHAc group, disaccharide trichloroacetimidate **65** was prepared. Glycosylation of aglycone **59** with **65** in the presence of $BF_3 \cdot Et_2O$ provided the desired product **66** in 61% yield as the sole anomer. Ultimately, global deprotection of **66** led to brasilicardin A (**48**) in 98% (for two steps).

**Total synthesis of brasilicardins A-D by Yoshimura and coworkers.** In 2018, Yoshimura and coworkers reported the asymmetric total synthesis of brasilicardin A-D (**48–51**) using readily available commercial materials (Scheme 11.7) [22]. The key intramolecular Michael addition in **67** was facilitated by NaHMDS, furnishing the desired cyclized product **68** in 93% yield as a single isomer. Following a series of transformations, the second intramolecular nitrile Michael addition in **69** provided the expected cyclization product **70** in 88% yield with 93 : 7 diastereoselectivity. After converting **70** into 1,1-dibromoalkene **71**, the treatment of **71** with $Me_2CuLi$ generated the vinylcopper species, which underwent the conjugate addition with the α,β-unsaturated ester moiety, delivering the tricyclic product **72** in 83% yield. Upon introduction of the amino acid chain, glycosylation of the aglycone **73** or **74** with disaccharide fluoride **75** under the promotion of $Cp_2HfCl_2$/AgOTf afforded the desired α-glycoside regio- and stereoselectively. Notably, the glycosylation approach employed by the Anada group was not applicable in the present substrates, as the C3-OH was left unprotected in this synthesis. Finally, removal of the *tert*-butyl ester with TFA, followed by global deprotection of the acetyl and Fmoc groups, accomplished the total synthesis of brasilicardin A (**48**, 43% for 2 steps) and B (**49**, 37% for 3 steps), respectively.

On another front, direct regioselective glycosylation of the C2-alcohol with a variety of rhamnosyl donors proved to be infeasible. Therefore, the C3-OH was selectively protected; subsequent glycosylation of the resultant **76** or **77** with the rhamnosyl *o*-cyclopropylethynylbenzoate **78** under the catalysis of $PPh_3AuNTf_2$ resulted in the desired α-glycoside **79** in high yield. Finally, TFA-mediated removal of the *tert*-butyl ester and subsequent global deprotection of the remaining acetyl, methoxyacetyl, and Fmoc groups with aqueous LiOH solution completed the total synthesis of brasilicardin C (**50**, 94% for two steps) and D (**51**, 92% for three steps), respectively.

## Efrotomycin

Efrotomycin (**80**, Figure 11.5), the most intricate member of the elfamycins, was initially isolated from *Nocardia lactamdurans* by a research group from Merck, Sharp & Dohme [23]. Apart from exhibiting efficacy against bacilli and dysentery, these compounds possessed significant growth-promoting properties and were primarily intended for veterinary use. Structurally, efrotomycin (**80**) encompasses 21 stereocenters and seven geometrical stereosites, comprising two *O*-glycoside bonds, an amide linkage, and several olefinic moieties. The principal challenges encountered during the total synthesis of efrotomycin include: (i) the stereoselective and efficient assembly of the oligosaccharide component and (ii) the stereocontrolled synthesis of the tetrahydrofuran motif with the appropriate substituents

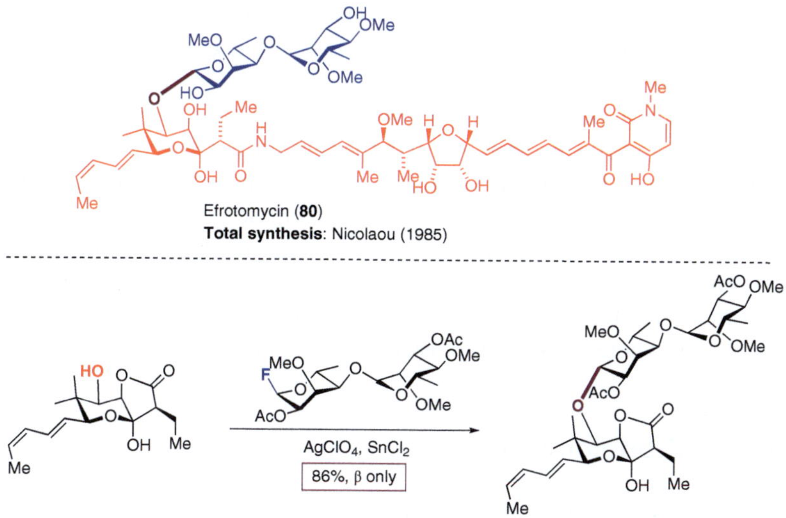


**Figure 11.5** Efrotomycin (**80**) and the key glycosylation step in the synthesis.

arranged in an all-*cis* fashion. In 1985, Nicolaou and coworkers reported the first total synthesis of efrotomycin (**80**), using a modular synthetic route that involved early-stage glycosylation with a disaccharide fluoride donor [24, 25].

## Peyssonoside A

Peyssonnoside A (**81**, Figure 11.6), a sulfated diterpene glycoside, was isolated from the red alga *Peyssonnelia* sp. by Kubanek and coworkers in 2019 [26]. It demonstrated promising activity against liver-stage *Plasmodium berghei* and moderate activity against methicillin-resistant *Staphylococcus aureus* (MRSA). Peyssonnoside A possesses several notable structural features, including (i) an

**Figure 11.6** Peyssonoside A (**81**) and the key glycosylation step in its total synthesis.

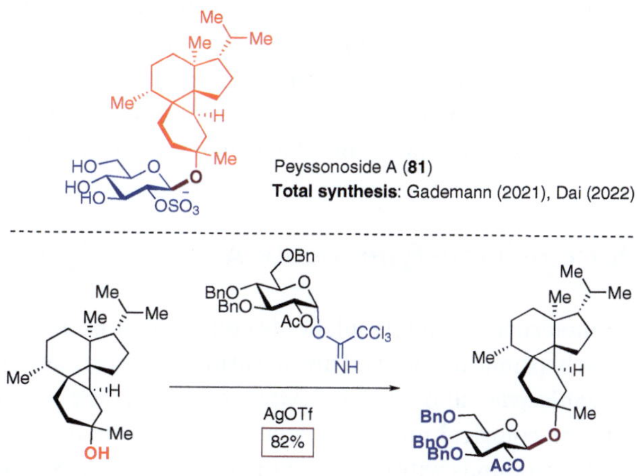

exceptional and unparalleled tetracyclic 5/6/3/6 carbon framework featuring an embedded pentasubstituted cyclopropane ring, (ii) the presence of six consecutive stereocenters and three consecutive quaternary stereocenters in close proximity. Moreover, the sulfated β-linked glycoside poses an additional synthetic challenge, due to the tertiary nature of the hydroxy group in the aglycone, which could result in elimination and orthoester byproducts. In 2021, Gademann and coworkers reported the first total synthesis of peyssonnosides A (**81**) in a concise, scalable, and highly diastereoselective manner, in which the saccharide was introduced at the late stage by employing glycosyl imidate as a donor, enabled by AgOTf [27]. In 2022, Dai and coworkers reported a 13-step enantioselective and stereoselective total synthesis of (−)-peyssonnoside A, where the same glycosylation method was used [28].

## Amycolamicin/Kibdelomycin

Amycolamicin (**82**, Figure 11.7) was first isolated from the soil actinomycete *Amycolatopsis* sp. MK575-fF5 by Igarashi and coworkers in 2009 [29], while kibdelomycin was isolated from *Kibdelosporangium* sp. (MA7385) by Singh and coworkers from Merck in 2011 [30] and was later found to be identical to amycolamicin. This compound exhibited potent and broad-spectrum antibacterial activities against both gram-positive bacteria and certain gram-negative pathogens. The chemical architecture of amycolamicin/kibdelomycin is unprecedented, featuring a *trans*-decalin, a tetramic acid, and two uncommon sugars. In 2021, Li and coworkers reported the first total synthesis of amycolamicin/kibdelomycin and provided evidence through careful titration that kibdelomycin is the salt form of amycolamicin [31]. The construction of the two unusual glycosidic bonds was achieved using the gold(I)-catalyzed glycosylation with glycosyl *ortho*-alkynyl benzoate donors. In 2022, Kuwahara and coworkers reported the second total synthesis of amycolamicin, employing a bicyclic *N*,*O*-acetal as donor for the introduction of the amycolose unit and an stereoconvergent *N*-acylation for introduction of the *N*-glycoside [32]. In the same year, Baran and coworkers reported their total synthesis of kibdelomycin, in which the amycolose unit was introduced by using trichloroethanyl amycoloside in the presence of TfOH with low diastereoselectivity [33]. The incorporation of the *N*-glycoside unit was achieved through a similar N-acylation protocol between β-ketothioester and aminal, facilitated by AgTFA. Later, Schobert et al. reported a formal synthesis of kibdelomycin [34].

## Strictosidine-Type Indole Alkaloid Glycosides

Monoterpenoid indole alkaloids constitute a prominent class of natural products, which primarily derived from various higher plant families such as *Rubiaceae*, *Loganiaceae*, and *Apocynaceae* species. These compounds exhibit significant pharmacological properties, making them valuable targets of research. One of the key compounds within this group is strictosidine (**83**, Figure 11.8), a monoterpenoid indole alkaloid glycoside, was first isolated from *Rhazya* plants species, specifically *Rhazya stricta* Decaisne and *Rhazya orientalis* in 1968 [35]. It features a densely packed dihydropyran ring, a glucosyl moiety, a bis(acetal) linkage, and nine stereocenters. Notably, strictosidine (**83**) has been identified as a common biosynthetic precursor to several monoterpenoid indole alkaloids, including cymoside (**84**), ophiorrhine A (**85**), and others. In 2020, Ishikawa and coworkers reported the first

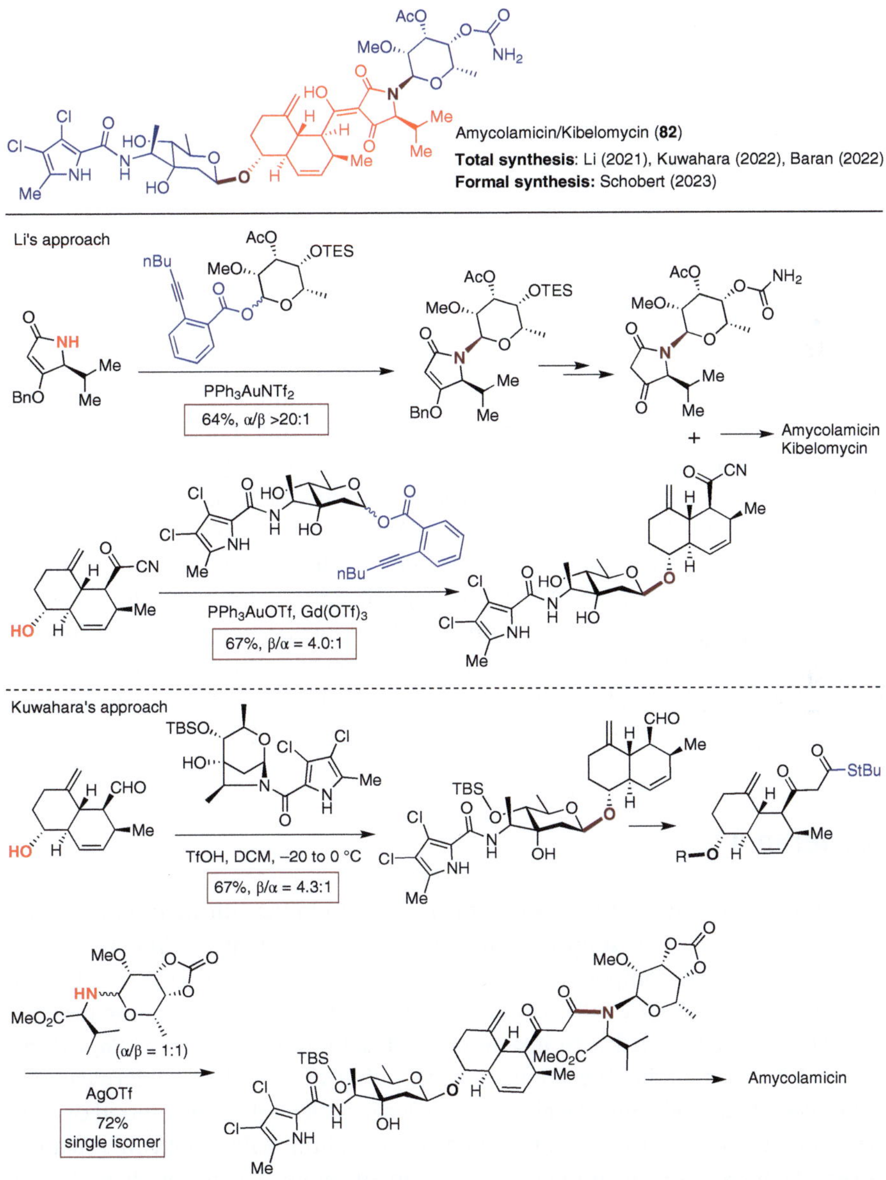


**Figure 11.7** Structure of amycolamicin/kibdelomycin (**82**) and the key glycosylation steps in the total synthesis.

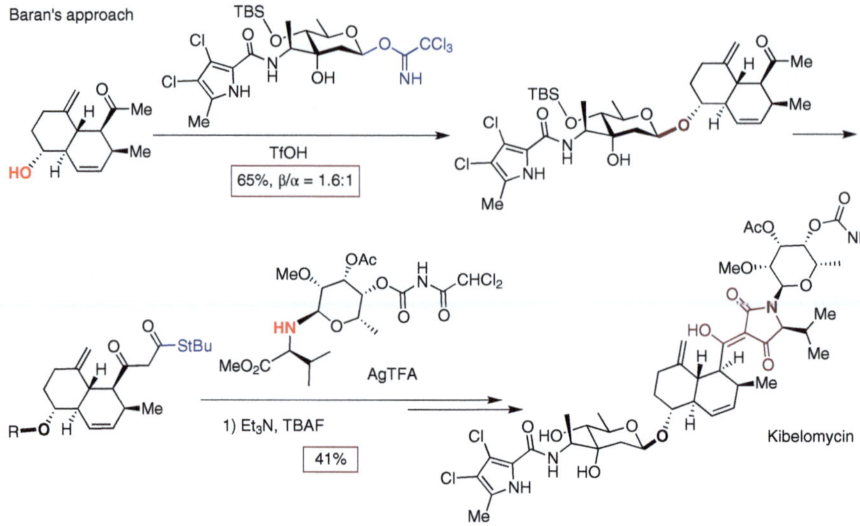


**Figure 11.7** (*Continued*)

total synthesis of strictosidine (**83**), with the glucose unit being installed with a glycosyl imidate donor [36]. In 2021, Garg and coworkers reported an alternative synthesis of strictosidine (**83**), featuring an oxa-Diels–Alder reaction utilizing a vinyl glycoside dienophile, along with an stereoselective enzymatic Pictet–Spengler reaction [37]. In 2020, Vincent and coworkers achieved the total synthesis of cymoside (**84**) through a bio-inspired oxidative cyclization strategy, in which the glucose unit was introduced at the late stage of synthesis [38, 39]. Recently, the Vincent group disclosed the total synthesis of ophiorrhine A (**85**) by employing a late-stage intramolecular Diels–Alder cycloaddition to form the intricate bridged and polycyclic system [40]. They utilized a similar glycosylated precursor as employed in the synthesis of strictosidine by the Ishikawa group.

## Cotylenin A

Cotylenin A (**86**, Figure 11.9) was initially isolated from the culture filtrate of a fungus (strain 501-7w) that stimulated the growth of cotyledons in Chinese cabbage seedlings [41]. This compound was able to induce differentiation in murine and human myeloid leukemia cells, as well as apoptosis in a wide range of human cancer cell lines when combined with interferon-α. Structurally, cotylenin A (**86**) exhibits a fused 5-8-5 carbocyclic ring system, incorporating an all-carbon quaternary stereogenic center, an acid-sensitive chiral allylic tertiary alcohol, a stereotetrad containing a *trans*-1,2-diol, and a four-substituted alkene. Additionally, cotylenin A features a distinct glucose-fused trioxabicyclo[2.2.1]heptane motif, adorned with methyl and epoxyethyl groups. In 2020, Nakada and coworkers reported a convergent enantioselective total synthesis of cotylenin A (**86**) [42], in which the sugar moiety was

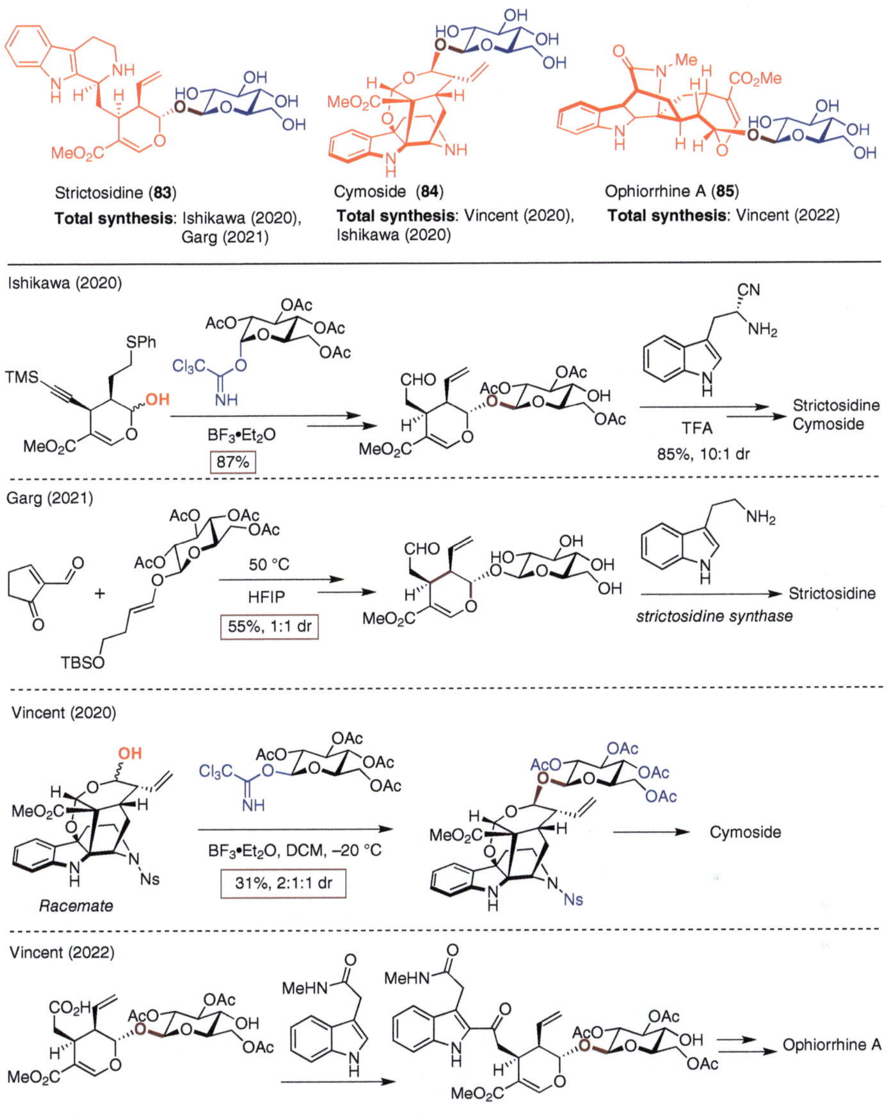


**Figure 11.8** Strictosidine-type indole alkaloid glycosides and the key steps in their synthesis.

introduced by using rhodium-catalyzed sulfonium ylide formation and subsequent Brønsted-acid-catalyzed glycosylation with a glycosyl sulfide as donor [43].

## Pyrolaside B

Pyrolaside B (**87**, Figure 11.10), a member of the phenol trimer class of natural products, was initially isolated from *Pyrola rotundifoliai* in 2005 [44]. This

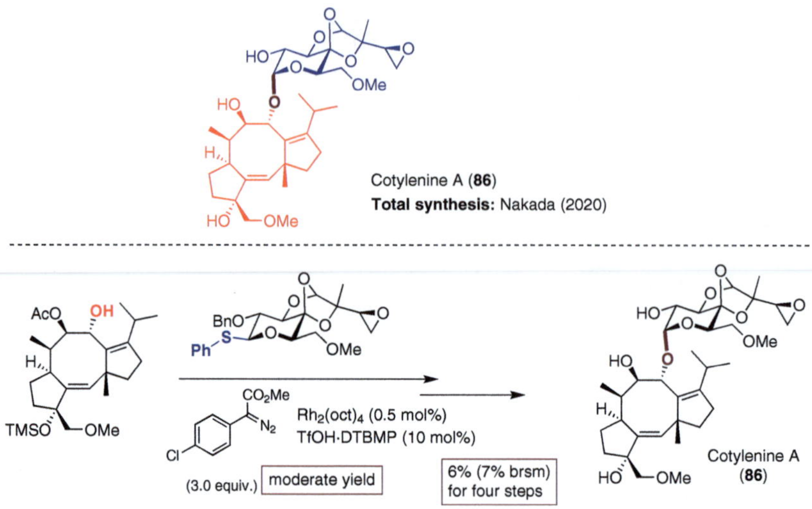


**Figure 11.9** Cotylenin A (**86**) and the key glycosylation step in the total synthesis.

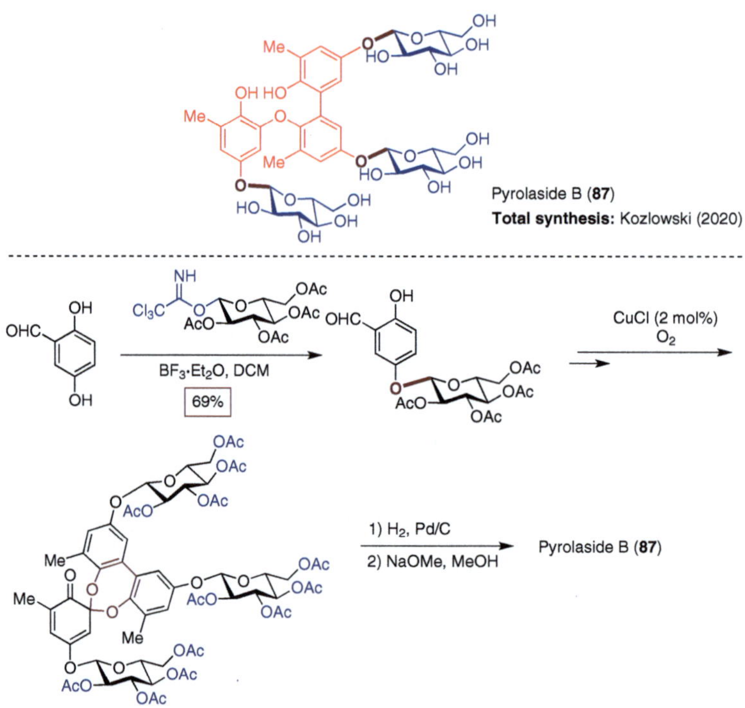


**Figure 11.10** Pyrolaside B (**87**) and the key steps in the synthesis.

compound demonstrated inhibitory activity against both *Micrococcus luteus* and *S. aureus*. In 2020, Kozlowski et al. reported a concise, five-step total synthesis of pyrolaside B (**87**) with an overall yield of 16% [45]. The authors developed a catalytic aerobic copper system for phenol trimerization, involving sequenced oxidative C–C and C–O coupling. Notably, the one-step construction of a spiroketal trimer intermediate played a critical role in the synthesis.

## References

**1** Sakuda, S., Isogai, A., Matsumoto, S., and Suzuki, A. (1987). *J. Antibiot.* 40: 296–300.
**2** Griffith, D.A. and Danishefsky, S.J. (1991). *J. Am. Chem. Soc.* 113: 5863–5864.
**3** Griffith, D.A. and Danishefsky, S.J. (1996). *J. Am. Chem. Soc.* 118: 9526–9538.
**4** Griffith, D.A. and Danishefsky, S.J. (1990). *J. Am. Chem. Soc.* 112: 5811–5819.
**5** Omura, S., Iwai, Y., Hirano, A. et al. (1977). *J. Antibiot.* 30: 275–282.
**6** Link, J.T., Raghavan, S., and Danishefsky, S.J. (1995). *J. Am. Chem. Soc.* 117: 552–553.
**7** Link, J.T., Raghavan, S., Gallant, M. et al. (1996). *J. Am. Chem. Soc.* 118: 2825–2842.
**8** Chambers, G.E., Sayan, A.E., and Brown, R.C.D. (2021). *Nat. Prod. Rep.* 38: 1794–1820.
**9** Wright, D.E. (1979). *Tetrahedron* 35: 1207–1237.
**10** Ganguly, A.K., Pramanik, B.N., Sarre, O.Z. et al. (1989). *Heterocycles* 28: 206–209.
**11** Nicolaou, K.C., van Delft, F.L., Conley, S.R. et al. (1997). *J. Am. Chem. Soc.* 119: 9057–9058.
**12** Nicolaou, K.C., Mitchell, H.J., Delft, F.L.V. et al. (1998). *Angew. Chem. Int. Ed.* 37: 1871–1874.
**13** Nicolaou, K.C., Rodríguez, R.M., Mitchell, H.J., and van Delft, F.L. (1998). *Angew. Chem. Int. Ed.* 37: 1874–1876.
**14** Nicolaou, K.C., Mitchell, H.J., Rodríguez, R.M. et al. (1999). *Angew. Chem. Int. Ed.* 38: 3345–3350.
**15** Nicolaou, K.C., Mitchell, H.J., Suzuki, H. et al. (1999). *Angew. Chem. Int. Ed.* 38: 3334–3339.
**16** Nicolaou, K.C., Rodríguez, R.M., Fylaktakidou, K.C. et al. (1999). *Angew. Chem. Int. Ed.* 38: 3340–3345.
**17** Nicolaou, K.C., Fylaktakidou, K.C., Mitchell, H.J. et al. (2000). *Chem. Eur. J.* 6: 3166–3185.
**18** Jaurand, G., Beau, J.-M., and Sinaÿ, P. (1982). *J. Chem. Soc., Chem. Commun.* 12: 701–703.
**19** Shigemori, H., Komaki, H., Yazawa, K. et al. (1998). *J. Org. Chem.* 63: 6900–6904.
**20** Anada, M., Hanari, T., Kakita, K. et al. (2017). *Org. Lett.* 19: 5581–5584.
**21** Jung, M.E. and Koch, P. (2011). *Org. Lett.* 13: 3710–3713.
**22** Yoshimura, F., Itoh, R., Torizuka, M. et al. (2018). *Angew. Chem. Int. Ed.* 57: 17161–17167.

23 Dewey, R.S., Arison, B.H., Hannah, J. et al. (1985). *J. Antibiot.* 38: 1691–1698.
24 Dolle, R.E. and Nicolaou, K.C. (1985). *J. Am. Chem. Soc.* 107: 1691–1694.
25 Dolle, R.E. and Nicolaou, K.C. (1985). *J. Am. Chem. Soc.* 107: 1695–1698.
26 Chhetri, B.K., Lavoie, S., Sweeney-Jones, A.M. et al. (2019). *J. Org. Chem.* 84: 8531–8541.
27 Chesnokov, G.A. and Gademann, K. (2021). *J. Am. Chem. Soc.* 143: 14083–14088.
28 Xu, B., Liu, C., and Dai, M. (2022). *J. Am. Chem. Soc.* 144: 19700–19703.
29 Igarashi, M., Sawa, R., and Homma, Y. (2010). Novel Compound Amycolamicin, Method for Production Thereof, and Use Thereof. Jpn Patent JP2009/058210; WO2010122669A1, filed/issued 24 April 2009.
30 Phillips, J.W., Goetz, M.A., Smith, S.K. et al. (2011). *Chem. Biol.* 18: 955–965.
31 Yang, S., Chen, C., Chen, J., and Li, C. (2021). *J. Am. Chem. Soc.* 143: 21258–21263.
32 Meguro, Y., Ito, J., Nakagawa, K., and Kuwahara, S. (2022). *J. Am. Chem. Soc.* 144: 5253–5257.
33 He, Y., Wang, C., Bi, D. et al. (2022). *Angew. Chem. Int. Ed.* 61: e202206183.
34 Schriefer, M.G., Treiber, L., and Schobert, R. (2023). *Chem. Sci.* 14: 3562–3568.
35 Smith, G.N. (1968). *Chem. Commun.* 912–914.
36 Sakamoto, J., Umeda, Y., Rakumitsu, K. et al. (2020). *Angew. Chem. Int. Ed.* 59: 13414–13422.
37 Anthony, S.M., Tona, V., Zou, Y. et al. (2021). *J. Am. Chem. Soc.* 143: 7471–7479.
38 Dou, Y., Kouklovsky, C., Gandon, V., and Vincent, G. (2020). *Angew. Chem. Int. Ed.* 59: 1527–1531.
39 Dou, Y., Kouklovsky, C., and Vincent, G. (2020). *Chem. Eur. J.* 26: 17190–17194.
40 Cao, W., Dou, Y., Kouklovsky, C., and Vincent, G. (2022). *Angew. Chem. Int. Ed.* 61: e202209135.
41 Sassa, T., Tojyo, T., and Munakata, K. (1970). *Nature* 227: 379–379.
42 Uwamori, M., Osada, R., Sugiyama, R. et al. (2020). *J. Am. Chem. Soc.* 142: 5556–5561.
43 Meng, L., Wu, P., Fang, J. et al. (2019). *J. Am. Chem. Soc.* 141: 11775–11780.
44 Chang, J. and Inui, T. (2005). *Chem. Pharm. Bull.* 53: 1051–1053.
45 Neuhaus, W.C. and Kozlowski, M.C. (2020). *Angew. Chem. Int. Ed.* 59: 7842–7847.

# Index



*Carbohydrate Chemistry in the Total Synthesis of Naturally Occurring Glycosides*, First Edition. Biao Yu and Xiaoyu Yang.
© 2024 WILEY-VCH GmbH. Published 2024 by WILEY-VCH GmbH.